Cross-Laminated Timber Design

Structural Properties, Standards, and Safety

Mustafa Mahamid

McGraw Hill

New York Chicago San Francisco
Athens London Madrid
Mexico City Milan New Delhi
Singapore Sydney Toronto

Library of Congress Control Number: 2020934159

McGraw Hill books are available at special quantity discounts to use as premiums and sales promotions or for use in corporate training programs. To contact a representative, please visit the Contact Us page at www.mhprofessional.com.

Cross-Laminated Timber Design: Structural Properties, Standards, and Safety

1 2 3 4 5 6 7 8 9 CD 25 24 23 22 21 20

ISBN 978-1-260-11799-8
MHID 1-260-11799-5

Sponsoring Editor
Ania Levinson

Editorial Supervisor
Donna M. Martone

Production Supervisor
Pamela Pelton

Acquisitions Coordinator
Elizabeth M. Houde

Project Manager
Apoorva Mathur,
Cenveo® Publisher Services

Copy Editor
Girish Sharma and Syed Haider,
Cenveo Publisher Services

Proofreader
Bruce Owens

Indexer
Edwin Durbin

Art Director, Cover
Jeff Weeks

Composition
Cenveo Publisher Services

Contents

Preface

Cross-laminated timber (CLT) is a relatively new system that started in Europe in the early 1990s. The wood design communities, the forest product industry, and researchers found in CLT an opportunity for increasing the use of wood in nonconventional and nontraditional applications. Following the European experience, forest product agencies in North America led by FPInnovations prepared peer-reviewed publications to provide immediate support for the design, construction, and manufacturing of CLT products, and provided technical information for implementing CLT systems in buildings codes and standards.

This book, in its first edition, provides the state of the art of recent developments in CLT design in its various disciplines. The book provides engineers practicing engineers and architects as well as students of these disciplines a comprehensive reference on the planning and design of CLT systems. It also gives the designer the information likely needed for all design phases. The book covers a general introduction to topics considered in design of CLT systems. These include codes and standards used in design of CLT systems; structural behavior, analysis, and design; structural design connections; hygrothermal performance of CLT assemblies; recommendations for design, construction, and maintenance; acoustics; fire safety for CLT projects; environmental aspects of wood as a construction material; and sustainability related to CLT.

The nine chapters of the book have been written by 11 contributors. They have presented their material in a ready-to-use form wherever possible. Therefore, derivations of formulas are omitted in all but a few instances, and many worked-out examples are given. Background information, descriptive matter, and explanatory material have been condensed or omitted. Because each chapter treats a subject that is broad enough to fill a book in itself, the contributors have had to select the material that, in their judgment, is likely to be most useful to the greatest number of users. References and sources of additional material are noted for most of the topics that could not be treated in sufficient detail. The editor is very grateful to the contributors for their tremendous efforts in writing, reviewing, and editing their work, and for their patience during the time it has taken to complete the first edition.

Mustafa Mahamid
University of Illinois at Chicago

Contributors

David Barber Arup

Denis Blount Arup

Jim L. Bowyer Professor Emeritus, Department of Bioproducts and Biosystems Engineering, University of Minnesota

Max Closen Dipl.-Ing (FH); MaSc.

Kathryn Fernholz Dovetail Partners, Inc.

Mustafa Mahamid PhD, SE, PE, P.Eng, F.SEI, F.ASCE, University of Illinois at Chicago

Matt Mahon LSTN Consultants

Lech Muszyński Wood Science and Engineering Department, Oregon State University

Keith Porter B. Eng, Dalhousie University

Mariapaola Riggio Wood Science and Engineering Department, Oregon State University

Borjen Yeh Ph.D., P.E., APA-The Engineered Wood Association

CHAPTER 1

Introduction to Cross-Laminated Timber

Mustafa Mahamid, PhD, SE, PE, P.Eng, F.SEI, F.ASCE
University of Illinois at Chicago

Cross-laminated timber (CLT) is an innovative wood product that was introduced in Austria and Germany in early 1990s and become a well-known engineered timber product of global interest.

CLT is usually composed of an uneven number of layers, usually three, five, or seven, as shown in Figs. 1.1 and 1.2, glued together on their wide face and sometimes on the narrow face as well. Nails or wooden dowels can be used as well to attach layers together, each layer is made of boards placed side by side, and the layers are arranged crosswise to each other at an angle of 90 degrees. Sometimes, and in special cases, two consecutive layers may be placed in the same direction to form a double layer that might be needed to increase the members' strength. CLT allows prefabricating full-size wall and floor elements as well as linear members that can support in- and out-of-plane loads. Among the various advantages of CLT include its lighter weight compared to other construction material, which results in smaller foundations, good thermal insulation, good sound insulation, and good performance when subjected to fire.

CLT applications are for residential and nonresidential structures; additional applications are mats that are used as temporary roads (see Fig. 1.3), temporary bridges, and crane supports during construction. CLT is a relatively new building system in North America and is a new class of timber products that is known as mass timber. CLT provides a competitive alternative to concrete, steel, and masonry in some building design applications. Masonry and concrete are heavy systems that have been used for single-family and multistory residential buildings around the world. Recent CLT projects in Europe and North America show that CLT can be competitive particularly in mid-rise and high-rise buildings.

This book covers structural, architectural, building science, environmental, and sustainability topics related to CLT; the book covers the following topics in detail:

- Codes and standards
- Structural design

FIGURE 1.1 Sample of CLT member—three-layer board arrangement.

FIGURE 1.2 Sample of CLT member—five-layer board arrangement. (*Courtesy of Sterling Lumber.*)

FIGURE 1.3 CLT mats used as temporary roads for construction purposes. (*Courtesy of Sterling Lumber.*)

- Connection design
- Hygrothermal performance of CLT assemblies: recommendations for design, construction, and maintenance
- Acoustics
- Fire
- Environmental aspects of wood as a construction material
- Sustainability

1.1 Codes and Standards

These CLT products can be used for structural and nonstructural applications. Although the production and design of CLT started in North America in 2008, these products have used in Europe for over 20 years. ANSI/APA PRG 320-2011 *Standard for Performance-Rated Cross Laminated Timber* [1] was the first North American CLT standard, and it was completed in December 2011. This standard, which was adopted by the 2015 *International Building Code* (IBC) [2], was subsequently revised and published as ANSI/APA PRG 320-2012 [3] in October 2012 and adopted by the 2015 *International Residential Code* (IRC) [4] in the United States and the 2014 CSA O86, *Engineering Design in Wood,* in Canada [5]. ANSI/APA PRG 320-2017 [6] was completed in October 2017 by the PRG 320 consensus-based canvas committee and approved by ANSI as the latest standard until early 2020. This version of the standard clarified and addressed issues that were emerging in CLT manufacturing in North America; ANSI/APA PRG 320-2017 [6] has been adopted by the 2018 IBC [7] and IRC [8]. The most recent ANSI/APA PRG 320 was published in January 2020 (ANSI/APA PRG 320 2019 [9]).

1.2 Structural Design

The structural design of CLT members represents a new generation of the design of wood structures. The structural use of large flat elements in wood construction was not known until few years ago with exception to thin panels. CLT is a new material for which code specification and regulations are still under development, and new provisions are being included in design codes slowly as research and knowledge evolve. CLT producers and suppliers as well as researchers in this field have done extensive research to come up with analysis and design methods and have laid out clear analysis and design methods to design such members. Despite the fact that CLT design is not fully regulated by design codes and jurisdictions, it has been used in structures through the process of certifying the products and through providing scientific evidence of the product to the local jurisdiction. The structural chapter introduces the available analysis and design methods of CLT members; the chapter is focused on the available methods that structural engineers can use and does not discuss how these methods were developed. For more details on modeling and derivation of these methods, the reader is referred to other publications.

1.3 Connection Design

Structural design consists of specifying the appropriate structural elements as well as joining them together to create structural systems. With the expansion of mass timber, there is a corresponding demand for long, high-strength fasteners that can be site installed with ease, speed, and precision. New lines of self-tapping structural screws have proven especially suitable for use with CLT for these reasons. Self-tapping screws will form the main focus of the connections chapter of this book. Connections must be strong enough to provide continuous load paths from applied gravity, wind, and seismic loads on the structure down to the foundation. The connection chapter provides extensive details on connections types, wood, connectors, and fastener limit states and details.

In addition to the required structural performance of CLT members in buildings, the book covers recommendations for design, construction and maintenance for hygrothermal, acoustics, fire, environmental aspects, and sustainability.

1.4 Hygrothermal Performance of CLT Assemblies: Recommendations for Design, Construction, and Maintenance

The hygrothermal performance of building assemblies is a result of their response to heat, air, and moisture transfer phenomena. The overall building service life, energy efficiency, and comfort and health of occupants require good control of the hygrothermal behavior of each assembly. In Chapter 5, the hygrothermal behavior of CLT assemblies is discussed with the aim of defining general design principles for durability. Practical recommendations for the design, construction, and maintenance of CLT structures are presented with the specific objective of maximizing service life and durability as related to hygrothermal performance. The heat, air, and moisture phenomena and relevant variables can be considered at different scales: at the material level, such as the hygrothermal parameters determined by the natural characteristics and composition of wood; at the product level, such as differences in the manufacturing process and product-specific characteristics of CLT panels; at the component/assembly level, such as the specific location of the assembly within the building and its performance requirements; and at the scale of the building and site, such as type of occupancy, climate zone, and construction type. In Chapter 5, these considerations are addressed in order of increasing scale, from the material and product level to the assembly level and to the scale of the whole building.

Building enclosures play an important role in the hygrothermal performance of a building and represent a buffer from the external environment. Principles and recommendations for the design of CLT enclosure assemblies regarding weather protection, moisture control, airtightness, and thermal control are presented in Chapter 5 with schematic examples. Design principles are presented that account for differential movement between parts of a mass-timber building. These details intend to prevent potential negative effects on structural integrity, enclosure performance, and serviceability.

Additionally, Chapter 5 presents an approach to durability, encompassing design, construction, and maintenance as well as techniques for the inspection and monitoring of CLT components over time.

1.5 Acoustics

The impact that acoustics may have on building occupants and on people in the environment surrounding the building should be considered in the design and construction of buildings. The study and practice of architectural acoustics is broadly intended to address these impacts. For the characterization of the various acoustical impacts, architectural acoustics can be subdivided into basic elements, including room acoustics, sound isolation, footfall/impact noise isolation, mechanical noise and vibration control, and environmental noise control.

Chapter 6 of the book provides an overview of the unique acoustic considerations to be made for the design of projects with CLT mass-timber structure. The chapter will begin with a broad overview of acoustics in buildings to frame the discussion of acoustic concepts. Then the chapter outlines the limited acoustic code requirements in the United States for relevant project types; the chapter also presents a comparison between CLT and other structural systems with regard to acoustic, CLT detailing and lab tests, and other analysis methods.

1.6 Fire

Chapter 7 of the book covers an extensive overview on fire as related to timber and CLT as well as building performance when exposed to fire. The chapter covers the basics of timber reaction to fire, wood structure and chemistry, wood pyrolysis and combustion, wood char, char rates, building stability during fire, expected building performance when exposed to fire, high-rise construction, and CLT fire resistance rating.

1.7 Environmental Aspects of Wood As a Construction Material

Chapter 8 covers the environmental aspects of wood as a construction material in comparison with other construction material, sustainability versus the increased use of wood, and durability of wood over time. These issues are addressed by examining the environmental implications of wood as a construction material with comparison to alternative materials based on a systematic assessment of a range of impact estimators. The assessments include single-family residential structure, multistory apartment building, and mid-rise office building.

The chapter also discusses the current state of North American forests and recent and historical trends in forest cover and growth-harvest relationships. Forest conditions and trends in other world regions are also examined.

1.8 Sustainability

Sustainability is covered in Chapter 9 since CLT is a wood product; it depends upon the availability of forest resources to be produced. All materials, including wood products, have environmental impacts and considerations of sustainability that are relevant when evaluating the use of CLT as well as other building materials. Chapter 9 discusses the concept of sustainability, what it means, and what is considered. The specifics of forest sustainability are also addressed, including historic factors, international conditions, and North American forest resources. The modern methods of forest protection, management, and restoration are described along with current concerns for forest health, carbon storage, sustainable livelihoods, and other benefits associated with sustainability. Finally, this chapter outlines the various ways that CLT and the use of CLT in construction can contribute to sustainability goals for forests, communities, and our built environment.

References

[1] APA—The Engineered Wood Association. Standard for Performance-Rated Cross Laminated Timber, ANSI/APA PRG 320. Tacoma, Washington, U.S.A. 2011.

[2] International Code Council. International Building Code. Country Club Hills, Illinois, U.S.A. 2015.

[3] APA—The Engineered Wood Association. Standard for Performance-Rated Cross Laminated Timber, ANSI/APA PRG 320. Tacoma, Washington, U.S.A. 2012.
[4] International Code Council. International Residential Code. Country Club Hills, Illinois, U.S.A. 2015.
[5] Engineering Design in Wood CSA O86, Canadian Standards Association, 2014.
[6] APA—The Engineered Wood Association. Standard for Performance-Rated Cross Laminated Timber, ANSI/APA PRG 320. Tacoma, Washington, U.S.A. 2017.
[7] International Code Council. International Building Code. Country Club Hills, Illinois, U.S.A. 2018.
[8] International Code Council. International Residential Code. Country Club Hills, Illinois, U.S.A. 2018.
[9] APA—The Engineered Wood Association. Standard for Performance-Rated Cross Laminated Timber, ANSI/APA PRG 320. Tacoma, Washington, U.S.A. 2020.

CHAPTER 2

Product Standard for Cross-Laminated Timber

Borjen ("B.J.") Yeh, Ph.D., P.E.
APA—The Engineered Wood Association

Introduction

Cross-laminated timber (CLT), as shown in Figs. 2.1 and 2.2, is a prefabricated engineered wood product made of at least three orthogonal layers of graded sawn lumber or structural composite lumber (SCL) that are laminated by gluing with structural adhesives to form a solid rectangular-shaped, straight, and plane timber intended for structural (roof, floor, or wall) applications. These CLT products can be used for structural and nonstructural applications. For the purpose of this chapter, these CLT products are intended for structural applications, such as those used in building construction, and are different from those used in nonstructural applications, such as the truck mats used in oil fields.

While these engineered wood products have been used in Europe for over 20 years, the production of structural CLT and design of CLT structural systems started in North America around 2008. Today, there are four major manufacturers of structural CLT in North America. They are DR Johnson Wood Innovations in Riddle, Oregon; Nordic Structures in Chibougamau, Quebec; Smartlam in Columbia Fall, Montana; and Structurlam Products in Penticton, British Columbia, as shown in Fig. 2.3. There are at least three additional CLT manufacturers that are expected to join the production in North America in 2018. Check with www.apawood.org/manufacturer-directory for the latest directory for CLT manufacturers certified by APA.

For the acceptance of new construction materials or systems in North America, such as CLT, a consensus-based product standard is essential to the manufacturers, designers, and regulatory bodies. In recognition of this need, APA—The Engineered Wood Association in the United States and FPInnovations in Canada initiated a joint standard development process in 2010. The intent was to develop a binational CLT standard for North America using the consensus standard development process of APA as a standards developer accredited by the ANSI. After months of intensive committee meetings and balloting, the

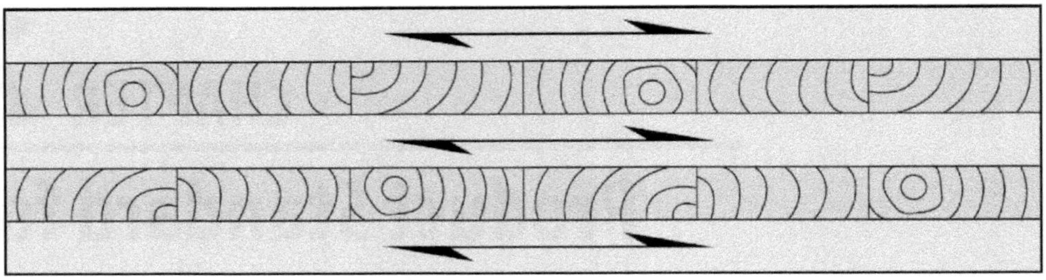

FIGURE 2.1 Cross section of a five-layer CLT Panel (arrows indicate the strength direction).

first North American CLT standard was completed as the ANSI/APA PRG 320-2011 *Standard for Performance-Rated Cross Laminated Timber* [1] in December 2011. This standard, which was adopted by the 2015 *International Building Code* (IBC) was subsequently revised and published as ANSI/APA PRG 320-2012 [2] in October 2012 and

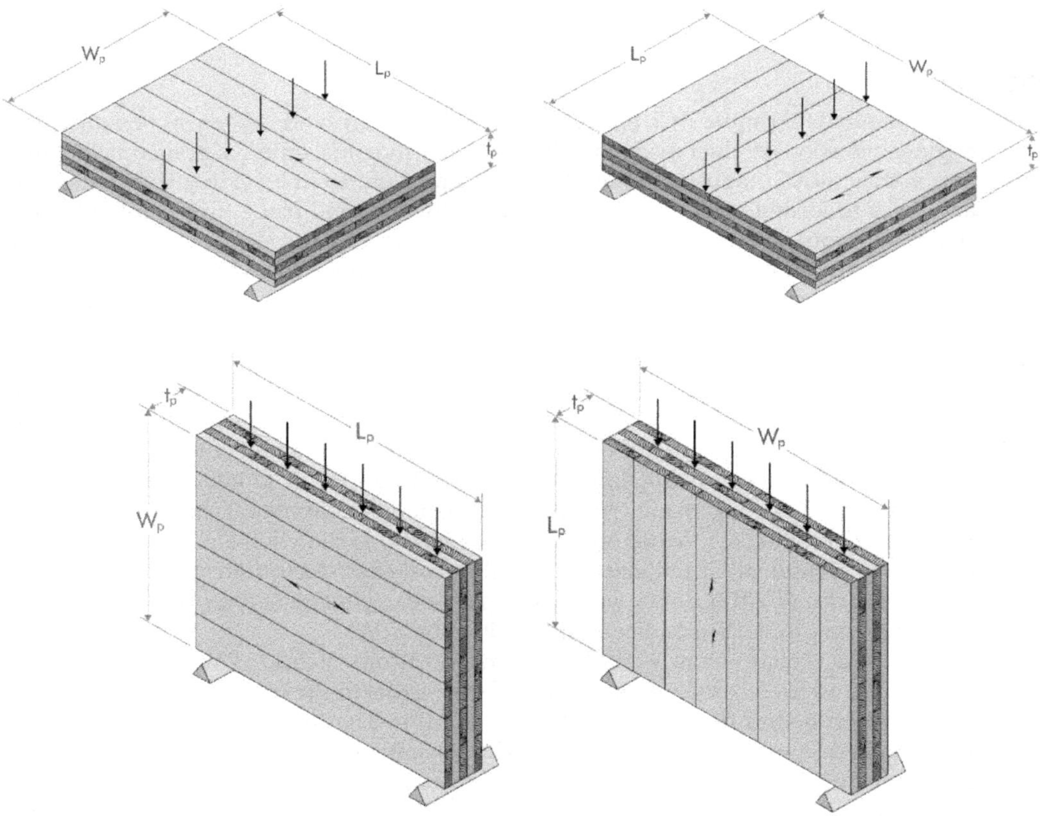

FIGURE 2.2 CLT orientations (top left: flatwise bending in the major strength direction; top right: flatwise bending in the minor strength direction; bottom left: edgewise bending in the major strength direction; bottom right: edgewise bending in the minor strength direction).

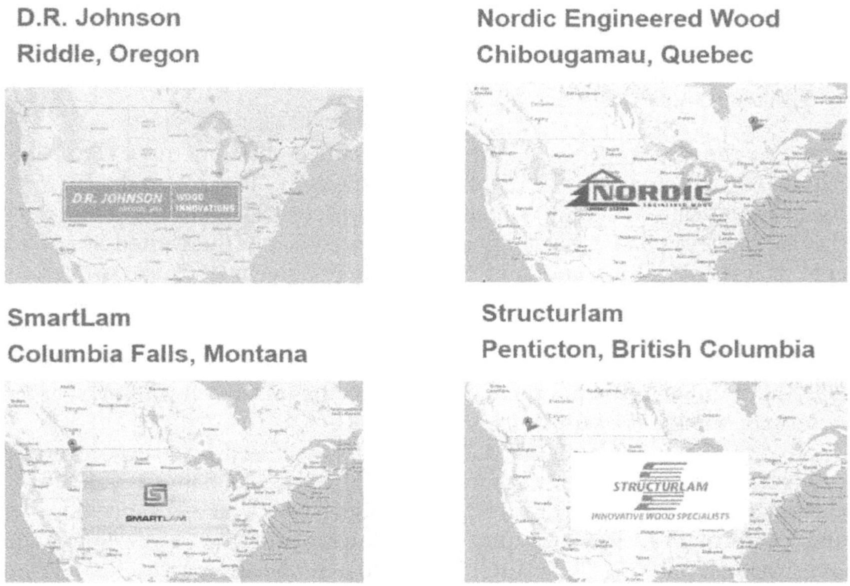

FIGURE 2.3 Current major structural CLT manufacturers in North America.

adopted by the 2015 *International Residential Code* (IRC) in the United States and the 2014 CSA O86, *Engineering Design in Wood*, in Canada.

In October 2017, a new ANSI/APA PRG 320-2017 was completed by the PRG 320 consensus-based canvas committee and approved by ANSI as the latest standard. ANSI/APA PRG 320-2017 [3], which have been adopted by the 2018 IBC and IRC, clarified and addressed issues that were emerging in CLT manufacturing in North America. This chapter provides detailed information for the background and key considerations based on ANSI/APA PRG 320-2017. The provisions directed quoted from ANSI/APA PRG 320-2017 are shown in *Italic* in the text below. Note that the referenced sections in the directed quoted provisions are those from ANSI/APA PRG 320-2017.

2.1 Scope of ANSI/APA PRG 320

The scope of ANSI/APA PRG 320, as stated in ANSI/APA PRG 320-2017, is as follows:

1. SCOPE
Cross-laminated timber (CLT) panels referenced in this standard are defined in 3.2 and shall be qualified and trademarked in accordance with this standard. This standard provides dimensions and tolerances, performance requirements, test methods, quality assurance, and trademarking for CLT panels.

CLT panels shall be used in dry service conditions, such as in most covered structures, where the average equilibrium moisture content of solid wood is less than 16 percent in the U.S., and is 15 percent or less over a year and does not exceed 19 percent in Canada. CLT panels qualified in accordance with the provisions of this standard are intended to resist the effects of moisture on structural performance as may occur due to construction delays or other conditions of similar severity. Products carrying a trademark of this standard shall be

> *used in accordance with the installation requirements prescribed in the recommendations provided by the CLT manufacturer, an approved agency, and/or its trade association. Finger joining, edge gluing, and face gluing between CLT panels, and camber of CLT panels are beyond the scope of this standard.*

Based on the stated scope, CLT products qualified and trademarked to ANSI/APA PRG 320 is limited to dry service conditions, such as in most covered structures where the mean equilibrium moisture content (EMC) of solid-sawn lumber is less than 16 percent (i.e., 65 percent relative humidity and 68°F or 20°C) in the United States and is 15 percent or less over a year and does not exceed 19 percent in Canada. Therefore, the CLT products manufactured to ANSI/APA PRG 320 may not be suitable for exterior applications where the products are exposed to the elements. Also, it should be noted that nail-laminated timber (NLT), dowel-laminated timber (DLT), finger-jointed or scarf-jointed CLT (in the billet form), or other CLT products manufactured without structural adhesive bonds are outside the scope of ANSI/APA PRG 320.

It is important to note that CLT products evaluated by a recognized inspection or product certification agency as meeting ANSI/APA PRG 320 is required by the U.S. and Canadian building codes to provide the designers with an assurance for product quality and performance. For example, the following are the requirements in the 2018 International Building Code (IBC) [4], the 2018 International Residential Code (IRC) [5], and the 2014 CSA O86, Engineering Design in Wood [6], which is referenced by the 2015 National Building Code (NBC) of Canada [7]. Therefore, CLT products that are not certified as in conformance to ANSI/APA PRG 320 are not in compliance with the IBC, IRC, and NBC, unless specifically approved by the engineer of record and the authority having jurisdiction (building regulators).

> ***(2018 IBC) 2303.1.4 Structural glued cross-laminated timber.***
> *Cross-laminated timbers shall be manufactured and identified in accordance with ANSI/APA PRG 320.*
>
> ***(2018 IRC) R502.1.6 Cross-laminated timber.***
> *Cross-laminated timber shall be manufactured and identified as required by ANSI/APA PRG 320.*
>
> ***(2014 CSA O86) 8.1 Scope***
> *The design values and methods given in Clause 8 apply only to panels of primary and custom CLT stress grades manufactured and certified in accordance with ANSI/APA PRG 320 and layups as defined in Clause 8.2. Panels with alternative CLT layups shall be designed in accordance with Clause 4.3.2.*

It is very important for the designer to understand that the acceptance of CLT products that have not demonstrated conformance to ANSI/APA PRG 320 is not as simple as a conversion of design properties published by the CLT suppliers, especially for those products imported from outside of North America. The CLT design standards in North America were developed based on an array of performance expectations stipulated in ANSI/APA PRG 320, such as heat durability, moisture durability, and fire performance, in addition to the compatibility of the design value derivation in North America. Accepting CLT products without demonstrated conformance to ANSI/APA PRG 320, as required by the U.S. and Canadian building codes, carries the responsibility of structural engineering and fire safety design, as well as the acceptance of product quality and durability in structural and fire performance.

2.2 Components for CLT

The major components for CLT include laminations, adhesives, and lamination joints (end joints, face joints, and edge joints if used). To manufacture a structural CLT, the quality of these components must be qualified and then quality controlled on an ongoing basis under an in-plant quality management system, which should cover the manufacturing processes and under a third-party independent product certification program.

2.2.1 Laminations

CLT is manufactured with laminations of dimension lumber or SCL, such as laminated veneer lumber (LVL), laminated strand lumber (LSL), or oriented strand lumber (OSL), which are bonded with structural adhesives through face joints, end joints, and/or edge joints. The requirements for lumber laminations in ANSI/APA PRG 320 are as follows:

6.1 Laminations—Lumber

6.1.1 Lumber species
Any softwood lumber species or species combinations recognized by American Lumber Standards Committee (ALSC) under PS 20 or Canadian Lumber Standards Accreditation Board (CLSAB) under CSA O141 with a minimum published specific gravity of 0.35, as published in the National Design Specification for Wood Construction (NDS) in the U.S. and CSA O86 in Canada, shall be permitted for use in CLT manufacturing provided that other requirements specified in this section are satisfied. The same lumber species or species combination shall be used within a single layer of CLT. Adjacent layers of CLT shall be permitted to be made of different species or species combinations.

6.1.2 Lumber grades
The minimum grade of lumber in the parallel layers of CLT shall be 1200f-1.2E MSR or visual grade No. 2. The minimum grade of lumber in the perpendicular layers of CLT shall be visual grade No. 3. Remanufactured lumber shall be considered as equivalent to solid-sawn lumber when qualified in accordance with Section 4.3.4 of ANSI A190.1 in the U.S. or SPS 1, 2, 4, or 6 in Canada. Proprietary lumber grades meeting or exceeding the mechanical properties of the lumber grades specified above shall be permitted for use provided that they are qualified in accordance with the requirements of an approved agency.

6.1.3 Lamination sizes

a. *Major Strength Direction—The net width of a lamination shall not be less than 1.75 times the lamination thickness for the parallel layers.*
b. *Minor Strength Direction—If the laminations in the perpendicular (cross) layers are not edge bonded, the net width of a lamination shall not be less than 3.5 times the lamination thickness for the perpendicular (cross) layers unless the interlaminar shear strength and creep are evaluated by testing in accordance with Section 8.5.5 and the principles of ASTM D6815, respectively.*
c. *Both Directions—The net thickness of a lamination for all layers at the time of gluing shall not be less than 5/8 inch (16 mm) or more than 2 inches (51 mm). In addition, the lamination thickness shall not vary within the same CLT layer.*

6.1.4 Moisture content
The moisture content of the lumber at the time of CLT manufacturing shall be 12 ± 3%. The moisture content of the SCL at the time of CLT manufacturing shall be 8 ± 3%.

6.2 Laminations—Structural Composite Lumber
SCL products meeting the requirements of ASTM D5456 and the equivalent specific gravity specified in 6.1.1 shall be permitted for use. SCL laminations must also meet the requirements of 6.1.3 through 6.1.6.

ANSI/APA PRG 320 utilizes the European experience in engineering theories and manufacturing processes of CLT and takes into consideration the characteristics of the North American lumber resource, manufacturing preference, and end-use expectations. For example, the standard permits the use of any softwood lumber species or species combinations recognized by the American Lumber Standards Committee (ALSC) under PS 20 [8] or the Canadian Lumber Standards Accreditation Board (CLSAB) under CSA O141 [9] with a minimum specific gravity (SG) of 0.35, as published in the National Design Specification for Wood Construction (NDS) [10] in the United States or the Engineering Design in Wood (CSA O86) [6] in Canada. One advantage of using standard-grade lumber is that such lumber will typically be marked as "HT" (heat treated), meaning that the resulting CLT product will also meet national and international phyto-sanitary requirements when the traceability (chain-of-custody) requirements of the lumber laminations can be properly demonstrated and certified. Note that CLT products made of hardwood lumber species are not part of ANSI/APA PRG 320 due to the lack of manufacturing experience and insignificance in commercial production volume for structural applications in North America today.

The minimum SG of 0.35 is intended as the lower bound for the CLT connection design since it is the near minimum value of commercially available wood species in North America, western woods in the United States, and northern species in Canada. To avoid differential mechanical and physical properties of lumber, the standard requires that the same lumber species or species combination be used within the same layer of the CLT while permitting adjacent layers of the CLT to be made of different species or species combinations. The standard also permits the use of SCL when qualified in accordance with ASTM D5456 [11]. In reality, however, it may be still years away before SCL would be used in CLT production because of apparent challenges in the face bonding of SCL to SCL or SCL to lumber. Due to the thickness variation and surface oxidation or inactivation of SCL, surface planing or sanding may be required for SCL before gluing. Another consideration is its cost competitiveness with lumber. Nonetheless, the advantage of SCL that can be produced in a long and wide billet form is one important reason that the ANSI/APA PRG 320 Committee elected to include SCL in the standard. Other attractive factors also include free of natural defects, such as wane, shake, and knots; more uniform stiffness and strength; and greater dimensional stability. The ANSI/APA PRG 320 Committee is working on more provisions that will be added to a future version of ANSI/APA PRG 320 to guide the use of SCL in CLT production.

Lumber grades in the parallel and perpendicular layers of CLT are required to be at least 1200f-1.2E MSR or visually graded No. 2 and visually graded No. 3, respectively. Remanufactured lumber is permitted as equivalent to solid-sawn lumber when qualified in accordance with ANSI A190.1 [12] in the United States or SPS 1, 2, 4, or 6 [13–16] in Canada. Proprietary lumber grades meeting or exceeding the mechanical properties of the lumber grades specified above are permitted provided that they are qualified in accordance with the requirements of an approved agency, which is defined in the standard as an independent inspection agency accredited under ISO/IEC 17020 [17] or an independent testing agency accredited under ISO/IEC 17025 [18] in the United States or a certification agency accredited under ISO/IEC 17065 [19] in Canada. This allows for a great flexibility in the utilization of forest resources in North America.

The net lamination thickness for all CLT layers at the time of gluing is required to be at least 5/8 in. (16 mm) but not thicker than 2 in. (51 mm) to facilitate face bonding. In addition, the lamination thickness is not permitted to vary within the same CLT layer except when it is within the lamination thickness tolerances—at the time of face bonding, variations in thickness across the width of a lamination is limited to ±0.008 in. (0.2 mm) or less, and the variation in thickness along the length of a lamination is limited to ±0.012 in. (0.3 mm). These maximum tolerances may need to be adjusted during qualification so as to produce acceptable face bond performance.

The net lamination width is required to be at least 1.75 times the lamination thickness for the parallel layers in the major strength direction of the CLT. This means that if 2× lumber (1-3/8 in. or 35 mm in net thickness after surfacing prior to gluing) is used in the parallel layers, the minimum net lamination width must be at least 2.4 in. (61 mm), i.e., 2 × 3 lumber. On the other hand, the net lamination width is required to be at least 3.5 times the lamination thickness for the perpendicular layers if the laminations in the perpendicular (cross) layers are not edge-bonded, unless the interlaminar shear strength and creep of the CLT are evaluated by testing. This means that if 2× lumber is used in the perpendicular layers, the net lamination width must be at least 4.8 in. (122 mm), i.e., 2 × 6 lumber.

This minimum lamination width in the perpendicular layers could become a problem for CLT manufacturers who may prefer to use 2 × 3 (net 1½ in. × 2½ in. or 38 mm × 63 mm) or 2 × 4 (net 1½ in. × 3½ in. or 38 mm × 89 mm) lumber. However, the Committee was concerned about the unbonded edge joints, which could leave gaps as potential stress risers. These, in turn, may reduce the effective interlaminar shear strength and stiffness and may result in excessive creep. Therefore, in this case, the manufacturers will have to either edge-glue the laminations or demonstrate the conformance to the standard by conducting interlaminar shear tests and ASTM D6815 [20] creep tests. It should be noted that this is an interim measure due to the lack of data at this point in time to address the concerns. As a result, it is expected that this provision may be revisited as more information becomes available.

The selection of lumber laminations represent a key step in CLT manufacturing, and the lumber grade should be selected in accordance to the CLT layup of the CLT panel. In addition, for a CLT appearance classification (discussed below), the outermost layer(s) may have specific visual characteristics for aesthetic purposes.

Most adhesives require that surfaces be planed prior to adhesive application and pressing to ensure a strong and durable gluebond. ANSI/APA PRG 320 recommends the following:

Note 5. It may be necessary to plane the lamination surface within 48 hours of face bonding for some wood species.

When the graded lumber is replaned just prior to bonding, depending on the amount of wood removed, this may alter the grade of the lumber, so a grade verification may be necessary. The use of rough-sawn lumber may seem to result in some saving because the lumber is required to be only planed once, and a lumber grading for visual or E-rating after planing may increase the net cost.

It should be noted that the packages of kiln-dried lumber are usually solid-stacked and dried to a moisture content (MC) of 19 percent or less at the time of surfacing, which may not be suitable for all CLT manufacturing processes. For example, some adhesives are sensitive to MC variations. ANSI/APA PRG 320 recommends that lumber having a MC of 12% ± 3% for CLT manufacturing to ensure proper bond quality of the product. If SCL is used, the target MC should be 8% ± 3% at the time of CLT manufacturing. It is recommended that the maximum difference in MC between adjacent pieces that are to be joined in CLT not exceed 5 percentage points. A handheld or online MC meter can be used to check the lumber MC.

ANSI/APA PRG 320 does not specifically address the wood temperature for CLT manufacturing with the expectation that this will be self-regulated by the adhesive manufacturer's specification. In general, wood temperature will affect the gluebond quality, and the adhesive manufacturer's recommendations should be followed. The ambient temperature in the manufacturing facility may also have an effect on some process parameters, such as the open assembly time and adhesive curing time. Therefore, it is recommended that the ambient temperature for the CLT manufacturing be at least 60°F (15°C).

In addition to the lumber MC and temperature, there are other lumber characteristics that may affect the quality of the adhesive bond. These either impact on the pressure that is effectively applied to the gluebond or simply reduce the available bonding surface. Lumber warp in the form of bow, crook, cup, and twist are examples of the former. Wane is a common example of the latter. Standard grades of framing lumber permit these characteristics to varying degrees. While these limits are acceptable for wood frame construction, some of these characteristics need to be restricted when manufacturing CLT in order to ensure formation of a good gluebond.

It is important that the impact of these characteristics, if permitted, be taken into account in the product manufacturing and expected gluebond performance. In ANSI/APA PRG 320, for example, this is addressed by grading to achieve an effective bond area of a minimum of 80 percent, as shown below:

> ***8.3 Qualification of Effective Bond Area***
>
> ***8.3.1 General***
>
> *The manufacturer shall establish visual grading rules for the bonded faces and limit the average glue skip to maintain an average effective bond area of 80% or more.*
>
> *The manufacturer's visual grading rules established to achieve the effective bond area shall include major visual characteristics based on characteristic measurements consistent with standard lumber grading rules.*

For example, wane will reduce the bonding area and concentrate the stresses in a CLT panel. However, wane cannot be ignored because it is a permitted characteristic in all lumber visual grades. The effect of wane can be accommodated by removing pieces with excessive amounts of wane and/or rearranging or reorienting pieces with wane.

2.2.2 Adhesives

Another critical component for CLT is the adhesives. The requirements for CLT adhesives in ANSI/APA PRG 320 are as follows:

> ***6.3 Adhesives***
>
> a. *In the U.S., adhesives used for CLT manufacturing shall meet the requirements of ANSI 405 with the exception that Section 2.1.6 of ANSI 405 (either ASTM D3434 or CSA O112.9) is not required. In addition, adhesives shall be evaluated for heat performance in accordance with Section 6.1.3.4 of DOC PS1.*
> b. *In Canada, adhesives shall meet the requirements of CSA O112.10, and Sections 2.1.3 and 3.3 (ASTM D7247 heat durability) of ANSI 405. In addition, adhesives shall be evaluated for heat performance in accordance with Section 6.1.3.4 of DOC PS1.*
> c. *For use in both the U.S. and Canada, adhesives shall meet both a and b in this section.*

The standard requires that the adhesives used for CLT manufacturing meet the requirements of ANSI 405 [21] with the exception that the extreme gluebond durability tests in ANSI 405 (either ASTM D3434 [22] or CSA O112.9 [23]), which are designed for adhesive qualification in exterior applications, is not required because CLT products manufactured to ANSI/APA PRG 320 are limited to dry service conditions, such as in most covered structures where the mean EMC of solid-sawn lumber is less than 16 percent (i.e., 65 percent relative humidity and 68°F or 20°C) in the United States and is 15 percent or less over a year and does not exceed 19 percent in Canada. Note that ANSI 405 includes ASTM D7247 heat durability tests [24]. CLT products qualified in accordance with the standard are intended to resist the effects of moisture on structural performance, as it may occur due to construction delays or other conditions of similar severity.

In Canada, CLT adhesives have to meet the requirements of CSA O112.10 [25] and ASTM D7247 heat durability, which are part of the requirements in ANSI 405. In addition, in both countries, CLT adhesives have to be evaluated for heat performance in accordance with PS1 [26]. The intent of the heat performance evaluation is to determine if an adhesive will exhibit heat delamination characteristics, which may increase the char rate of the CLT when exposed to fire in certain applications. If heat delamination occurs, the CLT manufacturer is expected to consult with the adhesive manufacturer and the approved agency to develop appropriate strategies in product manufacturing and/or end-use recommendations for the CLT fire design [27].

It is important to note that ANSI/APA PRG 320 does not currently have pass/fail criteria on adhesive heat delamination. This is because the standard CLT char rate, as stipulated in Chapter 16 of the NDS, has assumed the occurrence of adhesive heat delamination under fire exposure. However, in recent full-scale compartment fire tests, it was discovered that a certain type of CLT adhesive that causes adhesive heat delamination can result in a fire regrowth and second flashover, which is a concern by the fire service for mid- to high-rise tall wood buildings that may have a delayed firefighting if the automatic sprinklers are also malfunctioning or manually deactivated for some reason. Therefore, the ANSI/APA PRG 320 Committee is actively working with the International Code Council (ICC) Ad Hoc Committee on Tall Wood Buildings to revise the adhesive qualification requirements in ANSI/APA PRG 320 to prohibit adhesives that exhibit the heat delamination and fire regrowth behavior. In the meantime, the ANSI 405 Committee has also just approved the addition of CSA O177 [28] small-scale flame test to ANSI 405-2018 glulam adhesive standard. Since ANSI 405 is directly referenced in ANSI/APA PRG 320, the CSA O177 small-scale flame test will become a new requirement for all CLT adhesives when ANSI/APA PRG 320 adopts ANSI 405-2018.

Several types of structural adhesives have been successfully used in CLT production, as listed below:

- Phenolic types, such as phenol-resorcinol formaldehyde (PRF)
- Melamine types, such as melamine formaldehyde (MEL)
- Emulsion polymer isocyanate (EPI)
- One-component polyurethane (PUR)

PRF and MEL are well-known adhesives for structural use and commonly used for glulam manufacturing in North America. EPI adhesive is used for wood I-joist and lamination. PUR adhesive has been commonly used in Europe to produce CLT. It should be noted that not all formulations within an adhesive type will meet the requirements of the structural adhesive standard and that there may be considerable variation in working properties within each adhesive type. Documentation showing that the adhesive has met the appropriate standards is required for CLT product certification. In addition, the working properties of the adhesive needed by the manufacturing process should be considered and discussed with the adhesive supplier.

In addition to cost and working properties, each adhesive type may possess other attributes that may be important. For example, among the four adhesive types indicated above, PRF is dark brown, whereas MEL, EPI, and PUR are light colored. PUR is manufactured without the addition of solvents or formaldehyde and is moisture reactive. EPI is also free from formaldehyde. Due to the chemical reaction, PUR normally produces slight foaming during hardening.

2.2.3 Lamination Joints

Lamination joints include end joints, face joints, and edge joints. The requirements for lamination joints in ANSI/APA PRG 320 are as follows:

> ***6.4 Lamination Joints***
>
> ***6.4.1 General***
> *The lamination joints of CLT shall meet the requirements specified in this section.*
>
> ***6.4.2 End joints in laminations***
> *The strength, wood failure, and durability of lamination end joints shall be qualified in accordance with Section 12.1.3 of ANSI A190.1 and meet the requirements specified therein in the U.S., or shall be qualified in accordance with Section 9.5 of CSA O177 and meet the requirements specified therein in Canada.*
>
> ***6.4.3 Edge and face joints in laminations***
> *The wood failure and durability of the face and edge (when required for structural performance) joints shall be qualified in accordance with Section 12.1.2 of ANSI A190.1 and meet all requirements, except for the shear strength, specified in Sections 12.1.2(b) of that standard in the U.S., or shall be qualified in accordance with Sections 9.2 and 9.3 of CSA O177 and meet all requirements, except for the shear strength, specified therein in Canada.*

Adhesive-bonded edge joints between laminations in the same layer of CLT are not required in accordance with ANSI/APA PRG 320 unless CLT's structural and/or fire performance is qualified based on the use of adhesive-bonded edge joints. As previously mentioned, laminations with unbonded edge joints in the perpendicular layers are subject to the minimum width limitation of 3.5 times the lamination thickness. On the other hand, the end joints within the same lamination, as applicable (e.g., SCL layers may be provided in full width and full length), and the face joints between adjacent laminations must be qualified in accordance with the glulam standard, ANSI A190.1 in the United States and CSA O177 in Canada, with the exception that the interlaminar shear strength criteria do not apply due to the lower interlaminar shear strength when adjacent laminations are perpendicular. However, these provisions will be reviewed when more plant data are gathered and analyzed in the immediate future.

It should be also noted that the gap in the unbonded edge joint of the CLT is not specified in the current ANSI/APA PRG 320 even though the intent is to have the edge joint as tight as possible. In practicality, however, it is very difficult to have a completely tight unbonded edge joints. Therefore, the ANSI/APA PRG 320 Committee is currently working on a reasonable specification to ensure that the unbonded edge joints are reasonable close to each other. This specification is likely to be added to the future version of ANSI/APA PRG 320.

2.3 CLT Requirements

2.3.1 Dimensions and Dimensional Tolerances

Dimensions and dimensional tolerances of CLT, as specified in ANSI/APA PRG 320, are as follows:

> *5. PANEL DIMENSIONS AND DIMENSIONAL TOLERANCES*
>
> ***5.1 Thickness***
> *The thickness of CLT shall not exceed 20 inches (508 mm).*
>
> ***5.2 CLT Dimensional Tolerances***
> *Dimension tolerances permitted at the time of manufacturing shall be as follows:*
>
> *Thickness: ± 1/16 inch (1.6 mm) or 2% of the CLT thickness, whichever is greater*
> *Width: ± ⅛ inch (3.2 mm) of the CLT width*
> *Length: ± ¼ inch (6.4 mm) of the CLT length*
>
> *Textured or other face or edge finishes are permitted to alter the tolerances specified in this section. The designer shall compensate for any loss in cross-section and/or specified strength of such alterations.*
>
> ***5.3 Squareness***
> *Unless specified otherwise, the length of the two panel face diagonals measured between panel corners shall not differ by more than ⅛ inch (3.2 mm).*
>
> ***5.4 Straightness***
> *Unless specified otherwise, deviation of edges from a straight line between adjacent panel corners shall not exceed 1/16 inch (1.6 mm).*

The thickness of CLT is limited to 20 in. (508 mm) or less in ANSI/APA PRG 320. This is considered an upper limit that the CLT may be handled in production and transportation. In addition, dimension tolerances permitted in ANSI/APA PRG 320 are based on the measurements at the time of manufacturing. Textured or other face or edge finishes are permitted to alter the tolerances. However, the designers need to compensate for any loss in cross section and/or the specified strength due to such alterations.

The standard also specifies the CLT panel squareness, defined as the length of the two panel face diagonals measured between panel corners, to be within ⅛ in. (3.2 mm) or less. In addition, the CLT panel straightness, defined as the deviation of edges from a straight line between adjacent panel corners, is required to not exceed 1/16 in. (1.6 mm).

2.3.2 CLT Layups

As part of the standardization effort, seven CLT layups are stipulated in ANSI/APA PRG 320, while custom CLT products are also recognized, provided that the products are qualified by an approved agency in accordance with the qualification and mechanical test requirements specified in the standard. The CLT layups are presented in the form of structural capacities, such as bending strength (F_bS), bending stiffness (EI), interlaminar shear strength (V_s), and shear rigidity (GA), in both major and minor strength directions. This allows for the needed flexibility to CLT manufacturers for conformance to the product standard based on the available material resources and required design capacities.

The CLT layups were developed based on the following prescriptive lumber species and grades available in North America:

- E1: 1950f-1.7E Spruce-Pine-Fir MSR lumber in all longitudinal layers and No. 3 Spruce-Pine-Fir lumber in all transverse layers
- E2: 1650f-1.5E Douglas fir-Larch MSR lumber in all longitudinal layers and No. 3 Douglas fir-Larch lumber in all transverse layers
- E3: 1200f-1.2E Eastern Softwoods, Northern Species, or Western Woods MSR lumber in all longitudinal layers and No. 3 Eastern Softwoods, Northern Species, or Western Woods lumber in all transverse layers
- E4: 1950f-1.7E Southern Pine MSR lumber in all longitudinal layers and No. 3 Southern Pine lumber in all transverse layers
- V1: No. 2 Douglas fir-Larch lumber in all longitudinal layers and No. 3 Douglas fir-Larch lumber in all transverse layers
- V2: No. 1/No. 2 Spruce-Pine-Fir lumber in all longitudinal layers and No. 3 Spruce-Pine-Fir lumber in all transverse layers
- V3: No. 2 Southern Pine lumber in all longitudinal layers and No. 3 Southern Pine lumber in all transverse layers

The required characteristic strengths and moduli of elasticity for the laminations used to manufacture each layup of CLT are listed in Table 2.1. The corresponding allowable stress design (ASD) values for the United States and limit states design (LSD) design values for Canada are provided in Tables 2.2 and 2.3 (i.e., Tables A1 and A4 in ANSI/APA PRG 320), respectively. As seen from the list above, both mechanically graded lumber (for "E" classes) and visually graded lumber (for "V" classes) are included in this standard. Also included are three major species groups in North America: Douglas fir-Larch, Spruce-Pine-Fir, and Southern Pine (note that Layups E4 and V3 using Southern Pine lumber are not available in Canada). With the published lumber properties for each layup, the design capacities of the CLT were derived based on the "shear analogy" method developed in Europe [29] and the following assumptions:

- The modulus of elasticity of lumber in the perpendicular to grain direction, E_{90}, is 1/30 of the modulus of elasticity of lumber in the parallel to grain direction, E_0.
- The modulus of shear rigidity of lumber in the parallel to grain direction, G_0, is 1/16 of the modulus of elasticity of lumber in the parallel to grain direction, E_0.
- The modulus of shear rigidity of lumber in the perpendicular to grain direction, G_{90}, is 1/10 of the modulus of shear rigidity of lumber in the parallel to grain direction, G_0.

The design capacities for CLT are provided in the format of ASD for the United States, as shown in Table 2.4 (i.e., Table A2 of ANSI/APA PRG 320), and LSD for Canada, as shown in Table 2.5 (i.e., Table A4 in ANSI/APA PRG 320). The allowable bending and shear strengths can be readily converted to the characteristic bending and shear strengths (5th percentile with 75 percent confidence), respectively, by multiplying by an adjustment

CLT layup	Laminations used in major strength direction						Laminations used in minor strength direction					
	f_b (psi)	E (10^6 psi)	f_t (psi)	f_c (psi)	f_v (psi)	f_s (psi)	f_b (psi)	E (10^6 psi)	f_t (psi)	f_c (psi)	f_v (psi)	f_s (psi)
E1	4095	1.7	2885	3420	425	140	1050	1.2	525	1235	425	140
E2	3465	1.5	2140	3230	565	185	1100	1.4	680	1470	565	185
E3	2520	1.2	1260	2660	345	115	735	0.9	315	900	345	115
E4	4095	1.7	2885	3420	550	180	945	1.3	525	1375	550	180
V1	1890	1.6	1205	2565	565	185	1100	1.4	680	1470	565	185
V2	1835	1.4	945	2185	425	140	1050	1.2	525	1235	425	140
V3	1575	1.4	945	2375	550	180	945	1.3	525	1375	550	180

For SI: 1 psi = 0.006895 MPa.

[a]See Section 4 of ANSI/APA PRG 320-2017 for symbols.

[b]Tabulated values are test values and shall not be used for design. See Table 2.2 for design properties.

[c]Custom CLT layups that are not listed in this table shall be permitted in accordance with Section 7.2.1 of ANSI/APA PRG 320-2017.

[d]The characteristic values shall be determined as follows from the published reference design value unless otherwise justified by the approved agency:

$f_b = 2.1 \times$ published ASD reference bending stress (F_b),

$f_t = 2.1 \times$ published ASD reference tensile stress (F_t),

$f_c = 1.9 \times$ published ASD reference compressive stress parallel to grain (F_c),

$f_v = 3.15 \times$ published ASD reference shear stress (F_v).

TABLE 2.1 Required Characteristic Test Values[a–d] for Laminations Used in PRG 320 CLT

CLT layup	Laminations used in major strength direction						Laminations used in minor strength direction					
	F_b (psi)	E[d] (10^6 psi)	F_t (psi)	F_c (psi)	F_v (psi)	F_s (psi)	F_b (psi)	E[d] (10^6 psi)	F_t (psi)	F_c (psi)	F_v (psi)	F_s (psi)
E1	1950	1.7	1375	1800	135	45	500	1.2	250	650	135	45
E2	1650	1.5	1020	1700	180	60	525	1.4	325	775	180	60
E3	1200	1.2	600	1400	110	35	350	0.9	150	475	110	35
E4	1950	1.7	1375	1800	175	55	450	1.3	250	725	175	55
V1	900	1.6	575	1350	180	60	525	1.4	325	775	180	60
V2	875	1.4	450	1150	135	45	500	1.2	250	650	135	45
V3	750	1.4	450	1250	175	55	450	1.3	250	725	175	55

For SI: 1 psi = 0.006895 MPa.

[a]See Section 4 of ANSI/APA PRG 320-2017 for symbols.

[b]Tabulated values are ASD reference design values and not permitted to be increased for the lumber size and flat use adjustment factors in accordance with the NDS. The design values shall be used in conjunction with the section properties provided by the CLT manufacturer based on the actual layup used in manufacturing the CLT panel (see Table 2.4).

[c]Custom CLT layups that are not listed in this table shall be permitted in accordance with Section 7.2.1 of ANSI/APA PRG 320-2017.

[d]The tabulated E values are published E for lumber. For calculating the CLT design properties shown in Table 2.4, the transverse E of the lamination is assumed to be E/30, the longitudinal G of the lamination is assumed to be E/16, and the transverse G of the lamination is assumed to be longitudinal G/10.

TABLE 2.2 ASD Reference Design Values[a–c] for Laminations Used in PRG 320 CLT (for use in the United States)

CLT layup	Laminations used in major strength direction						Laminations used in minor strength direction					
	f_b (MPa)	$E^{(d)}$ (MPa)	f_t (MPa)	f_c (MPa)	f_v (MPa)	f_s (MPa)	f_b (MPa)	$E^{(d)}$ (MPa)	f_t (MPa)	f_c (MPa)	f_v (MPa)	f_s (MPa)
E1	28.2	11,700	15.4	19.3	1.5	0.50	7.0	9000	3.2	9.0	1.5	0.50
E2	23.9	10,300	11.4	18.1	1.9	0.63	4.6	10,000	2.1	7.3	1.9	0.63
E3	17.4	8300	6.7	15.1	1.3	0.43	4.5	6500	2.0	5.2	1.3	0.43
V1	10.0	11,000	5.8	14.0	1.9	0.63	4.6	10,000	2.1	7.3	1.9	0.63
V2	11.8	9500	5.5	11.5	1.5	0.50	7.0	9000	3.2	9.0	1.5	0.50

For SI: 1 MPa = 145 psi.

(a)See Section 4 of ANSI/APA PRG 320-2017 for symbols.

(b)Tabulated values are LSD design values and not permitted to be increased for the lumber size adjustment factor in accordance with CSA O86. The design values shall be used in conjunction with the section properties provided by the CLT manufacturer based on the actual layup used in manufacturing the CLT panel (see Table 2.5).

(c)Custom CLT layups that are not listed in this table shall be permitted in accordance with Section 7.2.1 of ANSI/APA PRG 320-2017.

(d)The tabulated E values are published E for lumber. For calculating the CLT design properties shown in Table 2.5, the transverse E of the lamination is assumed to be E/30, the longitudinal G of the lamination is assumed to be E/16, and the transverse G of the lamination is assumed to be longitudinal G/10.

TABLE 2.3 LSD Specified Strength and Modulus of Elasticity(a–c) for Laminations Used in PRG 320 CLT (for use in Canada)

factor of 2.1. The allowable bending stiffness and shear rigidity are based on the mean values, and no adjustments are required.

It should be noted that based on the recent full-scale CLT tests for thicker CLT (depths of seven layers or more), the standard includes a strength reduction factor of 0.85 for the calculated bending strengths in the major strength direction. It remains unclear at this point if such a factor can be attributed to the volume effect. In general, a shorter span-to-depth ratio is often associated with interlaminar shear failure during flexure. This may be addressed in the future version of the standard when additional data become available.

Custom CLT layups are permitted in ANSI/APA PRG 320 when accepted by an approved agency in accordance with the qualification and mechanical test requirements specified in the standard, as shown below:

> ***7.2.1 Custom CLT layups***
> *Custom CLT layups are permitted when approved by an approved agency in accordance with the qualification and mechanical test requirements (see 8.4 and 8.5) specified in this standard. In this case, a unique CLT layup designation shall be assigned by the approved agency if the custom product represents a significant product volume of the manufacturer to avoid duplication with an existing CLT layup designation that has been assigned to other manufacturers.*

The custom layups may include double outer layers or unbalanced layups when clearly identified for installation, as required by the manufacturer and the approved agency. However, the standard requires that a unique CLT grade designation be assigned by the approved agency if the custom product represents a significant product volume of the manufacturer to avoid duplication with an existing CLT grade designation that has been assigned to other manufacturers.

CLT Layup	CLT t_p (in.)	Lamination thickness (in.) in CLT layup							Major strength direction				Minor strength direction			
		=	⊥	=	⊥	=	⊥	=	$(F_bS)_{eff,f,0}$ (lbf-ft/ft of width)	$(EI)_{eff,f,0}$ (10^6 lbf-in.^{2}/ft of width)	$(GA)_{eff,f,0}$ (10^6 lbf/ft of width)	$V_{s,0}$ (lbf/ft of width)	$(F_bS)_{eff,f,90}$ (lbf-ft/ft of width)	$(EI)_{eff,f,90}$ (10^6 lbf-in.^{2}/ft of width)	$(GA)_{eff,f,90}$ (10^6 lbf/ft of width)	$V_{s,90}$ (lbf/ft of width)
E1	4 1/8	1 3/8	1 3/8	1 3/8					4525	115	0.46	1430	160	3.1	0.61	495
	6 7/8	1 3/8	1 3/8	1 3/8	1 3/8	1 3/8			10,400	440	0.92	1970	1370	81	1.2	1430
	9 5/8	1 3/8	1 3/8	1 3/8	1 3/8	1 3/8	1 3/8	1 3/8	18,375	1089	1.4	2490	3125	309	1.8	1960
E2	4 1/8	1 3/8	1 3/8	1 3/8					3825	102	0.53	1910	165	3.6	0.56	660
	6 7/8	1 3/8	1 3/8	1 3/8	1 3/8	1 3/8			8825	389	1.1	2625	1430	95	1.1	1910
	9 5/8	1 3/8	1 3/8	1 3/8	1 3/8	1 3/8	1 3/8	1 3/8	15,600	963	1.6	3325	3275	360	1.7	2625
E3	4 1/8	1 3/8	1 3/8	1 3/8					2800	81	0.35	1110	110	2.3	0.44	385
	6 7/8	1 3/8	1 3/8	1 3/8	1 3/8	1 3/8			6400	311	0.69	1530	955	61	0.87	1110
	9 5/8	1 3/8	1 3/8	1 3/8	1 3/8	1 3/8	1 3/8	1 3/8	11,325	769	1.0	1940	2180	232	1.3	1520
E4	4 1/8	1 3/8	1 3/8	1 3/8					4525	115	0.50	1750	140	3.4	0.62	605
	6 7/8	1 3/8	1 3/8	1 3/8	1 3/8	1 3/8			10,400	440	1.0	2410	1230	88	1.2	1750
	9 5/8	1 3/8	1 3/8	1 3/8	1 3/8	1 3/8	1 3/8	1 3/8	18,400	1089	1.5	3050	2800	335	1.9	2400
V1	4 1/8	1 3/8	1 3/8	1 3/8					2090	108	0.53	1910	165	3.6	0.59	660
	6 7/8	1 3/8	1 3/8	1 3/8	1 3/8	1 3/8			4800	415	1.1	2625	1430	95	1.2	1910
	9 5/8	1 3/8	1 3/8	1 3/8	1 3/8	1 3/8	1 3/8	1 3/8	8500	1027	1.6	3325	3275	360	1.8	2625
V2	4 1/8	1 3/8	1 3/8	1 3/8					2030	95	0.46	1430	160	3.1	0.52	495
	6 7/8	1 3/8	1 3/8	1 3/8	1 3/8	1 3/8			4675	363	0.91	1970	1370	81	1.0	1430
	9 5/8	1 3/8	1 3/8	1 3/8	1 3/8	1 3/8	1 3/8	1 3/8	8275	898	1.4	2490	3125	309	1.6	1960
V3	4 1/8	1 3/8	1 3/8	1 3/8					1740	95	0.49	1750	140	3.4	0.52	605
	6 7/8	1 3/8	1 3/8	1 3/8	1 3/8	1 3/8			4000	363	0.98	2420	1230	88	1.0	1750
	9 5/8	1 3/8	1 3/8	1 3/8	1 3/8	1 3/8	1 3/8	1 3/8	7100	899	1.5	3050	2800	335	1.6	2400

For SI: 1 in. = 25.4 mm; 1 ft = 304.8 mm; 1 lbf = 4.448 N.

[a]See Section 4 of ANSI/APA PRG 320-2017 for symbols.

[b]This table represents one of many possibilities that the CLT could be manufactured by varying lamination grades, thicknesses, orientations, and layer arrangements in the layup.

[c]Custom CLT layups that are not listed in this table shall be permitted in accordance with Section 7.2.1 of ANSI/APA PRG 320-2017.

TABLE 2.4 ASD Reference Design Values[a–c] for CLT Listed in Annex A (for use in the United States)

CLT layup	CLT t_p (mm)	Lamination thickness (mm) in CLT layup							Major strength direction				Minor strength direction			
		=	⊥	=	⊥	=	⊥	=	$(F_bS)_{eff,f,0}$ (10^6 N-mm/m of width)	$(EI)_{eff,f,0}$ (10^9 N-mm²/m of width)	$(GA)_{eff,f,0}$ (10^6 N/m of width)	$v_{s,0}$ (kN/m of width)	$(F_bS)_{eff,f,90}$ (10^6 N-mm/m of width)	$(EI)_{eff,f,90}$ (10^9 N-mm²/m of width)	$(GA)_{eff,f,90}$ (10^6 N/m of width)	$v_{s,90}$ (kN/m of width)
	105	35	35	35					42	1088	7.3	34	1.4	32	9.1	12
E1	175	35	35	35	35	35			98	4166	15	47	12	836	18	34
	245	35	35	35	35	35	35	35	172	10,306	22	59	28	3183	27	46
	105	35	35	35					36	958	8.0	43	0.94	36	8.2	15
E2	175	35	35	35	35	35			83	3674	16	59	8.1	929	16	42
	245	35	35	35	35	35	35	35	146	9097	24	74	19	3537	25	58
	105	35	35	35					26	772	5.3	29	0.92	23	6.4	10
E3	175	35	35	35	35	35			60	2956	11	40	8.0	604	13	29
	245	35	35	35	35	35	35	35	106	7313	16	50	18	2299	19	40
	105	35	35	35					15	1023	8.0	43	0.94	36	8.7	15
V1	175	35	35	35	35	35			35	3922	16	59	8.1	929	17	42
	245	35	35	35	35	35	35	35	61	9708	24	74	19	3537	26	58
	105	35	35	35					18	884	7.2	34	1.4	32	7.5	12
V2	175	35	35	35	35	35			41	3388	14	47	12	836	15	34
	245	35	35	35	35	35	35	35	72	8388	22	59	28	3183	23	46

For SI: 1 mm = 0.03937 in.; 1 m = 3.28 ft; 1 N = 0.2248 lbf.

[a]See Section 4 of ANSI/APA PRG 320-2017 for symbols.

[b]This table represents one of many possibilities that the CLT could be manufactured by varying lamination grades, thicknesses, orientations, and layer arrangements in the layup.

[c]Custom CLT layups that are not listed in this table shall be permitted in accordance with Section 7.2.1 of ANSI/APA PRG 320-2017.

TABLE 2.5 LSD Stiffness and Unfactored Resistances Values[a–c] for CLT Listed in Annex A (for use in Canada)

2.3.3 CLT Qualification

ANSI/APA PRG 320 stipulates the requirements for plant prequalification and structural performance qualification.

Plant Prequalification

The plant prequalification is intended to ensure that the CLT plant is qualified for the manufacturing factors, such as the assembly time, lumber MC, adhesive spread rate, clamping pressure, pressing time, and wood surface temperature, prior to the normal production. The plant prequalification can be conducted with full-thickness CLT panels of 24 in. (610 mm) or more in the major strength direction and 18 in. (457 mm) or more in the minor strength direction. Two replicated CLT panels are required to be manufactured for prequalification for each combination of factors considered. The two replicated CLT panels must not be extracted from a single full-size CLT panel.

The plant prequalification includes the evaluation of gluebond (block shear) and durability. Figure 2.4 shows the locations where the block shear and delamination specimens should be taken for the prequalification to ensure the dispersion of the specimens within a sampled CLT qualification panel. Results obtained from the prequalification are required to be documented and serve as the basis for manufacturing factors specified in the in-plant manufacturing standard.

The prequalification and qualification requirements are specified in Section 8 of ANSI/APA PRG 320, as shown below:

> ***8.1 Qualification Requirements***
> *Required qualification tests for CLT components, such as lumber, adhesives, and end, face, and edge joints are provided in Section 6 and summarized in Table 2. This section provides requirements for plant qualification and CLT qualification tests to meet the structural performance levels specified in Tables A2 and A4.*

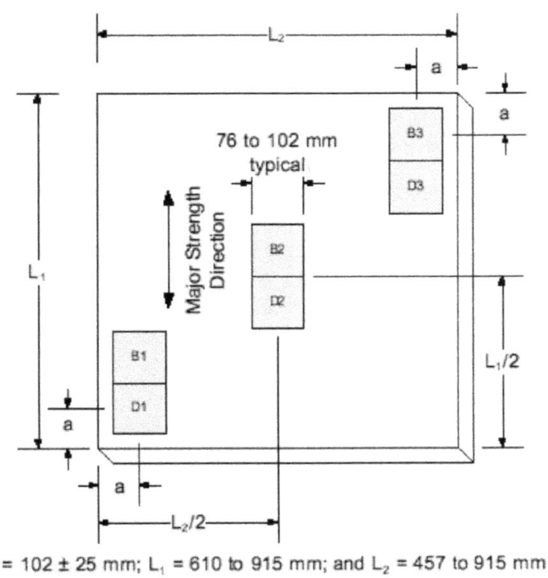

FIGURE 2.4 Block shear ("B") and delamination ("D") Specimen Locations.

8.2 Plant Prequalification

8.2.1 General

The CLT plant shall be prequalified for the manufacturing factors considered (see 8.2.2) using full-thickness qualification panels of 24 inches (610 mm) or more in the major strength direction and 18 inches (457 mm) or more in the minor strength direction (hereafter referred to as "Prequalification panels"). A minimum of two replicate CLT prequalification panels shall be manufactured for prequalification for each combination of factors considered in 8.2.2. The two replicate CLT prequalification panels shall not be extracted from a single full-size CLT panel.

Prequalification panels shall be prepared at the facility or at an alternative facility acceptable to the approved agency. All prequalification panels shall be:

a. Of the same approximate length and width at the time of pressing;

b. Pressed individually; and

c. Taken from approximately the geometric center of the larger panel, if applicable.

8.2.2 Fabrication of prequalification panels

Application of pressure to manufacture prequalification panels shall reflect the key characteristics of the manufacturing equipment, including the platen and glue spreader (as applicable) that is or will be used in the facility to be qualified. The applicability of the results shall be documented by the approved agency.

Factors considered for prequalification evaluation shall include assembly time, lumber moisture content, adhesive spread rate, clamping pressure, and wood surface temperature, as specified in the manufacturing standard of the plant and accepted by the approved agency.

8.2.3 Conditioning of prequalification panels

Prequalification panels shall be stored in an indoor environment for a minimum of 24 hours or until the adhesive has cured sufficiently to permit evaluation, whichever is longer.

8.2.4 Specimens

Six square/rectangular specimens (three for block shear tests, i.e., "B" specimens and three for delamination tests, i.e., "D" specimens) shall be extracted from each prequalification panel at the locations shown in Figure 2.4 and labeled to indicate the panel number and the specimen position within the panel. If the prequalification panel is larger than the specified prequalification panel size, the prequalification sampling area shall be 24 inches (610 mm) to 36 inches (910 mm) square located at the geometric center of the panel.

8.2.5 Test methods and requirements

The specimens obtained in accordance with 8.2.4 shall be qualified and meet the requirements specified in 6.4.3 of this standard.

8.3 Qualification of Effective Bond Area

8.3.1 General

The manufacturer shall establish visual grading rules for the bonded faces and limit the average glue skip to maintain an average effective bond area of 80% or more.

The manufacturer's visual grading rules established to achieve the effective bond area shall include major visual characteristics based on characteristic measurements consistent with standard lumber grading rules.

8.3.2 Sample selection and inspection

Samples shall be drawn from representative production of laminations meeting the manufacturer's visual grading rules and positioned in accordance with the in-plant manufacturing standard. The layer formed by the laminations shall be verified by the approved agency to provide an effective bond area of 80% or more over any randomly selected area not less than 48 inches (1,220 mm) square.

Structural Performance Qualification

To confirm the major CLT design properties, structural performance tests are required in accordance with ANSI/APA PRG 320, as shown below. The structural performance tests include bending strength, bending stiffness, and interlaminar shear in both major and minor strength directions. The sample size for bending stiffness must be sufficient for estimating the population mean within 5 percent precision with 75 percent confidence, or 10 specimens, whichever is greater. The sample size for bending strength and interlaminar shear must be sufficient for estimating the characteristic value with 75 percent confidence in accordance with ASTM D2915 [30].

The bending tests are required to be conducted flatwise (loads are applied perpendicular to the face layer of CLT) in accordance with the third-point load method of ASTM D198 [31] or ASTM D4761 [32] using the specimen width of not less than 12 in. (305 mm) and the on-center span of approximately 30 times the specimen depth. The Committee considered that a minimum specimen width of 12 in. (305 mm) is necessary to distinguish CLT from typical beam elements. However, it has been reported that for some CLT layups, the use of the span-to-depth ratio of 30 for bending tests in the minor strength direction may result in excessive deflection before the specimen reaches the peak load. Therefore, it is expected that this provision will be revisited in the near future. The weight of the CLT panel is permitted to be included in the determination of the CLT bending strength.

The interlaminar shear tests are required to be conducted flatwise in accordance with the center-point load method of ASTM D198 or ASTM D4761 using the specimen width of not less than 12 in. (305 mm) and the on-center span of 5 to 6 times the specimen depth. The bearing length must be sufficient to avoid bearing failure but not greater than the specimen depth. All specimens must be cut to length without overhangs, which are known to increase the interlaminar shear strength in shear tests.

7.2 Structural Performance Requirements

Structural performance shall be evaluated for each CLT layup unless otherwise noted in this section. CLT shall meet the minimum structural performance based on the properties shown in Table 1 multiplied by the section properties provided by the manufacturer and accepted by an approved agency when evaluated and confirmed by test results in accordance with 8.5. CLT panels manufactured with SCL layers, which do not meet Table 1 requirements, shall be qualified in accordance with 7.2.1.

8.4 Qualification for Structural Performance

Following plant prequalification, a representative sample of CLT panels shall be manufactured for qualification tests in accordance with 8.4.1 and 8.4.2. Depending on the number of layups intended for qualification, a qualification plan shall be developed and accepted by an approved agency in accordance with the principles prescribed in this section.

8.4.1 Required mechanical property qualification

The flatwise bending and flatwise shear properties of CLT layups in both major and minor strength directions shall be tested in accordance with 8.5.3 and 8.5.4 to confirm the design values shown in Table A2 for use in the US or Table A4 for use in Canada, or the design values approved by an approved agency.

8.4.2 Optional mechanical property qualification

When edgewise bending and edgewise shear properties are to be approved by an approved agency, qualification tests shall be conducted in accordance with 8.5.5 and 8.5.6, respectively.

8.5 Mechanical Property Qualification

The design values from required mechanical property qualification (8.4.1) and optional mechanical property qualification (8.4.2) shall be approved by an approved agency in accordance with this section.

8.5.1 Sampling

Test samples shall be representative of typical production and shall be sampled at the manufacturing facility by an approved agency using the layup intended for qualification. The sample size required for stiffness capacities shall be sufficient for estimating the population mean within 5% precision with 75% confidence, or 10 specimens, whichever is greater. In general, a sample size larger than 10 is needed when the coefficient of variation is greater than 13%. The sample size required for strength capacities shall be sufficient for estimating the characteristic value with 75% confidence in accordance with ASTM D2915.

8.5.2 Sample conditioning

CLT panel samples shall be stored in an indoor environment for a minimum of 24 hours or until the adhesive has cured sufficiently to permit evaluation, whichever is longer. The CLT samples at the time of mechanical tests shall have an average moisture content of not less than 8%.

8.5.3 Flatwise bending properties

Flatwise bending stiffness and bending moment capacity (resistance) shall be evaluated in accordance with 8.5.3.1 and 8.5.3.2

8.5.3.1 Flatwise bending test methods

Flatwise bending tests shall be conducted in both major and minor strength directions in accordance with the third-point load method of Sections 4 through 12 of ASTM D198 or Section 8 of ASTM D4761 using the specimen width of not less than 12 inches (305 mm) and the on-center span equal to approximately 30 times the specimen depth for the tests in the major strength direction and approximately 18 times the specimen depth for the tests in the minor strength direction. The weight of the CLT panel is permitted to be included in the determination of the flatwise bending moment capacity (resistance).

8.5.3.2 Flatwise bending qualification requirements

In the U.S. and Canada, the average flatwise bending stiffness determined from qualification tests shall equal or exceed the published flatwise bending stiffness $[(EI)_{eff,f,0}$ *or* $(EI)_{eff,f,90}]$.

In the U.S., the characteristic flatwise bending moment capacity determined from qualification tests shall equal or exceed the published ASD reference flatwise bending moment capacity $[(F_bS)_{eff,f,0}$ *or* $(F_bS)_{eff,f,90}]$ *times 2.1. In Canada, the characteristic flatwise bending moment resistance determined from qualification tests shall equal or exceed the published LSD flatwise bending resistance* $[(f_bS)_{eff,f,0}$ *or* $(f_bS)_{eff,f,90}]$ *divided by 0.96.*

8.5.4 Flatwise shear properties

Flatwise shear stiffness and capacity (resistance) shall be evaluated in accordance with 8.5.4.1 and 8.5.4.2.

8.5.4.1 Flatwise shear test methods

Flatwise shear stiffness tests shall be conducted in both major and minor strength directions in accordance with Sections 45 through 52 of ASTM D198.

Flatwise shear tests shall be conducted in both major and minor strength directions in accordance with the center-point load method of Sections 4 through 12 of ASTM D198 or Section 7 of ASTM D4761 using the specimen width of not less than 12 inches (305 mm) and the on-center span equal to 5 to 6 times the specimen depth. The bearing length shall be sufficient to avoid bearing failure, but not greater than the specimen depth. All specimens are to be cut to length with no overhangs allowed.

8.5.4.2 Flatwise shear qualification requirements

In the U.S. and Canada, the average flatwise shear stiffness determined from qualification tests shall equal or exceed the published shear stiffness in flatwise bending $[(GA)_{eff,f,0}$ *or* $(GA)_{eff,f,90}]$.

In the U.S., the characteristic flatwise shear capacity determined from qualification tests shall equal or exceed the published ASD reference flatwise shear capacity ($V_{s,0}$ *or* $V_{s,90}$) *times 2.1. In Canada, the characteristic flatwise shear resistance determined from qualification tests shall equal or exceed the published LSD flatwise shear resistance* ($v_{s,0}$ *or* $v_{s,90}$) *divided by 0.96.*

8.5.5 Edgewise bending properties

Edgewise bending stiffness and bending moment capacity (resistance) shall be permitted to be evaluated in accordance with 8.5.5.1 and 8.5.5.2. The volume, creep and load duration effects of edgewise bending capacity (resistance) shall be evaluated in accordance with the principles of Sections 7.4.1 and 7.4.2 of ASTM D5456.

8.5.5.1 Edgewise bending test methods

Bending tests shall be conducted edgewise in both major and minor strength directions in accordance with the third-point load method of Sections 4 through 12 of ASTM D198 or Section 6 of ASTM D4761 using the specimen depth of not less than 12 inches (305 mm) and the on-center span equal to approximately 18 times the specimen depth. The weight of the CLT panel is permitted to be included in the determination of the edgewise bending moment capacity (resistance).

8.5.5.2 Edgewise bending qualification requirements

In the U.S. and Canada, the average edgewise bending stiffness determined from qualification tests divided by the calculated moment of inertia ($I_{e,0}$ or $I_{e,90}$) shall equal or exceed the published edgewise bending modulus of elasticity ($E_{e,0}$ or $E_{e,90}$).

In the U.S., the characteristic edgewise bending moment capacity determined from qualification tests shall equal or exceed the published ASD reference edgewise bending stress ($F_{b,e,0}$ or $F_{b,e,90}$) multiplied by the calculated edgewise section modulus ($S_{e,0}$ or $S_{e,90}$) and an adjustment factor of 2.1. In Canada, the characteristic edgewise bending moment resistance determined from qualification tests shall equal or exceed the published LSD specified edgewise bending strength ($f_{b,e,0}$ or $f_{b,e,90}$) multiplied by the calculated edgewise section modulus ($S_{e,0}$ or $S_{e,90}$) and divided by an adjustment factor of 0.96.

8.5.6 Edgewise shear properties

Edgewise shear stiffness and capacity (resistance) shall be permitted to be evaluated in accordance with 8.5.6.1 and 8.5.6.2.

8.5.6.1 Edgewise shear test methods

Edgewise shear stiffness tests shall be conducted in both major and minor strength directions in accordance with Sections 45 through 52 of ASTM D198.

Edgewise shear capacity (resistance) tests shall be conducted in both major and minor strength directions in accordance with the full-scale test method specified in Annex A3 of ASTM D5456. The web thickness of the I-shaped cross section shall be the CLT thickness. The specimen shall contain at least one edge joint, as applicable, in the middle ⅓ of the specimen depth.

8.5.6.2 Edgewise shear qualification requirements

For use in the U.S. or Canada, the average edgewise shear stiffness determined from qualification tests divided by the gross thickness of CLT (t_p) shall equal or exceed the published modulus of rigidity (shear modulus) in edgewise bending ($G_{e,0}$ or $G_{e,90}$).

In the U.S., the characteristic edgewise shear capacity determined from qualification tests shall equal or exceed the published ASD reference edgewise shear capacity ($F_{v,e,0}\ t_p$ or $F_{v,e,90}\ t_p$) multiplied by an adjustment factor of 2.1. In Canada, the characteristic edgewise shear resistance determined from qualification tests shall equal or exceed the published LSD edgewise shear resistance ($f_{v,e,0}\ t_p$ or $f_{v,e,90}\ t_p$) divided by an adjustment factor of 0.96.

It is important to note that the CLT mechanical properties that are published in Table 2.4 (i.e., Table A2 in ANSI/APA PRG 320) and Table 2.5 (i.e., Table A2 in ANSI/APA PRG 320) are mandatorily required to be qualified. This includes most flatwise mechanical properties. As previously mentioned, the CLT flatwise bending properties can be reliably predicted through the principle of engineering mechanics. On the other hand, there are no engineering models that can be used to reliably predict most CLT edgewise mechanical properties. Therefore, the qualification of CLT edgewise mechanical properties are optional in ANSI/APA PRG 320; that is, they are required only when the CLT manufacturer elects to publish their *proprietary* edgewise bending properties through the approval of the approved agency. In this case, ANSI/APA PRG 320 provides standard test methods and data analysis requirements so that the proprietary CLT edgewise bending design values from different CLT manufacturers can be tested, analyzed, and published in a consistent manner. This should help avoid confusion to the designers when comparing proprietary CLT edgewise bending design values.

Process Change Qualification

When process changes occur in production, qualification tests are required, depending on the extent of the changes and their impacts to the CLT performance. ANSI/APA PRG 320 lists some key changes and the required responses, as summarized below:

Process change	Response
• Press equipment • Adhesive formulation class • Addition or substitution of species from a different species group • Changes to the visual grading rules that reduce the effective bond area or the effectiveness of the applied pressure (e.g., warp permitted)	Plant prequalification and structural reevaluation
• Other changes to the manufacturing process or component quality not listed above • Adhesive composition (e.g., fillers and extenders)	Plant prequalification
• Increase in panel width or length of more than 20%	Structural reevaluation

2.3.4 Appearance Classification

It should be noted that the appearance classification of CLT, just like glulam, does not affect the mechanical properties of the product. Therefore, it is called the "classification" instead of the "grade," which may have the implication of structural performance in convention. In ANSI/APA PRG 320, there are no mandatory appearance classifications for CLT. The Committee elected to leave the CLT appearance classifications to an agreement between the buyer and seller. However, nonmandatory classifications based largely on selected glulam appearance classifications in ANSI A190.1 are included in the appendix, which covers the Architectural and Industrial Appearance Classifications, as shown below. Depending on the market demand, the appearance classifications may be standardized in the future as needed.

A1. Architectural Appearance Classification

An appearance classification normally suitable for applications where appearance is an important, but not overriding consideration. Specific characteristics of this classification are as follows:

- *In exposed surfaces, all knot holes and voids measuring over ¾ inch (19 mm) are filled with a wood-tone filler or clear wood inserts selected for similarity with the grain and color of the adjacent wood.*
- *The face layers exposed to view are free of loose knots and open knot holes are filled.*
- *Knot holes do not exceed ¾ inch (19 mm) when measured in the direction of the lamination length with the exception that a void may be longer than ¾ inch (19 mm) if its area is not greater than ½ in.^{2} (323 mm^2).*
- *Voids greater than 1/16 inch (1.6 mm) wide created by edge joints appearing on the face layers exposed to view are filled.*
- *Exposed surfaces are surfaced smooth with no misses permitted.*

A2. Industrial Appearance Classification
An appearance classification normally suitable for use in concealed applications where appearance is not of primary concern. Specific characteristics of this grade are as follows:

- *Voids appearing on the edges of laminations need not be filled.*
- *Loose knots and knot holes appearing on the face layers exposed to view are not filled.*
- *Members are surfaced on face layers only and the appearance requirements apply only to these layers.*
- *Occasional misses, low laminations or wane (limited to the lumber grade) are permitted on the surface layers and are not limited in length.*

2.4 CLT Manufacturing Process

2.4.1 CLT Online Quality Control, Surface Sanding, and Cutting

An industrial sanding machine designed for wood composite products such as plywood may be used to sand one CLT panel at a time to the target thickness with a tolerance of +0.004 in. (0.1 mm). Tighter tolerances may be specified by building project. After sanding, CLT panels are then conveyed to a machining station where a multiaxis numerically controlled machine cuts out openings for windows and doors, splices, and other required parts. Cutting is performed under strictly controlled conditions for maximum accuracy. Minor repairs are carried out manually at this stage of the manufacturing process.

2.4.2 Product Certification, Marking, Packaging, and Shipping

Product certification and marking ensure that the correct product is specified, delivered, and installed. It is also an important part of product conformity assessment by providing the information to allow designers, contractors, and the authority having jurisdiction to check the authenticity of the product. CLT products represented as conforming to the ANSI/APA PRG 320 standard are required to bear the stamp of an approved agency which either inspects the manufacturer or has tested a random sampling of the finished products in the shipment being certified for conformance with the standard.

CLT products conforming to ANSI/APA PRG 320 are required to be identified with marks in accordance with the following provisions:

8.8 Certification and Trademarking

8.8.1 Certification
CLT products represented as conforming to this standard shall bear the stamp of an approved agency which (1) either inspects the manufacturer or (2) has tested a random sampling of the finished products in the shipment being certified for conformance with this standard.

8.8.2 Product marking
CLT products represented as conforming to this standard shall be identified with marks containing the following information:

a. *CLT grade qualified in accordance with this standard;*
b. *The CLT thickness or identification;*
c. *The mill name or identification number;*
d. *The approved agency name or logo;*
e. *The symbol of "ANSI/APA PRG 320" signifying conformance to this standard;*
f. *Any manufacturer's designations which shall be separated from the grade-marks or trademarks of the approved agency by not less than 6 inches (152 mm); and*
g. *"Top" stamp on the top face of custom CLT panels used for roof or floor if manufactured with an unbalanced layup.*

> ***8.8.3 Frequency of marking***
> *Non-custom and other required marks in this section shall be placed on standard products at intervals of 8 feet (2.4 m) or less in order that each piece cut from a longer piece will have at least one of each of the required marks.*
>
> ***8.8.4 Custom products***
> *For products manufactured to meet specific job specifications (custom products), the marking shall be permitted to contain information less than that specified in 8.8.2. However, custom products shall bear at least one mark containing a required identification. When long CLT products shipped to a job are to be cut later into several members for use in the structure, the frequency of marking required in 8.8.3 shall be followed.*
>
> ***8.8.5 Voiding marks***
> *CLT products originally marked as conforming to this standard but subsequently rejected as not conforming thereto shall have any reference to the standard obliterated or voided by the manufacturer.*
>
> *Note 16. This can be performed by blocking out the stamp with permanent black ink or light sanding.*

Note from Section 8.8.3 of ANSI/APA PRG 320 that the non-custom and other required marks must be placed on standard products at intervals of 8 ft (2.4 m) or less in order that each piece cut from a longer piece will have at least one of each of the required marks. For products manufactured to meet specific job specifications (custom products), the marking may contain information less than that specified for standard CLT products. However, custom products must bear at least one mark containing a required identification, as indicated in Section 8.8.4 of ANSI/APA PRG 320. Also noted from Section 8.8.4 of ANSI/APA PRG 320 is that when long CLT products shipped to a job site are to be cut later into several members for use in the structure, the frequency of marking must be applied at intervals of 8 ft (2.4 m) or less.

Additional markings on the panels may show the main direction loading of the panels in the structure and, possibly, the zones designed to receive connectors. Section 8.8.5 of ANSI/APA PRG 320 addresses the voiding of marks for non-confirming products. Because CLT panels are intended for use under dry service conditions, the panels should be protected from weather during transportation, storage, and construction on the job site.

2.5 Quality Assurance

Quality assurance is required by ANSI/APA PRG 320 to ensure the CLT product quality through detecting changes in properties that may adversely affect the CLT performance. In this regard, an ongoing evaluation of the manufacturing process, including end, face, and edge (if used) joints in laminations; effective bonding area; lamination grade limitations; and the finished production inspection, is required to be conducted by the CLT manufacturer to confirm that the product quality remains in satisfactory compliance to the product specification requirements. The production must be held pending results of the quality assurance testing on representative samples. In addition, the product quality assurance must be audited by an independent inspection or certification agency on a regular basis in accordance with the building code requirements.

As there are a number of process-related issues that would affect the integrity of the bondline, there should be a process in place to qualify a plant to ensure that it has the means to assess and control the quality of the input components and the final product. Industrial mass production of CLT panels requires an in-plant process control program, as shown in ANSI/APA PRG 320 as follows:

9. QUALITY ASSURANCE

9.1 Objectives

This section is intended for use with CLT products that have qualified for trademarking under this standard. The purpose of this section is to assure product quality by detecting changes in properties that may adversely affect the CLT performance. In all cases, the criteria to which the CLT products are tested shall be provided in the Mill Specification or equivalent document.

9.2 Process Control

Ongoing evaluation of the process properties listed in this section shall be performed to confirm that the CLT quality remains in satisfactory compliance to the product specification requirements. Sampling methods and quality assurance testing shall be documented in an in-plant manufacturing standard and accepted by the approved agency. All processes and test records relevant to the production shall be retained based on the manufacturer's record retention policy and are subject to audit by the approved agency. Production shall be held pending results of the quality assurance testing on representative samples.

9.3 End, Face, and Edge Joints in Laminations

The lamination end joints, face joints, and edge joints (when applicable) shall be sampled and tested for ongoing quality assurance in accordance with Table 3 of ANSI A190.1 and meet the strength (required for end joints only), wood failure, and durability requirements specified therein in the U.S., or shall be sampled and tested in accordance with Section 7 of CSA O122 and meet the strength (required for end joints only), wood failure, and durability requirements specified therein in Canada. Special considerations for face bonding of the CLT panel as a whole are provided in 9.3.1 through 9.3.3 of this standard.

9.3.1 Effective bonding area

On-grade lumber shall be laid up to maintain an effective bonding area of not less than 80% on surfaces to be bonded for each bondline.

Note 17. To maintain an effective bond area, laminations in cross-plies may need to be oriented such that the bark and pith faces of adjacent pieces are generally alternated.

9.3.2 Lamination grade limits

Grade limits intended to limit the amount of lamination warp that will not be corrected upon application of pressure shall be qualified in accordance with 8.3.

9.3.3 Glue skip in the face bondline

The average glue skip in a face bondline shall not exceed the level established to maintain the effective bonding area specified in 9.3.1. Glue skips are not assessed as delamination unless the inclusion of such skips does not invalidate the delamination requirements.

9.4 Finished Production Inspection

All production shall be inspected visually, and/or by measurements or testing for conformance to this standard with the following attributes:

a. Dimensions (width, depth, and length);
b. Shape, including straightness and squareness;
c. Type, quality and location of structural bondlines;
d. Appearance classification;
e. Layup, including lumber species and grades, placement, and orientation;
f. Moisture content; and
g. Application of the appropriate marks.

9.5 Minor Variations

A product is considered conforming to this standard when minor variations of a limited extent in noncritical locations exist, or when structural damage or defects have been repaired and, in the judgment of a qualified person, the product is structurally adequate for the use intended. The identity of the product and the nature of the minor variation shall be documented and provided to the designer of record upon request. A qualified person is one who is familiar with the job specifications and applicable design requirements and has first-hand knowledge of the manufacturing process.

ANSI/APA PRG 320 uses delamination testing as a means to assess quality and moisture durability of the bondline. In the delamination test, a specimen obtained from a production panel is saturated with water and then dried to evaluate the adhesive bondline's ability to resist the wood shrinkage and swelling stresses. The delamination test also assesses somewhat the ability of the adhesive to withstand moisture degradation. In the delamination test, separation in the wood adjacent to the bondline, as opposed to separation in the adhesive, is not considered delamination. The limits on the amount of delamination are based on the glulam standards, as indicated in Section 9.3 of ANSI/APA PRG 320. When all these requirements are met, the manufacturing process is deemed to be producing CLT with bondlines of acceptable quality.

Preliminary tests carried out at FPInnovations suggest that wood failure results from block shear specimens tested under vacuum-pressure-dry conditions could be used to assess the bond quality. For additional information on this topic, refer to the report on block shear testing of CLT [33].

The quality assurance process for CLT also includes the inspection of finished products, as shown in Section 9.4 of ANSI/APA PRG 320. When there are minor and noncritical variations in product quality, ANSI/APA PRG 320 permits a qualified person, usually the technical or quality director of the CLT plant, to make judgment for product acceptance provided that the requirements specified in Section 9.5 of ANSI/APA PRG 320 are followed. This practice is similar to glulam manufacturing in accordance with ANSI A190.1.

2.6 Conclusion

In North America, a limited number of CLT production lines have been commercialized since 2008. With the publication of ANSI/APA PRG 320 since 2011, the code-recognized CLT design standards have also been published in the 2015 ANSI/AWC NDS and the 2014 CSA O86. As a result, several timber structures have also been constructed in the United States and Canada using CLT products manufactured in North America. From a product certification perspective, APA as well as other ISO/IEC 17065-accredited certification agencies in North America are trademarking CLT products in accordance with ANSI/APA PRG 320 to provide the designers with construction materials that are consistent in quality and recognized by the building codes. Therefore, the designers can focus on the architectural and structural designs without the concern of material supplies and quality. This is a very significant step toward the wide acceptance of the relatively new construction products, such as CLT, in North America.

It should be noted that as CLT remains a relatively new engineered wood product in North America, and not only the product standard, ANSI/APA PRG 320, but also the design standards, NDS and CSA O86, are evolving. Therefore, the revision to ANSI/APA PRG 320 is expected to occur in a relatively short period in time until the CLT production becomes relatively stable in the immediate future. In the meantime, the building codes and regulations in North America are also making changes in the building height and area and fire protection requirements to accommodate this new form of mass timber products. For example, ANSI/APA PRG 320-2018 was published in February 2018 to address new adhesive qualification requirements to avoid a fire regrowth phenomenon.

After that, the ANSI/APA PRG 320 Committee will focus on additional details in the use of SCL in CLT production and the gluebond durability requirements for CLT.

At the same time, the 2020 NBC in Canada and the 2021 IBC in the United States are expected to expand the building height and area limitations for mass timber. New sub-building types under Type IV Construction have been proposed by the ICC Ad Hoc Committee on Tall Wood Buildings for the 2021 IBC to allow mass timber, such as CLT, to be constructed up to 18 stories in height, depending on the level of fire protection provided for the building. Similar proposals for tall wood buildings are also being proposed for the 2020 NBC in Canada. When completed, these revisions in the building codes, design standards, and product standard will pave the way for expanded market and production for CLT in North America in the next few years.

ISO Technical Committee (TC) 165 on Timber Structures, in the meantime, is near completion of the development of an ISO standard for CLT under the leadership of Dr. Yeh of APA and Mr. Kretschmann of the ALSC. This ISO standard is intended to harmonize the CLT standards from North America and Europe as an international standard, which will encourage the use of CLT in building construction globally.

References

[1] APA—The Engineered Wood Association. *Standard for Performance-Rated Cross Laminated Timber*, ANSI/APA PRG 320. Tacoma, Washington D.C., U.S.A. 2011.

[2] APA—The Engineered Wood Association. *Standard for Performance-Rated Cross Laminated Timber*, ANSI/APA PRG 320. Tacoma, Washington D.C., U.S.A. 2012.

[3] APA—The Engineered Wood Association. *Standard for Performance-Rated Cross Laminated Timber*, ANSI/APA PRG 320. Tacoma, Washington D.C., U.S.A. 2017.

[4] International Code Council. *International Building Code*. Country Club Hills, IL, U.S.A. 2018.

[5] International Code Council. *International Residential Code*. Country Club Hills, IL, U.S.A. 2018.

[6] Canadian Standards Association. *Engineering Design in Wood*. CSA O86. Toronto, Ontario, Canada. 2014 (Reprinted 2016).

[7] National Research Council. *National Building Code of Canada*. Ottawa, Canada. 2015.

[8] National Institute of Standards and Technology. *American Softwood Lumber Standard*. Voluntary Product Standard PS 20. Washington D.C., U.S.A. 2010.

[9] Canadian Standards Association. *Softwood Lumber*. CSA O141. Toronto, Ontario, Canada. 2005.

[10] American Wood Council. *National Design Specification for Wood Construction*. ANSI/AWC NDS. Leesburg, VA, U.S.A. 2012.

[11] ASTM International. *Standard Specification for Evaluation of Structural Composite Lumber Products*. ASTM D5456. West Conshohocken, PA, U.S.A. 2014b.

[12] APA—The Engineered Wood Association. *Structural Glued Laminated Timber*. ANSI A190.1. Tacoma, Washington D.C., U.S.A. 2017.

[13] National Lumber Grades Authority. *Special Products Standard for Fingerjoined Structural Lumber*. SPS 1. New Westminster, BC, Canada. 2017.

[14] National Lumber Grades Authority. *Special Products Standard for Machine Graded Lumber*. SPS 2. New Westminster, BC, Canada. 2017.

[15] National Lumber Grades Authority. *Special Products Standard for Fingerjoined Machine Graded Lumber*. SPS 4. New Westminster, British Columbia, Canada. 2014.

[16] National Lumber Grades Authority. *Special Products Standard for Structural Face-Glued Lumber*. SPS 6. New Westminster, BC, Canada. 2015.

[17] ISO. *General Criteria for the Operation of Various Types of Bodies Performing Inspection*. ISO/IEC 17020. Geneva, Switzerland. 2012.

[18] ISO. *General Requirements for the Competence of Testing and Calibration Laboratories*. ISO/IEC 17025. Geneva, Switzerland. 2005.

[19] ISO. *Conformity Assessment—Requirements for Bodies Certifying Products, Processes, and Services*. ISO/IEC 17065. Geneva, Switzerland. 2012.

[20] ASTM International. *Standard Specification for Evaluation of Duration of Load and Creep Effects of Wood and Wood-Based Products*. ASTM D6815. West Conshohocken, PA, U.S.A. 2009 (R2015).
[21] APA—The Engineered Wood Association. *Standard for Adhesives for Use in Structural Glued Laminated Timber*. ANSI 405. Tacoma, Washington D.C., U.S.A. 2018.
[22] ASTM International. *Standard Test Method for Multiple-Cycle Accelerated Aging Test (Automatic Boil Test) for Exterior Wet Use Wood Adhesives*. ASTM D3434. West Conshohocken, PA, U.S.A. 2000 (R2013).
[23] Canadian Standards Association. *Evaluation of Adhesives for Structural Wood Products (Exterior Exposure)*. CSA O112.9. Toronto, Ontario, Canada. 2010 (R2014).
[24] ASTM International. *Standard Test Method for Evaluating the Shear Strength of Adhesive Bonds in Laminated Wood Products at Elevated Temperatures*. ASTM D7247. West Conshohocken, PA, U.S.A. 2016.
[25] Canadian Standards Association. *Evaluation of Adhesives for Structural Wood Products (Limited Moisture Exposure)*. CSA O112.10. Toronto, Ontario, Canada. 2008.
[26] National Institute of Standards and Technology. *Structural Plywood*. Voluntary Product Standard PS 1. Washington D.C., U.S.A. 2009.
[27] Frangi, A., M. Fontana, E. Hugi, and R. Jobstl. *Experimental Analysis of Cross-Laminated Timber Panels in Fire. Fire Safety Journal* 44: 1078–1108. 2009.
[28] Canadian Standards Association. *Qualification Code for the Manufacturers of Structural Glued-Laminated Timber*. CSA O177. Toronto, Ontario, Canada. 2006 (R2015).
[29] Gagnon, S. and M. Popovski. Structural Design of Cross-Laminated Timber Elements. In: Chapter 3, *CLT Handbook*. FPInnovations. Quebec, Quebec, Canada. 2011.
[30] ASTM International. *Standard Practice for Sampling and Data-Analysis for Structural Wood and Wood-Based Products*. ASTM D2915. West Conshohocken, PA, U.S.A. 2010.
[31] ASTM International. *Test Methods of Static Tests of Lumber in Structural Sizes*. ASTM D198. West Conshohocken, Pennsylvania, U.S.A. 2015.
[32] ASTM International. *Standard Test Methods for Mechanical Properties of Lumber and Wood-Based Structural Material*. ASTM D4761. West Conshohocken, PA, U.S.A. 2013.
[33] Casilla, R., C. Pirvu, B. J. Wang, and C. Lum. *Block Shear Testing of CLT Panels—An Exploratory Study*. FPInnovations. Vancouver, BC, Canada. 2011.

CHAPTER 3

Structural Behavior, Analysis, and Design of Cross-Laminated Timber

Mustafa Mahamid, PhD, SE, PE, P.Eng, F.ASCE, F.SEI
University of Illinois at Chicago

Introduction

The structural design of cross-laminated timber (CLT) members represents a new generation of the design of wood structures. The structural use of large flat elements in wood construction was not known until early 1990s in Europe, and it is relatively new in North America with the exception of thin panels. CLT is a new material for which code specification and regulations are still under development, and new provisions are being included in design codes slowly as research and knowledge evolve. CLT producers and suppliers as well as researchers in this field have done extensive research to come up with analysis and design methods; some methods have been laid out. Despite the fact that CLT design is not fully regulated by design codes and jurisdictions, it has been used in structures through the process of certifying of the products and through providing scientific evidence of CLT product to the local jurisdiction.

This chapter introduces the available analysis and design methods of CLT members; these methods are based on theory, modified theories, and procedures and extensive research performed by researchers and CLT suppliers. The chapter is focused on the available methods that structural engineers can use; the chapter does not discuss how these methods were developed; for more details on modeling and derivation of these methods, the reader is referred to other publications. Sections 3.1–3.4 of this chapter are based on analysis and design procedures presented in Bernasconi [1] and European Standard EN 1995-1-1 2004/A1 2008 Eurocode 5 [2]. The verification of limit states in the first four sections of this chapter could be performed using the Eurocode 5 [3]. Sections 3.5 to 3.10 are based on the methods presented in the CLT Handbook, Canadian and U.S. editions [3, 4].

The chapter introduces the basics for understanding the behavior and for the analysis and design of CLT members; for in-depth information on how these methods were derived and used, the reader is referred to literature and references.

Other structural related topics, such as codes and standards, connections, and fire, are discussed in other chapters of this book.

3.1 Introduction to Structural Analysis of CLT

The analysis of CLT members is based on basic structural mechanics theories that are known to structural engineers. The method presented here presents the analysis of any CLT member for any product that is formed by structural gluing of several layers of wood boards of the same thickness and possibly of different thicknesses and of the same wood species and possibly of different wood species and grades without reference to any specific CLT product.

The method applies to CLT members that are composed of solid wood glued together in cross layers. Other materials that are formed by cross layers but not glued together are not covered in this chapter. Figure 3.1 shows the bending test of CLT panel composed of southern yellow pine boards.

3.2 Flexural Members

3.2.1 Structural Analysis

The determination of the internal stresses in a CLT flexural member involves the determination of the bending and shear stresses in a single plan. The analysis principle presented below is used to check the adequacy of the flexural and the shear strength of the CLT member regardless if bending is one direction or in two directions.

FIGURE 3.1 Bending test of CLT panel composed of southern yellow pine boards. (*Courtesy of Brindley Engineering Corporation.*)

3.2.2 Flexural Stresses

The distribution of bending stresses in a CLT member must take into account the layered structure of the member and the different material properties of the individual layers. The calculation of the flexural stiffness of the CLT member must take into account the modulus of elasticity in each direction: in the longitudinal direction E_0 and in the transverse direction E_{90}. Due to the big difference between the moduli of elasticity, E_0 and E_{90}, for the same material (wood species), the transverse direction can be neglected. Figure 3.2 shows an example of a CLT member arrangement. The figure shows the different moduli of elasticity, geometric reference, and bending stress distribution across the depth of the member.

The flexural stiffness of a CLT member per unit width, b, of the element is given by Eq. (3.1):

$$K = EJ = \sum (J_i \cdot E_i) + \sum (A_i \cdot a_i^2 \cdot E_i) \tag{3.1}$$

where K = flexural stiffness in the considered direction
J_i = moment of inertia of a single layer
E_i = modulus of elasticity of a single layer
A_i = area of the single layer
a_i = distance from the center of gravity of the section
t_i = thickness of a single layer

The distribution of bending stresses shown in Fig. 3.2 can be determined by Eq. (3.2):

$$\sigma = \frac{M}{K} \cdot a \cdot E_i \tag{3.2}$$

where M = bending moment
a = distance from the center of gravity to the layer under consideration
K = obtained from Eq. (3.1)

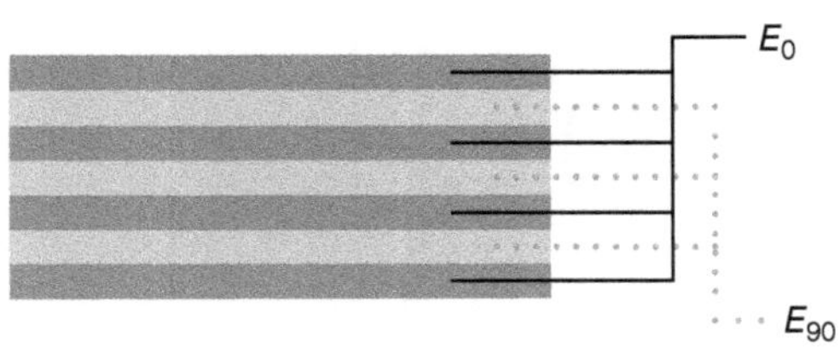

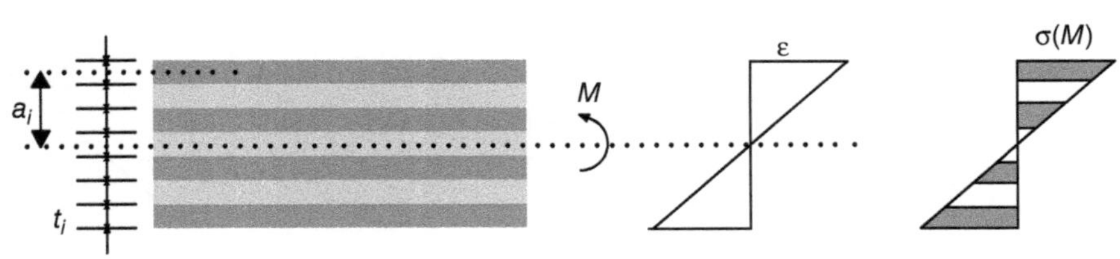

FIGURE 3.2 Bending stress distribution and geometric reference of a CLT member.

In the case of a section with layers that are composed of material of the same strength, the maximum flexural stress in tension is given by

$$\sigma = \frac{M}{W^*} \tag{3.3}$$

where $W^* = \frac{K}{h/2} \cdot E_0$

h = depth of the CLT member
E_0 = modulus of elasticity of the external layers

For bending in the perpendicular direction of the CLT member, which is referred to as the weak direction, the two outer layers can be neglected, and the same principle can be applied. CLT member cross section with orientation in the weak axis is shown in Fig. 3.3.

The calculated bending stress, $\sigma_{m,d}$, should be verified with calculated bending strength as applicable per the design standard:

$$\sigma_{m,d} \leq \text{Calculated flexural strength of the CLT section as applicable per the design standard} \tag{3.4}$$

The flexural resistance values are obtained from the characteristic strength values and by applying the appropriate modification factors and safety factors as applicable by the design standard.

The flexural strength of the CLT member can be considered as the flexural strength corresponding to the material of the wood layer considered.

The Eurocodes consider system coefficients which allow increasing the strength of the member by redistribution of stresses to the layers oriented in the same direction

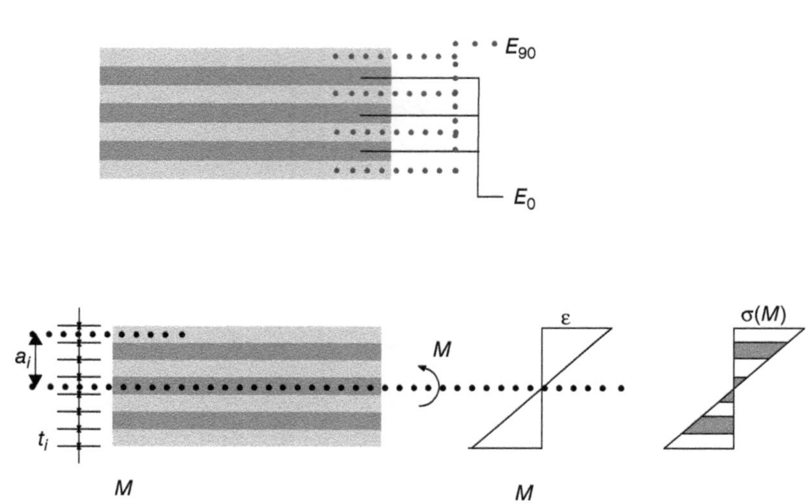

FIGURE 3.3 Weak direction of the CLT member; distribution of stresses due to the bending moment, *M*; and geometric references.

which will alleviate stresses in the weak direction. The system coefficient is defined in the Eurocodes [1, 2] as

$$k_{sys} = 1 + 0.025 \cdot n$$

$$k_{sys,max} = 1.1$$

where n = number of parallel stressed layers and $n > 1$.

The number of effective layers for the determination of k_{sys} should be defined on the basis of the width of the CLT member and on the maximum width of the individual layer. Refer to [1, 2] for more details on the system coefficient and its applications in calculating the strength of a CLT member.

For biaxial bending, the analysis is performed in the two directions independently. The maximum flexural tensile stress parallel to grain should be considered in the considered stress direction. The stress distribution in both directions is shown in Fig. 3.4.

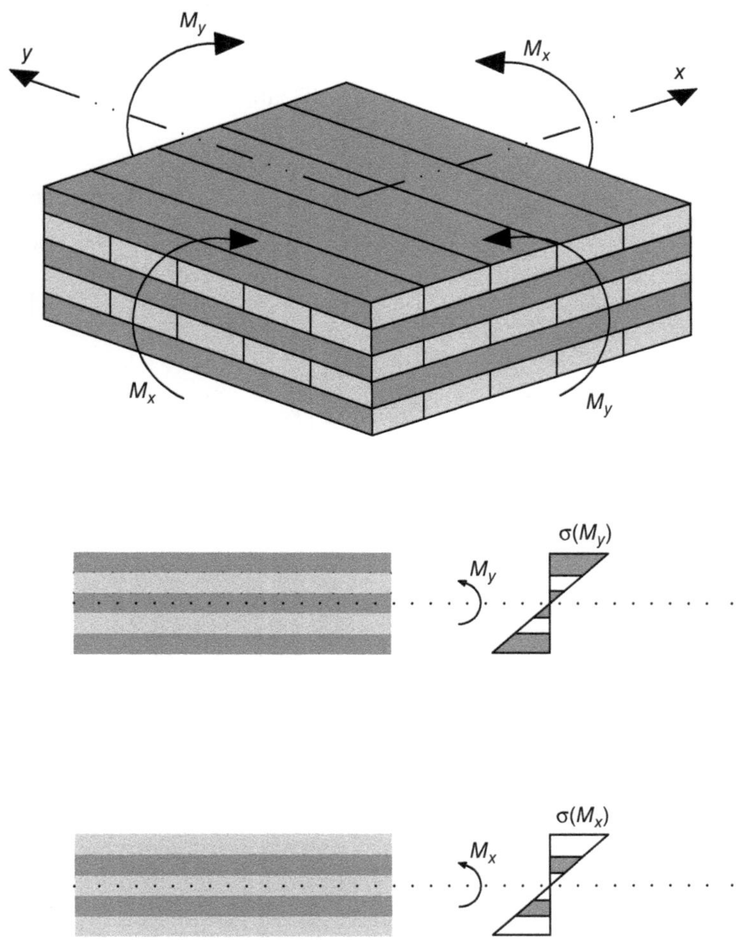

FIGURE 3.4 Biaxial bending in the plane of the plate.

The maximum stress $\sigma_{mx,max,d}$ and $\sigma_{my,max,d}$ in the x and y directions respectively, obtained in each direction should be verified with the allowable per Eqs. (3.5) and (3.6):

$$\sigma_{mx,max,d} \leq \text{Calculated flexural strength of the CLT section as applicable per the design standard} \tag{3.5}$$

$$\sigma_{my,max,d} \leq \text{Calculated flexural strength of the CLT section as applicable per the design standard} \tag{3.6}$$

The flexural resistance values are obtained from the characteristic strength values and by applying the appropriate modification factors and safety factors as applicable by the design standard.

The interaction between the stresses in the two directions, which may have an effect on the ultimate resistance, has not been determined yet.

In the case of axial forces present in the CLT member, the stresses are determined based on the layers with fibers in the same direction as the axial force and neglecting the transverse fibers. The axial stresses in the direction considered should be combined with flexural tensile stresses in the same direction. The interaction between the axial stresses and flexural tensile stresses should be checked in both directions. Figure 3.5 shows the biaxial flexural and the axial forces on a CLT panel.

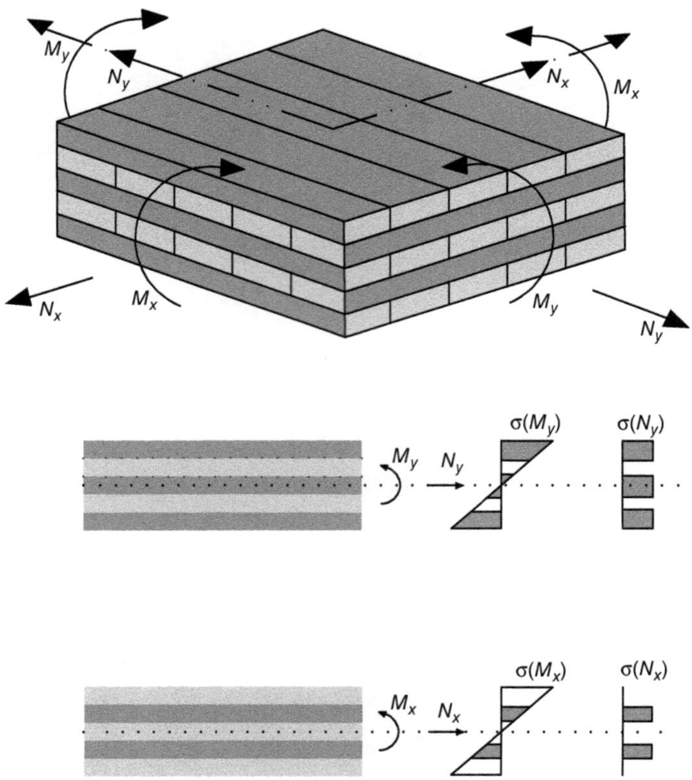

FIGURE 3.5 Biaxial flexural and axial forces on a CLT panel cross section and their stress distribution.

The interaction checks are presented in the following equations:

$$\frac{\sigma_{\text{mx,max},d}}{\text{calculated flexural strength}}+\frac{\sigma_{\text{cx,max},d}}{\text{calculated axial compressive strength}}\leq 1 \quad (3.7)$$

$$\frac{\sigma_{\text{my,max},d}}{\text{calculated flexural strength}}+\frac{\sigma_{\text{cy,max},d}}{\text{calculated axial compressive strength}}\leq 1 \quad (3.8)$$

or

$$\frac{\sigma_{\text{mx,max},d}}{\text{calculated flexural strength}}+\frac{\sigma_{\text{tx,max},d}}{\text{calculated axial tensile strength}}\leq 1 \quad (3.9)$$

$$\frac{\sigma_{\text{my,max},d}}{\text{calculated flexural strength}}+\frac{\sigma_{\text{ty,max},d}}{\text{calculated axial tensile strength}}\leq 1 \quad (3.10)$$

where, $\sigma_{\text{cx,max},d}$ and $\sigma_{\text{cy,max},d}$ are maximum compression stresses in the x- and y-directions respectively in the x and y directions respectively, $\sigma_{\text{tx,max},d}$ and $\sigma_{\text{ty,max},d}$ are maximum tension stresses in the x- and y-directions respectively.

The calculated strength is relative to the layer considered. The flexural and axial resistance values are obtained from the characteristic strength values and by applying the appropriate modification factors and safety factors as applicable by the design standard.

3.2.3 Shear

The principle used in flexural stress calculations is followed for shear stress calculations. The determination of shear stresses should take into consideration the different material characteristics of the individual layers. However, in the determination of shear stresses, the cross layers should be considered, as they are essential in transferring the shear stresses between the parallel layers in the direction considered. Figure 3.6 shows bending and shear stress distributions in a CLT member cross section oriented in the two different directions.

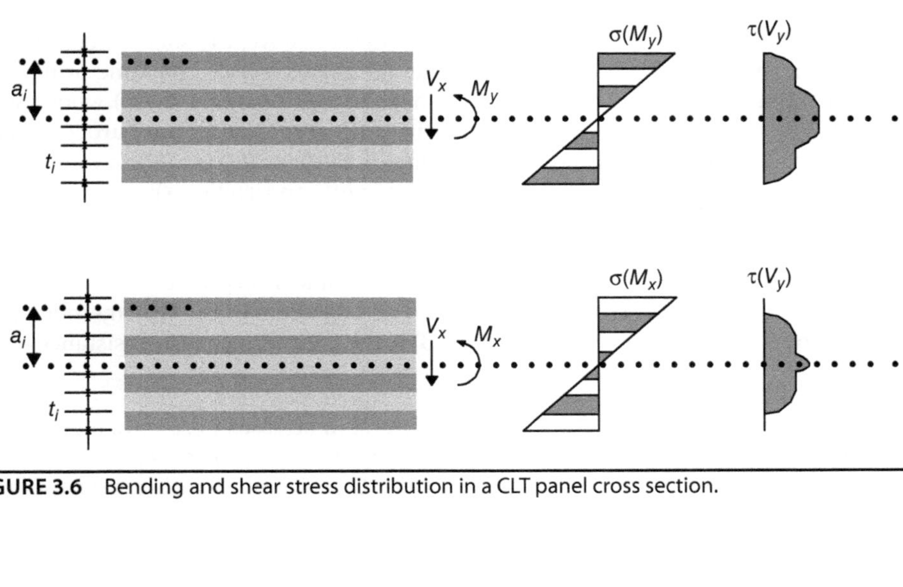

FIGURE 3.6 Bending and shear stress distribution in a CLT panel cross section.

The shear stress distribution is given by Eq. (3.11):

$$\tau(a) = \frac{V \cdot S}{K \cdot b} = \frac{V \cdot \int E \cdot a \cdot dA}{K \cdot b} \tag{3.11}$$

where a = distance from the center of gravity to each layer
b = width of the considered element
S = static moment of the layer
K = obtained from Eq. (3.1)

The shear stress distribution shows that the stress can reach the maximum value in the transverse layers of the section considered. The shear limit state must take into account the fact that the shear strength depends on the orientation of the fibers in the considered element. The shear strength check should be performed in both directions, not only in the longitudinal (strong) direction, where it is usually used, but also in the transverse (weak) layer, where the shear strength is much smaller. In the transverse direction, the material fails in tension perpendicular to gain to the main stress direction (parallel to grain) at a +/− 45-degree angle with respect to the axis of the element considered. The resistance to this type of failure is known as "rolling shear" and referred to with the index "r" and sometimes "s".

The shear stress distribution described above is approximated because the shear deformation of the individual layers is not considered here. This is justified by the fact that the thickness is sufficiently small with the element's width-to-thickness ratio greater than 10.

The check for shear given in Eq. (3.12) for the layers in the direction considered, and the shear check for the transverse direction with respect to the direction considered is given in Eq. (3.13):

$$\tau_d \leq \text{calculated shear strength in the direction considered} \tag{3.12}$$

$$\tau_d \leq \text{rolling shear strength} = \text{calculated shear strength in the transverse direction to the direction considered} \tag{3.13}$$

where, t_d is the calculated shear stress either in direction considered or in the transverse direction to direction considered.

According to the calculation model mentioned above, the shear strength of the CLT element should be taken as that of the glue-laminated wood. The same principle applies to the determination to the transverse shear strength (rolling shear) whose strength is twice the tensile strength perpendicular to grain according to the Eurocodes and other sources. The same model allows a 25% increase to the transverse shear strength.

The model for calculating the shear strength and rolling shear strength for CLT members is indicated in the model above.

The shear resistance values are obtained from the characteristic strength values and by applying the appropriate modification factors and safety factors.

Depending on the composition and the arrangement of the CLT member, the lowest value obtained from Eqs. (3.12) and (3.13) is the controlling shear resistance.

3.2.4 Bearing Stresses

CLT members may be subjected to compression stresses perpendicular to grain that may cause crushing of the wood if the bearing strength is exceeded. The bearing stresses develop at the contact areas of the CLT member with their supports and if

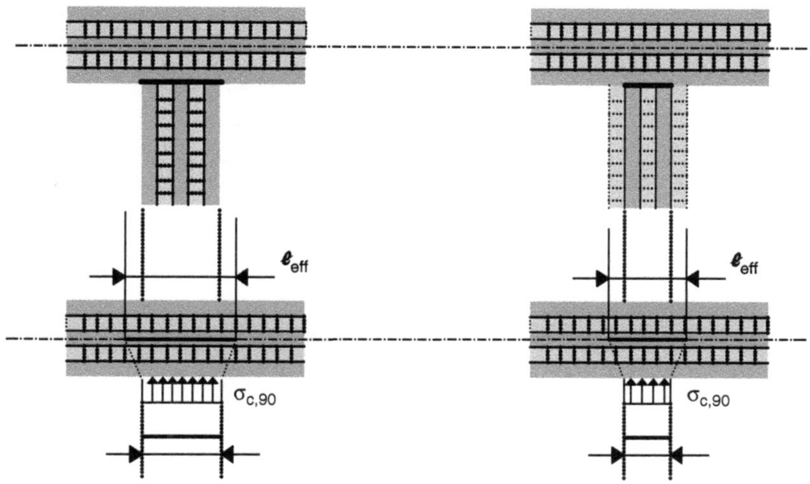

FIGURE 3.7 Compression perpendicular to the grain at the supports.

they are supporting concentrated loads, such as the case of a horizontal CLT panel supported by a wall. Figure 3.7 shows example of compression perpendicular to grain at the supports.

The determination of compression stress perpendicular to grain depends on the contact surface area between the supporting and the supported member. In Fig. 3.6, the contact area is defined by the width of the vertical element (sum of the thickness of individual layers). In the case of homogeneous support, made from other material for example, the actual surface area is considered. The check for compression stress perpendicular to grain, $\sigma_{c,90,d}$, is shown in Eq. (3.14):

$$\sigma_{c,90,d} \leq \text{calculated compression strength perpendicular to grain} \tag{3.14}$$

The compression strength perpendicular to grain resistance values are obtained from the characteristic, strength values and by applying the appropriate modification factors and safety factors.

3.2.5 Serviceability Check and Deflection Calculations

To determine the deflection and the serviceability limit state of a CLT member for a distributed load over its surface area, the flexural and the shear rigidities (stiffnesses) of the CLT member should be determined. The flexural rigidity (flexural stiffness) has already been defined by Eq. (3.1). The shear rigidity (shear stiffness) of a CLT member is based on the contribution of the individual layers and the difference in their stiffnesses. The shear modulus, G, depends on the orientation of stress and grain. G_0 is defined as the shear modulus for the longitudinal shear stress, and G_{90} is defined as the shear modulus for the transverse shear stress. Longitudinal and transverse shear stresses as well as the appropriate shear modulus for each case are shown in Fig. 3.8.

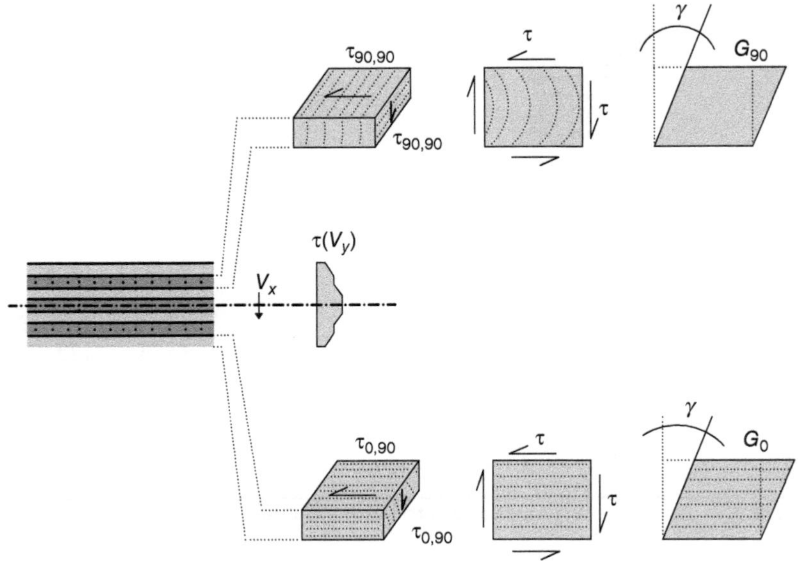

FIGURE 3.8 Shear stress in different layers with the shear modulus G_0 and G_{90}.

The shear stiffness of the CLT section is obtained from Eq. (3.15) [1]:

$$S_v = GA' = \kappa \cdot \sum (G_i \cdot b_i \cdot t_i) = \kappa \cdot \sum (G_i \cdot A_i) \tag{3.15}$$

where S_v = shear stiffness in the considered direction
G_i = single layer shear modulus
A_i = area of a single layer
b_i = width of the element considered
t_i = thickness of the single layer
κ = correction coefficient

The correction factor, κ, allows for the consideration of nonuniform characteristics of the whole thickness, and it depends on the geometry and the characteristics of the individual layers. The κ factor can be determined using Eq. (3.16) [1]:

$$\kappa = \frac{1}{\sum (G_i \cdot A_i) \cdot \frac{1}{K^2} \cdot \int \frac{S^2(z)}{G_i(z) \cdot b} dz} \tag{3.16}$$

where $S(z)$ = static moment
z = vertical axis (material thickness)

The value of the coefficient κ varies and depends on the number of layers, the ratio between G_0 and G_{90}, and the thickness of the material. The analysis of the variations of this coefficient shows that for a constant thickness of the individual layers and for a ratio between G_0 and G_{10} equal to 10, the following values of κ are obtained (Table 3.1) [1].

	Number of layers					
	3	5	7	9	11	13
κ	0.21	0.24	0.26	0.27	0.27	0.27

TABLE 3.1 κ Coefficient vs. Number of Layers

Depending on the composition of the CLT member with layers of different thicknesses and with an odd number of layers in which there are more layers oriented in the strong direction more than others, it is shown that the κ coefficient value is between 0.20 and 0.30.

As per the European standards [2] and the CLT Handbook [3, 4], G_{90} is obtained from Eq. (3.17):

$$G_{90} = \frac{G_0}{10} \tag{3.17}$$

3.2.6 Wall Elements

Introduction

The wall element, from a structural point of view, is a slab in a vertical position to which the plate effect is added when subjected to bending, which is due to wind loads, for example, applied on the building exterior.

For the plate structural effect, the theory mentioned above remains valid, and the structural behavior of the slab (wall) must be analyzed in detail before sizing and performing the necessary design checks.

The internal forces in a plate element are shown in Fig. 3.9; the internal forces n_x and n_y represent the axial stresses, and n_{xy} represents the shear stress in the plane of the CLT member. The unit for n_x, n_y, and n_{xy} is force per unit width (kN/m or lb/ft). Forces in lb or kN are expressed in capital letters, such as N_x, N_y, and N_{xy}.

Different cases for strength and stiffness are analyzed and discussed in the following sections.

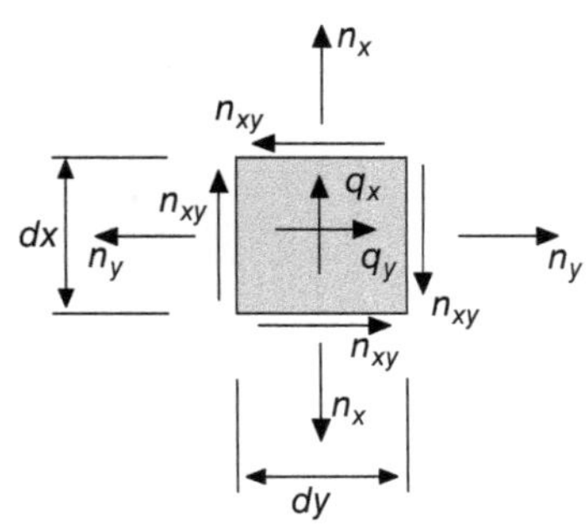

FIGURE 3.9 Plate element with the forces acting on it.

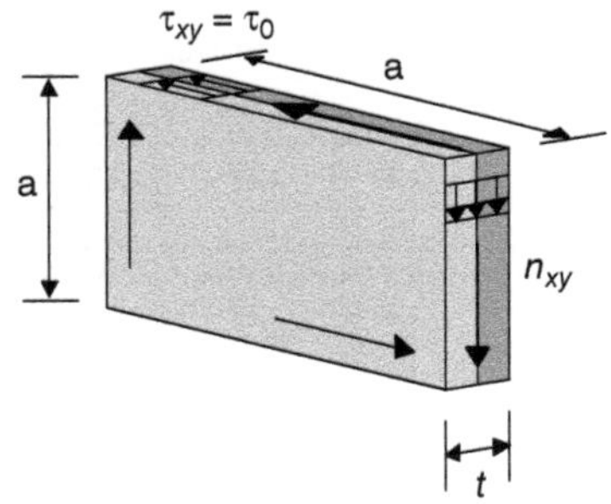

FIGURE 3.10 Shear stresses on a plate element.

Shear Stress in the Plane of the Wall

The shear stresses in a plate element, in general, can be determined from Eq. (3.18); shear stresses in a plate element are shown in Fig. 3.10:

$$\tau_{xy} = \tau_0 = \frac{n_{xy}}{t} = \frac{V_{xy}}{t \cdot a} \tag{3.18}$$

where n_{xy} is the force per unit length, V_{xy} is the force, t is the thickness of the element, and a is the length of the element. The stresses due to n_{xy} force create a tension stress in the CLT element as a result of the multilayer structure of the material.

In a CLT member, the material is not homogeneous across the material thickness, as it is made up of several cross layers that have different material properties in the different directions. The structural analysis starts from the principle that different cross layers are connected to each other only through the contact surface between the horizontal and vertical edges and that the transfer of the forces between one layer and another is achieved by the glue. The shear stresses in a two orthogonal layers are shown in Fig. 3.11.

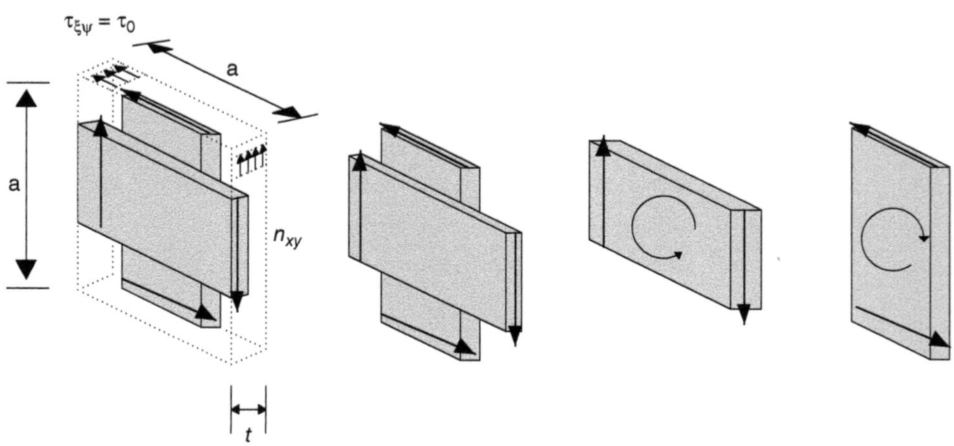

FIGURE 3.11 Stress on a shear element composed of two orthogonal layers.

The shear stress in the two-board element shown in the figure is composed of shear stress and twisting moment at the contact surfaces. The shear stresses and the stresses resulting from the twisting moments represent the state of stress caused by the n_{xy} force in a CLT member. These two components are also defined as limit states: shear and torsion limit states.

The stresses resulting from shear and torsion limit states should be determined and checked separately since they are two different limit states and correspond to two different resistance factors. The stresses resulting from shear and torsion limit states can be determined based on one basic and ideal element consisting of two orthogonal layers that are part of the thickness of a member and based on neglecting the lack of symmetry in the plane of the plate. The general cases for shear stresses in a member discussed above are shown in Fig. 3.12.

For the general case, the nominal shear stress shown in Fig. 3.12 can be obtained from Eq. (3.19) and is based on ideal element without cracks:

$$\tau_0 = \frac{V_{xy}}{t \cdot a} \tag{3.19}$$

For the real case, cases (b) and (c) of Fig. 3.12 are considered, and the shear force V can be determined from Eq. (3.20):

$$V = t \cdot a \cdot \tau_0 \tag{3.20}$$

The tension due to torsion is obtained from the twisting moment, [Eq. (3.21)], divided by the polar moment of inertia, [Eq. (3.22)], resulting in the shear stress in Eq. (3.23):

$$M_t = t \cdot a^2 \cdot \tau_0 \tag{3.21}$$

$$W_p = \frac{a^3}{3} \tag{3.22}$$

$$\tau_T = \frac{M_t}{W_p} = 3 \cdot \tau_0 \cdot \frac{t}{a} \tag{3.23}$$

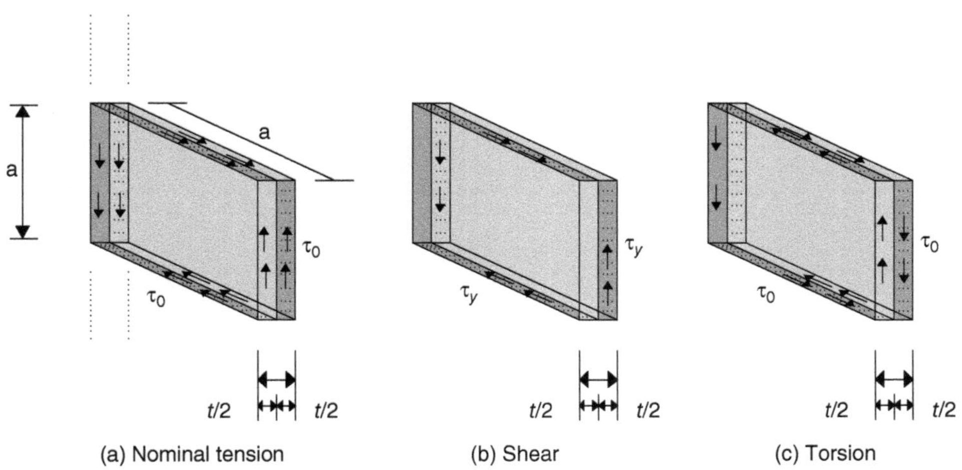

FIGURE 3.12 General cases of shear stress.

Now the actual thickness of the CLT member should be defined and applied to the calculations, that is, to apply it to a real element with a finite number of layers where symmetry planes are not always present as opposed to an ideal element and to what is shown in Fig. 3.12. The actual thickness definition is considered as the glued surfaces given that equilibrium is satisfied. The thicknesses of the individual layers as well as the total thickness of the member should be transformed into an ideal thickness, t^*, which will be used as the basis for the calculations for the actual shear and torsional stresses.

The ideal thicknesses are defined in the following way with reference to the glued surfaces:

- For the outer layers: the minimum value between the thickness of the inner layer and the double of the thickness of the outer layer to the glued surface considered: for example, for surface A: $t^*_A = \min.\ (2t_1, t_2)$
- For the inner layers: the minimum value of the thicknesses of the layers adjacent to the surface glued: for example, for the surface B: $t^*_B = \min.\ (t_3, t_4)$

The individual layer thicknesses and surfaces A and B are shown in Fig. 3.13.

The total ideal thickness of the element is given by Eq. (3.24):

$$t^*_{\text{tot}} = \sum_1^n t^*_i \tag{3.24}$$

where t^*_i = ideal thickness of the individual layers (in relation to the glued surface)
n = number of glued surfaces

It will be noted that the ideal total calculated thickness is smaller than the actual thickness of the plate. The strength check must be performed for the two cases separately as follows:

$$\tau_{v,d} \leq \text{calculated shear strength due to shear as applicable per design standard} \tag{3.25}$$

$$\tau_{T,d} \leq \text{calculated shear strength due to torsion as applicable per design standard} \tag{3.26}$$

where, $\tau_{v,d}$ and $\tau_{T,d}$ are the calculated shear stress due to shear and due to torsion respectively.

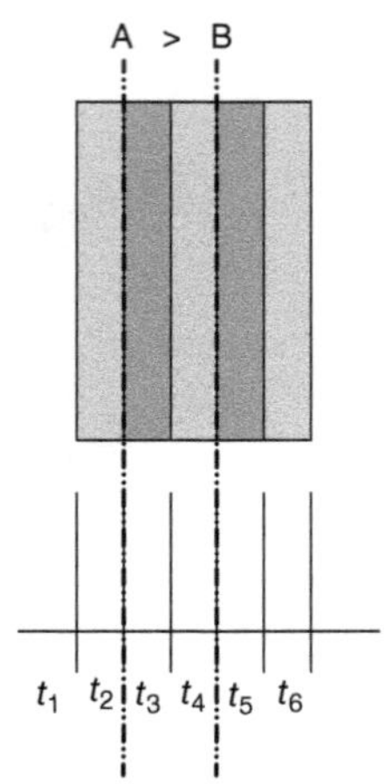

FIGURE 3.13 Ideal thicknesses of the layers for calculating the shear strength.

For the strength check, it is first necessary to define the determining force, referred to the glued surfaces:

$$n_{xy,i} = n_{xy} \cdot \frac{t_i^*}{\sum_1^n t_i^*} = n_{xy} \cdot \frac{t_i^*}{t_{tot}^*} \tag{3.27}$$

It is therefore necessary to determine the ideal reference shear stress using the ideal thickness of the plate:

$$\tau_{0,i} = \frac{n_{xy,j}}{t_i^*} = n_{xy} \cdot \frac{t_i^*}{\sum_1^n t_i^*} \cdot \frac{1}{t_i^*} = n_{xy} \cdot \frac{1}{\sum_1^n t_i^*} = n_{xy} \cdot \frac{1}{t_{tot}^*} \tag{3.28}$$

It is therefore possible to perform the shear strength check for the two limit states separately:

$$\tau_{v,d} = 2 \cdot \tau_0^* \leq \text{calculated shear strength due to shear as applicable per design standard} \tag{3.29}$$

$$\tau_{T,d} = 2 \cdot \tau_0^* \cdot t_i^* / a \leq \text{calculated shear strength due to torsion as applicable per design standard} \tag{3.30}$$

Now it is possible to proceed with the shear strength check of the CLT wall members based on the model of complete mechanical and structural behavior.

The shear stress is determined from Eq. (3.31):

$$\tau_{v,d} = \frac{3}{2} \cdot \frac{n_{xy,d}}{t_{min}} \leq \text{calculated shear strength due to shear as applicable per design standard} \tag{3.31}$$

where t_{min} = minimum sum of the layer thicknesses in the same direction.

The difference is essentially given by the coefficient 3/2, in which the shear stress distribution is assumed to be similar to that of a beam under bending. Other than this basic difference, similar results are obtained for elements formed by layers of boards of identical thickness.

The shear due to torsion check is determined from Eq. (3.32):

$$\tau_{T,d} = \frac{F_d \cdot h}{\sum J_p} \cdot \frac{a}{2} \leq \text{calculated shear strength due to torsion as applicable per design standard} \tag{3.32}$$

where F_d = force acting on the wall
a = dimension of the contact surface between the layers
h = height of the wall
J_p = polar moment of inertia of the same surface

The shear resistance values are obtained from the characteristic strength values and by applying the appropriate modification factors and safety factors as applicable by the design standard.

Shear Stiffness and Deformation in the Plane of the Plate (Slab and Wall)

The shear stiffness of the wall element must take into account the limit states defined in the previous section and their interaction. The total deformation is therefore given by the sum of the component given by tension (due to shear) with the component given by torsion. For the tension component, the deformation is given by Eq. (3.33):

$$\gamma_1 = \frac{\tau_0}{G_{0,\text{mean}}} \tag{3.33}$$

The torsion component is given by Eq. (3.34):

$$\Phi = \gamma_2 = \frac{M_T}{G_T \cdot J_P} \cdot \frac{t}{2} = \frac{\tau_0 \cdot t \cdot a^2}{G_t \cdot \frac{a^4}{6}} \cdot \frac{t}{2} = \frac{6 \cdot \tau_0}{G_{0,\text{mean}}} * \left(\frac{t}{a}\right)^2 \tag{3.34}$$

where G_t is taken as equal to half of $G_{0,\text{mean}}$, that is, of the value of the shear modulus of the boards used for the production of CLT.

The shear stiffness for the plate is important for practical applications of CLT and is expressed as

$$D_{xy} = G^* \cdot t_{\text{CLT}} \tag{3.35}$$

where D_{xy} = shear stiffness of the CLT plate
G^* = CLT plate shear modulus
t_{CLT} = plate thickness

The determination of D_{xy} depends on several factors, including the geometry of the CLT member (a and t). Furthermore, the approach represented here includes several simplifications and approximations, which can be corrected with the appropriate correction coefficients. Given $G_{0,\text{mean}}$, obtained from the characteristics of the material, the stiffness of the CLT plate can be obtained from Eq. (3.36):

$$\frac{G^*}{G_{0,\text{mean}}} = \frac{1}{1 + 6 \cdot \alpha_T \cdot \left(\frac{t}{a}\right)^2} \tag{3.36}$$

where

$$\alpha_T = 0.32 \cdot \left(\frac{t}{a}\right)^{0.77} \tag{3.37}$$

The correction factor α_T represents a numerical correction coefficient.

The study of the stiffness of the CLT plate showed that the approximation is good enough for practical applications.

3.2.7 Axial Forces in the Plane of the Plate—Forces N_x and N_y (n_x and n_y)

The structural behavior, analysis, and design for axial forces acting in the plane of the plate are based on the theory and behavior of a plate element. Axial forces and stresses acting on a CLT plate are shown in Fig. 3.14.

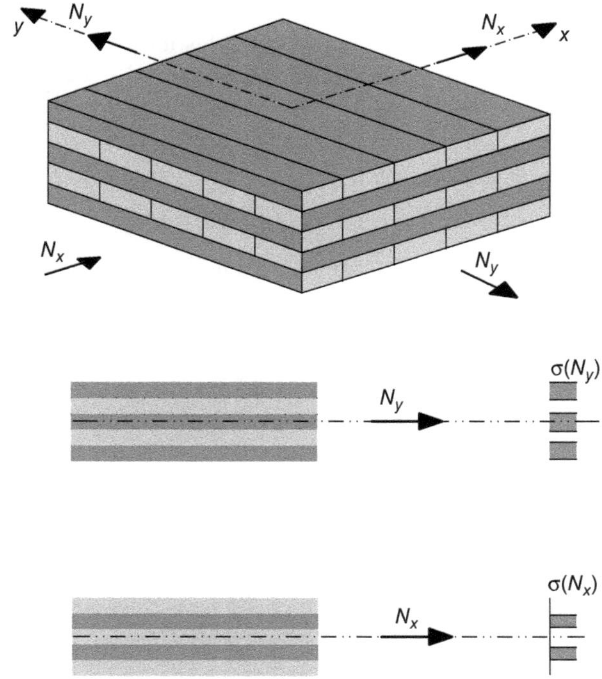

FIGURE 3.14 Stress due to axial forces N_x and N_y (or n_x and n_y if using force per unit length).

The tension strength of the CLT member must take into account the characteristics and the material properties of the different layers of the CLT element. Following the same analogy for bending members, considering the significant difference between the modulus of elasticity, *E*, of the different layers in the considered direction and the transverse direction, it is certainly possible to neglect the transverse layers and to perform the calculations considering the layers in the stress direction only. Figure 3.15 shows the stresses considered in each direction. In the *y*-direction, the top, middle, and bottom layers are considered the layers in the stress direction (strong modulus of elasticity in the

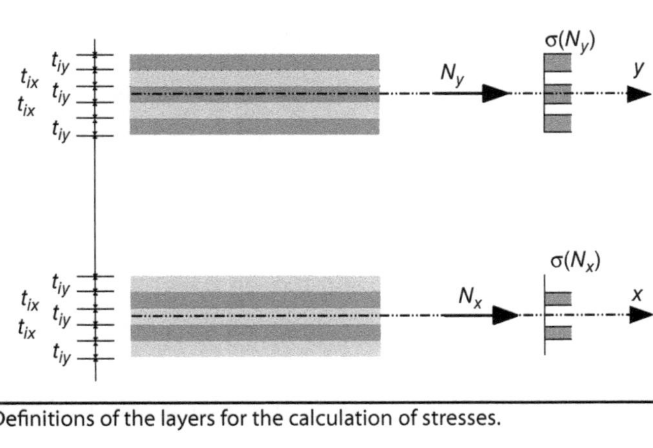

FIGURE 3.15 Definitions of the layers for the calculation of stresses.

y-direction). In the x-direction, the second and the fourth layers are considered the layers in the stress direction (strong modulus of elasticity in the x-direction).

The effective area for each direction considered is calculated as per Eqs. (3.38) and (3.39):

$$A_x = \sum b \cdot t_{ix} = t_x \cdot b \tag{3.38}$$

$$A_y = \sum b \cdot t_{iy} = t_y \cdot b \tag{3.39}$$

where t_{ix} = thickness of individual layers oriented in the x-direction
t_{iy} = thickness of individual layers oriented in the y-direction
t_x = total thickness of the layers oriented in the x-direction
t_y = total thickness of the layers oriented in the y-direction
b = width of the element
A_x = effective area in the x-direction
A_y = effective area in the y-direction

The tension stresses are calculated using Eqs. (3.40) and (3.41):

$$\sigma_x = \frac{n_x}{t_x} = \frac{N_x}{A_x} \tag{3.40}$$

$$\sigma_y = \frac{n_y}{t_y} = \frac{N_y}{A_y} \tag{3.41}$$

where N_x = axial force in the x-direction
n_x = axial force per unit of width in the x-direction
N_y = axial force in the y-direction
n_y = axial force per unit of width in the y-direction

The axial strength check is performed by comparing the axial stress with the design axial capacity of the member; the calculated axial stress must be less or equal to the design capacity. The design capacity depends on the capacity of the material, taking into account that the tensile strength values are different from the compressive strength values:

$$\sigma_{x,c,d} = \frac{n_x}{t_x} \leq \text{calculated compression strength in the } x\text{-direction parallel to grain} \tag{3.42}$$

$$\sigma_{y,c,d} = \frac{n_y}{t_y} \leq \text{calculated compression strength in the } y\text{-direction parallel to grain} \tag{3.43}$$

or

$$\sigma_{x,c,d} = \frac{n_x}{t_x} \leq \text{calculated tension strength in the } x\text{-direction parallel to grain} \tag{3.44}$$

$$\sigma_{y,c,d} = \frac{n_y}{t_y} \leq \text{calculated tension strength in the } y\text{-direction parallel to grain} \tag{3.45}$$

where, $\sigma_{x,c,d}$ and $\sigma_{y,c,d}$ are the calculated axial stress due to axial load in the x- and y-direction respectively.

Axial strains can be calculated in the two directions; strains could be tensile or compressive strains, depending on the axial force applied. Strains in the x- and y-directions are obtained from Eqs. (3.46) and (3.47):

$$\varepsilon_x = \frac{n_x}{D_x} = \frac{N_x}{D_x \cdot b} = \frac{N_x}{E_{0,\text{mean}} \cdot A_x} \tag{3.46}$$

$$\varepsilon_y = \frac{n_y}{D_y} = \frac{N_y}{D_y \cdot b} = \frac{N_y}{E_{0,\text{mean}} \cdot A_y} \tag{3.47}$$

where

$$D_x = E_{0,\text{mean}} \cdot t_x \tag{3.48}$$

$$D_y = E_{0,\text{mean}} \cdot t_y \tag{3.49}$$

3.3 Combined Loads

Members subjected to compressive forces such as walls may be subjected to instability and therefore may experience buckling. Such members may be subjected to transverse load, such as wind load, which creates a bending moment on the wall which increases the potential for buckling. Design of such members should consider the effect of both axial forces and bending moments.

The combined load should be verified with allowable limits as applicable by the design standards. The verification of the ultimate limit state must consider the combination of the stresses due to the moment and to the axial force as shown and discussed in Sec. 3.2.2.

3.4 Complex Elements

The procedures and analyses methods presented in the previous sections deal with basic structural elements that are subjected to one force or a combination of forces. The verification of member capacities is performed by applying the fundamental principles of wood design and the behavior of CLT members. The basic principles presented in the sections above apply to the most common cases in practice; however, when complex and uncommon cases are encountered, such as members with irregular geometry (for example, walls with openings), a more detailed analysis is required, such as finite-element analysis. The determination of internal stresses is obtained in these cases by performing numerical modeling of the structural element by defining the different strength and stiffness characteristics of the material in the two different directions of the element.

When the stresses are obtained from analysis, the design verifications can be performed as presented in the sections above.

3.5 Analytical Methods and Design Procedures for CLT Members

Several methods have been originally introduced and adopted to determine the mechanical properties of CLT in Europe. These methods were either experimental or analytical. Extensive tests were conducted to evaluate the flexural properties of floor elements with specific span-to-depth ratios and specific material and layer arrangements. However, the issue with the experimental approach is that more tests should be performed every time there is a change in the type of material, layer arrangement, manufacturing, or any other parameter.

Therefore, the analytical approach is the desired approach once verified with experimental results. The stiffness and strength properties for the CLT element are determined from the properties of each layer in the CLT element.

One of the common analytical methods for the analysis and design of CLT members is known as the "gamma method" or (the γ-method); this method is based on the "Mechanically Jointed Beams Theory," which is available in Annex B of Eurocode 5 [2]. Due to the difference in the properties of each layer, "effective stiffness" is used as the property for the CLT element. The gamma method introduced the "connection efficiency factor" (γ_i) to account for the shear deformation in the perpendicular layer; $\gamma = 1$ is used for a completely glued member, and $\gamma = 0$ is used when there is no connection at all. The gamma method provides a closed-form solution for a simply supported beam or panel with sinusoidal load distribution; however, the difference between exact solution and uniformly distributed loads and point loads is small and acceptable for practical purposes [5].

Another method used to determine the design properties of a CLT member is the "k-method," in which Blass and Fellmoser [6] have applied the "composite theory" to predict the design properties of CLT members. This method is acceptable for high span-to-depth ratio members, and it does not account for shear deformation in the individual layers.

The most recent method is known as the "shear analogy method" [7]. This method is applicable to solid panels with cross layers, takes into account shear deformation, and is not limited to certain number of layers. This method is considered adequate for CLT members and as the most accurate compared to other methods in the literature.

The lumber material properties for each layer used in the calculations of the design properties should be based on published design standards and should be consistent the design philosophy used (NDS in the United States, CSA O86 in Canada, etc.).

The design procedures presented in this chapter are applicable to CLT manufactured with glued surfaces (face glue). The chapter does not cover nailed or doweled CLT products.

3.5.1 Mechanical Properties of CLT Elements

For CLT elements used in floor and wall systems, CLT members are typically composed of individual board members; the thickness of these members varies between ⅜ and 2 in. (~10 and 50 mm), and the width varies between 3 and 10 in. (~80 and 240 mm). The individual boards used are either visually or machine stress rated with average moisture content of 12% ± 2%. The individual lumber members used for CLT manufacturing should be graded by a recognized standard grading agency. Boards graded using in-house quality control standards maybe used as long as they are validated by a certification agency.

3.5.2 Rolling Shear Modulus and Shear Deformation—Loads Perpendicular to the Plane

One of the key parameters that may control the design of CLT members is rolling shear strength and stiffness. The load-bearing behavior of the CLT member is affected by the material itself and by its constructive anisotropy [8]. Bejtka and Lam (2008) confirmed this finding at the University of British Columbia using CLT panels built with Canadian lodgepole pine laminates [9]. Fellmoser and Blass showed that stress distribution in the CLT panel layers and the effective bending stiffness of the panel depend on the rolling shear modulus of the cross layers [10]. The parameters that affect the rolling shear modulus are many such as moisture content, cross-layer density, laminate thickness, type of species, annual ring orientation, and size and geometry of the board's cross section.

The literature has recommendations for rolling shear modulus, G_R. As per Mestek et al. (2008) [8], G_R is assumed to be 1/10 of the shear modulus parallel to the grain of the boards, G_0 ($G_R \approx G_0/10$).

From literature and experience, the shear modulus G of wood products is generally assumed to be between 1/12 and 1/20 of the true modulus of elasticity, that is, $E_{true}/G_0 \approx$ 12–20. For softwood lumber, this ratio may be assumed to be 16 [3]. Figure 3.16 illustrates the rolling shear deformation behavior of a five-layer CLT cross section [3].

3.5.3 Shear Deformation Due to Loads Perpendicular to Plane

Shear deformation of CLT panels subjected to uniform loaded may be neglected for a span-to-depth ratio (l/d) greater than 20 [8]. Other sources in the literature suggest a minimum span-to-depth ratio of 30 before neglecting shear deformation. As per the CLT Handbook, Canadian edition, [3] for slabs with a span-to-depth ratio of 30, the contribution of shear deformation was about 11 percent, and 22 percent for slab with a span-to-depth ratio of 20.

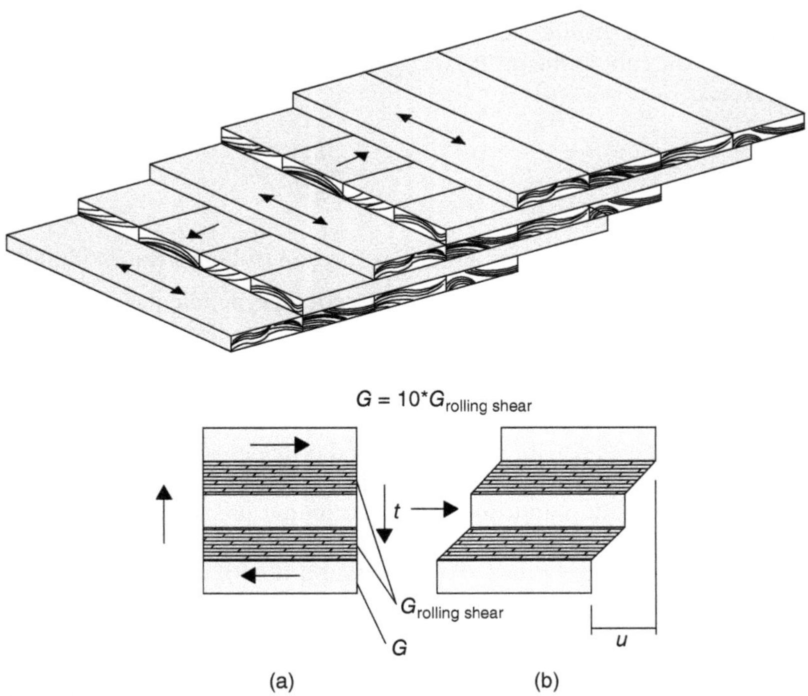

FIGURE 3.16 Rolling shear deformation of a five-layer CLT member, (*a*) undeformed (*b*) deformed.

3.5.4 Analytical Design Methods for CLT Elements Used in Floor Systems

Several methods have been developed to determine the basic mechanical properties of CLT elements. This section provides details about the common design methods used for CLT elements. Since CLT panels are relatively light building material compared to concrete and steel, the design could be controlled by serviceability requirements such as deflection, vibration, and creep instead of strength requirements such as bending and shear strengths.

Mechanically Jointed Beams Theory (Gamma Method)

The mechanically jointed beams theory, or the gamma method (γ-method), was developed in 1955 by Karl Möhler [3]. This method, used by some CLT manufacturers, was developed for beams with *I* or *T* cross-section shapes connected with mechanical fasteners. The effective stiffness of the beam (EI_{eff}) depends on the section properties of the beam and the connection efficiency between the different parts of the beam cross section using the connection efficiency factor gamma (γ), which depends on slip of the fasteners between the different parts of the cross section. Gamma (γ) is 0 if there is no mechanical connection and is 1 for rigid connection such as glued sections.

Since CLT members have no mechanical joints (surfaces are glued), the conventional γ-method is modified to make it applicable to CLT members. The boards on the longitudinal direction (strong direction) are considered as the load-carrying elements, and the rolling shear stiffness of the cross layers is considered as stiffness connecting the longitudinal layers. In other words, the rolling shear stiffness or deformation (Fig. 3.16) is considered as mechanical fasteners. Equation (3.50) shows the modification of the γ-method to make it applicable to CLT members. In this equation, s/K_i, which is the ratio for fasteners at each interface "*i*" for determining the connection efficiency factor, is replaced by the rolling shear slip, which is the shear deformation between the load carrying layers:

$$\frac{s}{K_i} = \frac{\bar{h}_i}{G_R \cdot b} \tag{3.50}$$

where G_R = shear modulus perpendicular to grain (rolling shear modulus)

$\bar{h}_i$ = thickness of board layers in direction perpendicular to the stress direction

b = width of the panel (1 ft or 1 m)

s = spacing between mechanical fasteners (not applicable to glued CLT members)

K_i = slip modulus of mechanical fasteners (not applicable to glued CLT members)

As mentioned above, the gamma method provides a closed-form solution for a simply supported beam or panel with sinusoidal load distribution; however, the difference between exact solution and uniformly distributed loads and point loads is small and acceptable for practical purposes [5].

The gamma method is based on the assumption of using a simply supported beam with a span equals l (length between the zero moments) to determine the γ-factor. In the case of a cantilever with a cantilever length l_c, two times the cantilever length is used in the calculation of effective bending stiffness EI_{eff}. In the case of continuous spans, there are two methods that can be followed to determine the effective bending stiffness EI_{eff}: a simplified method and an iterative method. The simplified method suggests using $0.8l$ as the distance between the inflection points, whereas in the iterative method, an initial value between the inflection points is used for the first iteration, say, $0.8l$. Then a computer program is used to determine the inflection points for a beam with EI_{eff}, then the new length between the inflection points is obtained and EI_{eff} recalculated. Other iterations can be performed similarly, usually after few iterations, and a stable solution is achieved for EI_{eff}.

Bending Strength and Stiffness—Loads Perpendicular to the Plane (Floor and Roof Systems)

The evaluation and approval of CLT members in Europe is performed by a combination of experimental and analytical approaches. The tests are based on EN and DIN standards. The approach presented herein is based on defining a relationship between the bending strength, obtained from experiments, and the effective section modulus of the CLT member. The section modulus is calculated using Eq. (3.51a):

$$S_{eff} = \frac{2 \cdot I_{eff}}{h_{tot}} = \frac{I_{eff}}{0.5 \cdot h_{tot}} \tag{3.51a}$$

where S_{eff} = effective section modulus
I_{eff} = effective moment of inertia
h_{tot} = total depth of the panel

$$EI_{eff} = \sum_{i=1}^{n} (E_i I_i + \gamma_i E_i A_i a_i^2) \tag{3.51b}$$

where $0 < \gamma \leq 1$
$\gamma = 1$ for rigid connection and $\gamma = 0$ for no connection. Typically, γ may vary from 0.85 to 0.99.

Per the mechanically jointed beams theory, as per Appendix B of Eurocode 5, the maximum bending stress in the panel can be obtained from Eq. (3.52):

$$\sigma_{max} = \sigma_{global} + \sigma_{local} \tag{3.52}$$

where σ_{local} = stress in the outside layer as a result of bending of that layer
σ_{global} = axial stress developed in the outside layer due to bending

The global stress and the local stress can be obtained from Eqs. (3.53) and (3.54):

$$\sigma_{global} = \frac{\gamma_1 E_1 a_1 M}{(EI)_{eff}} \tag{3.53}$$

$$\sigma_{local} = \frac{0.5 E_1 h_1 M}{(EI)_{eff}} \tag{3.54}$$

where a_1 = the distance between the centroid of the first layer and the centroid of the panel cross section
h_1 = the thickness of the first (outermost) layer

Substituting Eqs. (3.53) and (3.54) in Eq. (3.53), the maximum stress is obtained from Eq. (3.55):

$$\sigma_{\max} = \frac{\gamma_1 E_1 a_1 M}{(EI)_{\text{eff}}} + \frac{0.5 E_1 h_1 M}{(EI)_{\text{eff}}} \tag{3.55}$$

or

$$\sigma_{\max} = \frac{ME_1}{(EI)_{\text{eff}}} \cdot (\gamma_1 a_1 + 0.5h_1) \tag{3.56}$$

When the modulus of elasticity in all longitudinal layers is the same, $E_1 = E_2 = E_3 = E$, Eq. (3.56) becomes

$$\sigma_{\max} = \frac{M}{I_{\text{eff}}} \cdot (\gamma_1 a_1 + 0.5h_1) \tag{3.57}$$

The calculated maximum bending stress in Eq. (3.57) should be compared with allowable stress provided by the applicable design standard.

Shear Strength—Loads Perpendicular to the Plane (Floor and Roof Systems)

The shear strength of structural glued members is usually determined from experiments. CUAP 03.04/06 provides details on test procedures that should be followed to determine the shear strength of glued structural members. The shear strength can be calculated using Eq. (3.58):

$$\tau = \frac{1.5 \cdot V}{A_{\text{gross}}} \tag{3.58}$$

where τ = maximum shear strength (psi or MPa)
V = maximum shear force (lb or N)
A_{gross} = gross cross-sectional area of the panel = $b \cdot h_{\text{tot}}$ (in.^{2} or mm^2)

Using simple beam theory and the theory of mechanically jointed beams, the maximum shear stresses occur where the normal stresses are equal to zero; the shear stress can be obtained by Eq. (3.59):

$$\tau = \frac{V \cdot (EQ)}{(EI)_{\text{eff}} \cdot b} \tag{3.59}$$

τ = maximum shear strength (ksi or MPa)
V = maximum shear force (lb or N)
Q = static moment of area for the cross-section (in.^{2} or mm^2)
b = width of the cross-section perpendicular to the shear flow (in. or mm), usually 1 ft or 1000 mm

Considering a CLT panel with five layers as shown in Fig. 3.17, the static moment of area Q for the section above the centroidal axis can be calculated using Eq. (3.60):

$$(EQ) = \gamma_1 \cdot E_1 \cdot a_1 \cdot A_1 + E_1' \cdot A_1' \cdot a_1' + \gamma_2 \cdot E_2 \cdot \frac{A_2}{2} \cdot \frac{h_2}{4} \tag{3.60}$$

The calculated maximum shear stress in Eq. (3.59) should be compared with calculated shear strength capacity provided as applicable by design standards.

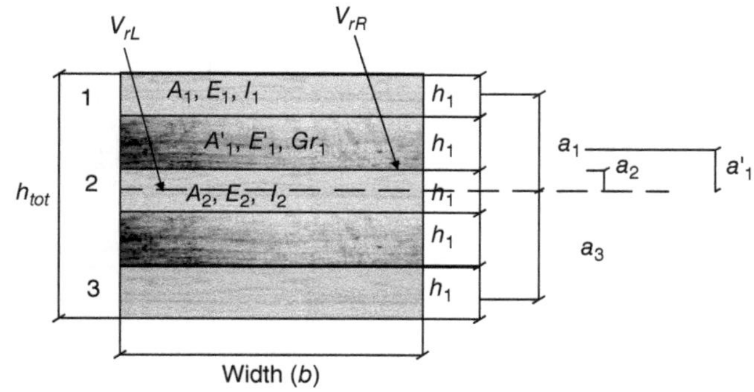

FIGURE 3.17 Cross section of CLT panel with five layers.

Composite Theory: k-Method

The composite theory method has been used in the plywood industry. This method does not take into account the plies stressed in the perpendicular direction in the determination of the bending properties, which means $E_{90} = 0$.

The method has been modified for design of CLT members to calculate EI_{eff} and assumes the following:

1. Linear stress-strain relationship is assumed with plane cross section remaining plane.
2. Shear deformation is not considered; therefore, the method may be used for high span-to-depth ratio $l/h \geq 30$.
3. Strength and stiffness properties are based on all layers, layers are loaded parallel to grain, and layers are loaded perpendicular to grain; the stiffness of layers stressed perpendicular to grain is calculated using the stiffness of the layers stressed parallel to grain divided by 30: $E_{90} = E_0/30$.
4. The composition factors, k_i, depend on the loading configuration and are presented in Table 3.2.

Using Table 3.2, the effective bending strength and stiffness for solid panels are presented in Table 3.3 [6].

The maximum bending stress is calculated using Eq. (3.61):

$$\sigma_{max} = \frac{M}{S} \tag{3.61}$$

The calculated maximum shear stress in Eq. (3.61) should be compared with allowable stress capacity provided by the applicable design standard.

Shear Analogy Method (by Kreuzinger)

The shear analogy method is known as the most accurate for the design of CLT members [6]. In this method, the different moduli of elasticity and shear moduli of elasticity of every single layer are considered for nearly any system with any number of layers and any span-to-depth ratio. In this method also, the shear deformation is not neglected.

The multilayer cross section of the CLT element is separated into two virtual elements: beam A and beam B. Beam A represents the inherent flexural strength of the individual layers along their own neutral axes, and beam B represents the "Steiner"

Load configuration	k_i
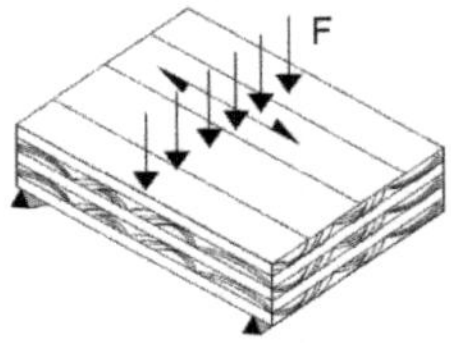	$k_1 = 1 - \left(1 - \frac{E_{90}}{E_0}\right) \cdot \frac{a_{m-2}^3 - a_{m-4}^3 + \cdots \pm a_1^3}{a_m^3}$
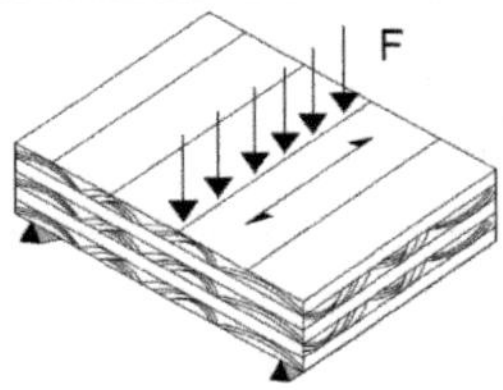	$k_2 = \frac{E_{90}}{E_0} + \left(1 - \frac{E_{90}}{E_0}\right) \cdot \frac{a_{m-2}^3 - a_{m-4}^3 + \cdots \pm a_1^3}{a_m^3}$
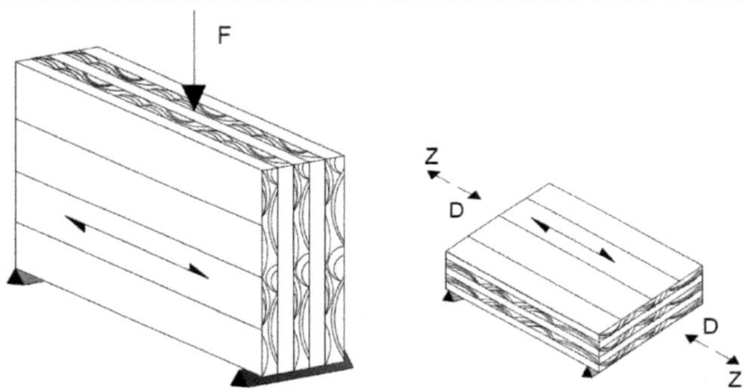	$k_3 = 1 - \left(1 - \frac{E_{90}}{E_0}\right) \cdot \frac{a_{m-2} - a_{m-4} + \cdots \pm a_1}{a_m}$
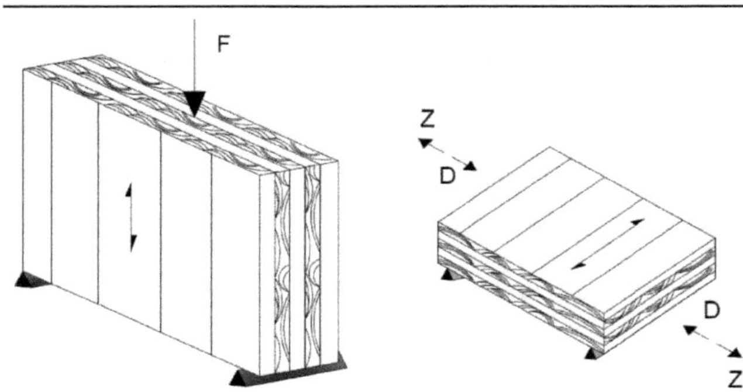	$k_4 = \frac{E_{90}}{E_0} + \left(1 - \frac{E_{90}}{E_0}\right) \cdot \frac{a_{m-2} - a_{m-4} + \cdots \pm a_1}{a_m}$

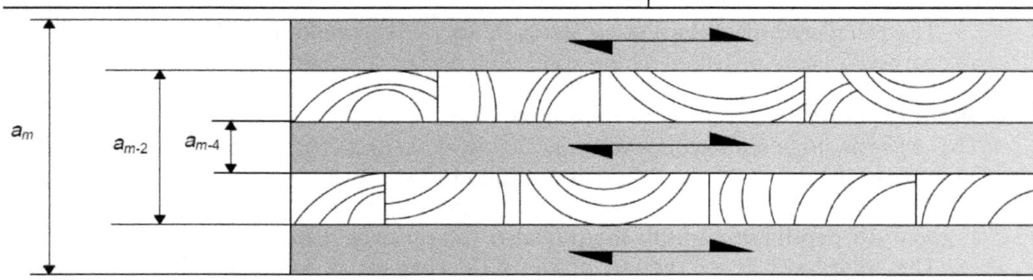

TABLE 3.2 Composition Factors, *k*, for Solid Wood Panels with Cross Layers [6].

Loading	To the grain of the outer skins	Effective strength value	Effective stiffness value
Perpendicular to the plane loading			
Bending	Parallel	$f_{b,0,eff} = f_{b,0}k_1$	$E_{b,0,eff} = E_0k_1$
	Perpendicular	$f_{b,90,eff} = f_{b,0}k_2a_m/a_{m-2}$	$E_{b,90,eff} = E_0k_2$
In-plane loading			
Bending	Parallel	$f_{b,0,eff} = f_{b,0}k_3$	$E_{b,0,eff} = E_0k_3$
	Perpendicular	$f_{b,90,eff} = f_{b,0}k_4$	$E_{b,90,eff} = E_0k_4$
Tension	Parallel	$f_{t,0,eff} = f_{t,0}k_3$	$E_{t,0,eff} = E_0k_3$
	Perpendicular	$f_{t,90,eff} = f_{t,0}k_4$	$E_{t,90,eff} = E_0k_4$
Compression	Parallel	$f_{c,0,eff} = f_{t,0}k_3$	$E_{c,0,eff} = E_0k_3$
	Perpendicular	$f_{c,90,eff} = f_{t,0}k_4$	$E_{c,90,eff} = E_0k_4$

TABLE 3.3 Effective Values of Strength and Stiffness for Solid Wood Panels with Cross Layers [6].

points part of the flexural strength, the flexible shear strength of the panel, as well as the flexibility of all connections. The beams are coupled with an infinitely rigid connection so that both beams A and B deflect together an equal amount of deflection. Figure 3.18 shows the entire cross section with its elements using the shear analogy method.

The bending stiffness of beam A, which depends on the inherent stiffness of the individual layers of the cross section, is given in Eq. (3.62):

$$B_A = \sum_{i=1}^{n} E_i \cdot I_i = \sum_{i=1}^{n} E_i \cdot b_i \cdot \frac{h_i^3}{12} \tag{3.62}$$

where $B_A = (EI)_A$

b_i = width of each individual layer, usually taken as 1 ft or 1 m for CLT panels

h_i = thickness of each individual layer

The bending stiffness of beam B is the sum of the Steiner points of all individual layers and is obtained using Steiner's theorem Eq. (3.63):

$$B_B = \sum_{i=1}^{n} E_i \cdot A_i \cdot z_i^2 \tag{3.63}$$

where $B_B = (EI)_B$

z_i = the distance between the centroid of each layer and the neutral axis

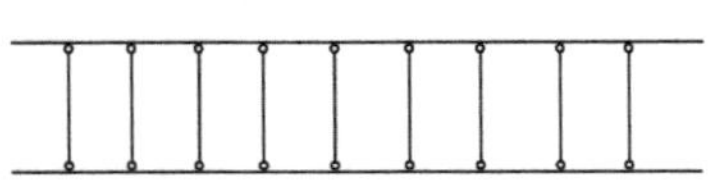

FIGURE 3.18 Beam differentiation using the shear analogy method.

Beam B also has shear stiffness and the stiffness of the flexible connections, if they exist. The shear stiffness of beam B, S_B, which is also expressed as $(GA)_B$, is obtained from Eq. (3.64):

$$\frac{1}{S_B}=\frac{1}{a^2}\cdot\left[\sum_{i=1}^{n-1}\frac{1}{k_i}+\frac{h_1}{2\cdot G_1\cdot b_1}+\sum_{i=2}^{n-1}\frac{h_i}{G_i\cdot b_i}+\frac{h_n}{2\cdot G_n\cdot b_n}\right] \tag{3.64}$$

where

$$k_i=\frac{K_i}{s_i} \tag{3.65}$$

where k_i = the slip of the fastener between the beams.

In B_A (Eq. 3.64) and B_B (Eq. 3.65) above, E_0 is used for the longitudinal layers, and $E_{90} = E_0/30$ is used for the cross layers.

Additionally, the shear modulus for the longitudinal layer is assumed to be G, where the shear modulus for the cross layer is assumed to be G_R, used for the rolling shear.

It is important to note that continuity of deflections between beams A and B must be valid at every point, that is, $\Delta_A = \Delta_B$. The shears and moment of the individual layers can be obtained using Eqs. (3.66) and (3.67), respectively:

$$V_{A,i}=\frac{E_i\cdot I_i}{B_A}\cdot V_A \tag{3.66}$$

$$M_{A,i}=\frac{E_i\cdot I_i}{B_A}\cdot M_A \tag{3.67}$$

where $V_{A,i}$ and $M_{A,i}$ are for each individual layer of beam A.

The shear and bending stresses for each individual layer, $\tau_{A,i}$ and $\sigma_{A,i}$, of beam A are calculated from Eqs. (3.68) and (3.69), respectively:

$$\tau_{A,i}=\frac{E_i\cdot I_i}{B_A}\cdot 1.5\cdot\frac{V_A}{b\cdot h_i} \tag{3.68}$$

$$\sigma_{A,i}=\pm\frac{M_{A,i}}{I_i}\cdot\frac{h_i}{2} \tag{3.69}$$

The shear and bending stresses for each layer are shown in Fig. 3.19.

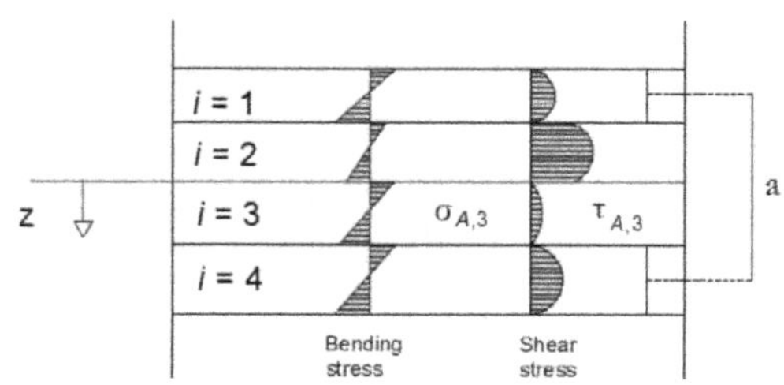

FIGURE 3.19 Bending and shear stresses in beam A using the shear analogy method [7].

For beam B, the axial forces, $N_{B,i}$, and normal stresses, $\sigma_{B,i}$, for each individual layer are calculated using Eqs. (3.70) and (3.71), respectively:

$$N_{B,i} = \frac{E_i \cdot A_i \cdot z_i}{B_B} \cdot M_B \tag{3.70}$$

$$\sigma_{B,i} = \frac{N_{B,i}}{b_i \cdot h_i} = \frac{E_i \cdot z_i}{B_B} \cdot M_B \tag{3.71}$$

The shear stresses at the interface of the two layers, $\tau_{Bi,i+1}$, is calculated by using Eq. (3.72):

$$\tau_{Bi,i+1} = \frac{V_B}{B_B} \cdot \sum_{j=i+1}^{n} E_j \cdot A_j \cdot z_j \tag{3.72}$$

The distribution of bending stresses and shear stresses for beam B is shown in Fig. 3.20.

The final stress distribution from beam A and beam B stresses is shown in Fig. 3.21. Note that the shear stress distribution in Fig. 3.21 includes the effect of the connector, which does not exist in CLT members.

The maximum deflection in the middle of a CLT member subjected to a uniformly distributed load can be obtained from Eq. (3.73). As shown in the equation, both flexural and shear deformations are taken into consideration:

$$u_{\max} = \frac{5}{384} \cdot \frac{qL^4}{(EI)_{\text{eff}}} + \frac{1}{8} \cdot \frac{qL^2 k}{(GA)_{\text{eff}}} \tag{3.73}$$

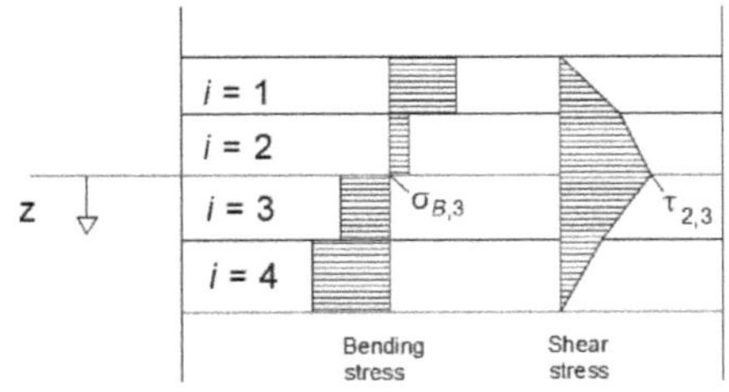

FIGURE 3.20 Normal and shear stresses in beam B using the shear analogy method [7].

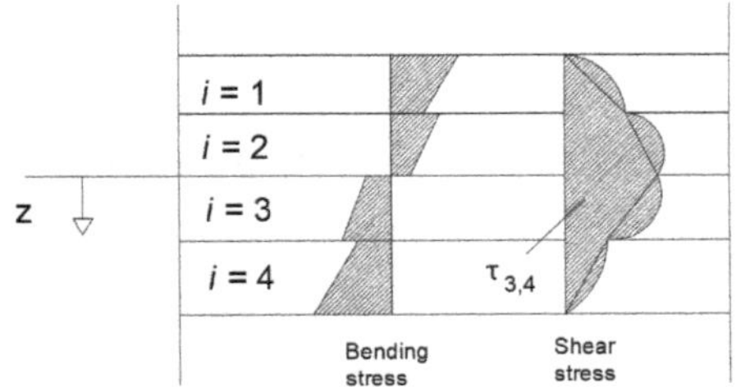

FIGURE 3.21 Final stress distribution obtained from the superposition of the results from beams A and B [7].

Equation (3.73) can also be expressed in Eq. (3.74):

$$u_{\max} = \frac{5}{384} \cdot \frac{qL^4}{(EI)_{\text{eff}}} \cdot \left(1 + \frac{48(EI)_{\text{eff}} k}{5(GA)_{\text{eff}} L^2}\right) \tag{3.74}$$

Equation (3.74) can also be expressed in Eq. (3.75):

$$u_{\max} = \frac{5}{384} \cdot \frac{qL^4}{(EI)_{\text{eff}}} \cdot (\alpha + \beta) \tag{3.75}$$

where $\alpha = 1.0$ and

$$\beta = \frac{48(EI)_{\text{eff}} \kappa}{5(GA)_{\text{eff}} L^2} \tag{3.76}$$

where κ = shear coefficient equals 1.2 [11].

The effective bending stiffness can be obtained from Eq. (3.77):

$$(EI)_{\text{eff}} = B_A + B_B = \sum_{i=1}^{n} E_i \cdot b_i \cdot \frac{h_i^3}{12} + \sum_{i=1}^{n} E_i \cdot A_i \cdot z_i^2 \tag{3.77}$$

The effective shear stiffness can be obtained from Eq. (3.78):

$$(GA)_{\text{eff}} = \frac{a^2}{\left[\left(\frac{h_1}{2 \cdot G_1 \cdot b}\right) + \left(\sum_{i=2}^{n-1} \frac{h_i}{G_i \cdot b_i}\right) + \left(\frac{h_n}{2 \cdot G_n \cdot b}\right)\right]} \tag{3.78}$$

For a concentrated load applied at the middle of the span of a CLT member, the maximum deflection equation can be calculated from Eq. (3.79):

$$u_{\max} = \frac{1}{48} \cdot \frac{PL^3}{(EI)_{\text{eff}}} + \frac{1}{4} \cdot \frac{PL}{\frac{(GA)_{\text{eff}}}{\kappa}} = \frac{1}{48} \cdot \frac{PL^3}{(EI)_{\text{eff}}} \cdot \left(1 + \frac{12 \cdot (EI)_{\text{eff}} \kappa}{(GA)_{\text{eff}} L^2}\right) \tag{3.79}$$

Equation (3.79) can be expressed as shown in Eq. (3.80):

$$u_{\max} = \frac{1}{48} \cdot \frac{PL^3}{(EI)_{\text{eff}}} \cdot (\alpha + \beta) \tag{3.80}$$

where $\alpha = 1.0$ and

$$\beta = \frac{48(EI)_{\text{eff}} \kappa}{(GA)_{\text{eff}} L^2} \tag{3.81}$$

where κ = shear coefficient equals 1.2 [11].

Simplified Design Methods for Calculation Bending and Shear Strengths for Slabs Subjected to Out-of-Plane Loading

In this section, simplified design equations are presented to calculate the capacity of CLT panels used for slab applications and subjected to out-of-plane loading. The bending stress can be obtained from Eq. (3.82):

$$\sigma = M \cdot y \cdot \frac{(E_i)}{(EI)_{\text{eff}}} \tag{3.82}$$

The maximum stress occurs at the extreme fibers of the cross section; in this case, for $y = \frac{h_{tot}}{2}$, the maximum bending stress is obtained from Eq. (3.83):

$$\sigma = M \cdot 0.5 \cdot h_{tot} \cdot \frac{(E_i)}{(EI)_{eff}} \tag{3.83}$$

The calculated bending stress in Eq. (3.83) should be compared with allowable stress provided by the applicable design standard.

The shear stress can be calculated using other methods available in the literature that are based on simple beam theory. The shear stress can be calculated using Eq. (3.84):

$$\tau_v = \frac{1.5 \cdot V}{c \cdot A_{gross}} \tag{3.84}$$

where c = a reduction factor and is calculated using Eq. (3.85):

$$c = \frac{I_{eff}}{I_{gross}} \tag{3.85}$$

According to Blass and Fellmoser [6], the maximum shear strength of CLT panels can be calculated as

$$\tau_v = \frac{V \gamma_i Q_i}{b \cdot I_{eff}} \tag{3.86}$$

where Q = static moment of area, calculated in a way similar to Eq. (3.60).

Regular Two-Way Slab System Loaded Perpendicular to the Slab Plane

In a typical building project, floor or roof CLT panels may be supported by walls, columns, or a combination of both. Depending on the support type used, various edge conditions for the panel may exist. For example, the plate could be simply supported on two parallel edges and connected to other plates along the other edge or simply supported on two parallel edges and free along the other two edges. Similar to any slab system, CLT slabs can be supported on three or four sides.

The CLT slab should be investigated carefully in regard to its support conditions since the support conditions affect the relative stiffness of the plates at the supports. The possible support conditions include slab hinged along two edges and free along two edges, slab hinged along three or four edges, and slab supported on columns.

The analysis and design of CLT slabs is similar to a certain extent to that of reinforced-concrete slab. Reinforced-concrete two-way slab systems consist of rectangular panels supporting uniform gravity loads (dead load, live loads, snow load). Two-way slab reinforced-concrete systems shall meet the following:

1. The longer to shorter span ratio (center to center) is less than or equal to 2.
2. For slabs with beams or walls between supports, the relative effective stiffness of beams or walls in the two directions shall be between 0.2 and 5.
3. The column offsets are not greater than 20 percent of the span length, in the direction of the offset, from either axis between the centerlines of successive columns.
4. Since resistance is in the two orthogonal dimensions, the slab reinforcement is placed in the two orthogonal directions of slab; this is similar to CLT panel behavior.

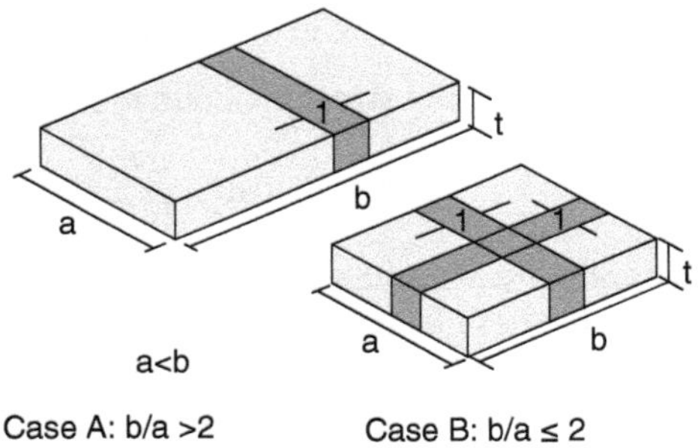

FIGURE 3.22 Slab panel loaded perpendicular to the plane: one-way and two-way slab concepts.

Two plates in bending are shown in Fig. 3.22; in these plates, "a" represents the shorter span of the panel, and "b" represents the longer direction span. Case A in the figure represents a rectangular plate with a b/a ratio greater than 2, and Case B represents a rectangular plate with a b/a ratio less than or equal to 2.

According to plate theory [11] and the assumptions for two-way slab design, it is suggested that panels supported on four sides with $b/a > 2$ be designed in one direction, which is the short direction, in this case, the span $L = a$. For panels supported on four sides with $b/a \leq 2$, it is suggested to design the panel as a two-way slab using $L_1 = a$ and $L_2 = b$ [11].

The analysis and design of CLT panels in two-way action is complex and should take into consideration if parameters such as support conditions, relative stiffness at the support, the moduli of elasticity in both directions, rolling shear in both directions, and so on. Therefore, to avoid this complexity in design, it is suggested that two-way CLT slab systems be designed as one-way systems, which will result in a more conservative result. Five-layer panels are recommended to be used if the two-way action is used in design, and the verification should be performed using a three-layer section in the center of the panel without the outer layers.

Bending Strength and Bending Stiffness—Loads Parallel to the Plane (Diaphragms)

Diaphragms are horizontal structural elements that are important to the design of the structure's lateral system. The lateral forces from earthquakes and winds are transferred to the vertical lateral system through floor and roof diaphragms and then to foundations. The in-plane stiffness and strength of the diaphragms affect the load distribution in the walls, which affects the design. A clear and detailed diaphragm design procedure is not available so far; several studies have been performed recently that should provide a clear approach for the design of diaphragms. So far, CLT diaphragms can be designed using the International Building Code (IBC 2012, 2015, 2018) [12-14]. Flexible and rigid diaphragms options should be evaluated, and the more critical ones should be considered.

Cantilevered and Statically Indeterminate CLT Elements

As mentioned previously, in the gamma method, CLT members are generally designed as simply supported members with a span equals l (length between the zero moments). In the case of a cantilever with a cantilever length l_c, two times the cantilever length is used in the calculation of effective bending stiffness EI_{eff}. In the case of continuous spans, there are two methods that can be followed to determine the effective bending stiffness EI_{eff}: a simplified method and an iterative method. The simplified method suggests using 0.8l as the distance between the inflection points, whereas in the iterative method, an initial value between the inflection points is used for the first iteration, say, 0.8l. Then a computer program is used to determine the inflection points for a beam with EI_{eff}, then the new length between the inflection points is obtained and EI_{eff} recalculated. Other iterations can be performed similarly, usually after few iterations, and a stable solution is achieved for EI_{eff}.

CLT Slab Supported by a Column/Post

The capacity of the CLT member in compression perpendicular to grain should be verified using the loaded area of the column/post using the applicable design standard.

Additional Stresses

Cross-layer boards that are not glued or not edge-bonded may experience additional stresses perpendicular to grain due to the rotation around their longitudinal axis. The tensile stresses perpendicular to grain and the rolling shear stresses should be verified by testing if the ratio between the width of the board and its thickness is less than 4.

Design Methods for CLT Elements Used in Wall Systems

CLT wall panels maybe subjected to vertical in-plane loads, lateral in-plane loads, and lateral out-of-plane loads. Wall panels subjected to out-of-plane loading only, such as wind loads, are designed similarly to slabs subjected to vertical out-of-plane loads. For wall panels subjected to in-plane lateral loads (like diaphragms), design procedure has not been provided with enough clarity and details. It is important to indicate that, for wall panels subjected to gravity loads, the outer layers should be arranged so that the grain direction is parallel to the gravity load.

CLT Wall Panels Under Axial In-Plane Loads and Out-of-Plane Loads

The analysis and design of CLT wall panels subjected to combined loading can be performed using several methods; these methods are described in the sections below.

3.6 Mechanically Jointed Columns Theory (Eurocode 5)

Annex 5 of Eurocode 5 includes the details of mechanically jointed columns. Using effective section properties' assumptions and using this method, the effective slenderness ratio, λ_{eff}, can be calculated using Eq. (3.87):

$$\lambda_{eff} = l \cdot \sqrt{\frac{A_{tot}}{I_{eff}}} \tag{3.87}$$

where A_{tot} = total area of the cross-sectional area of the panel
l = height of the wall, i.e., buckling length
I_{eff} = effective moment of inertia and is given as

$$I_{eff} = \frac{(EI)_{eff}}{E_{mean}} \tag{3.88}$$

where EI_{eff} defined in Section 3.5.4
E_{mean} = the modulus of elasticity of the boards acting parallel to the axial load, the vertical layers in this case

Using the calculated λ_{eff}, the compressive axial strength equation is used based on the range of λ_{eff} using the applicable design standard.

3.7 Design Procedures for CLT Elements Used as Beams and Lintels

The design of beams and lintels made of CLT can be designed using the procedure presented below. Figure 3.23 shows CLT member subjected to in-plane loading. Three-layer or seven-layer CLT can be used as well.

3.7.1 Bending Strength Calculations for In-Plane Loads

The bending stress can be calculated as shown in Eq. (3.89):

$$\sigma = M \cdot y \cdot \frac{(E_{\text{mean}})}{(EI)_{\text{eff}}} \tag{3.89}$$

The maximum stress occurs at $y = H/2$ and is shown in Eq. (3.90);

$$\sigma_{\text{max}} = M \cdot 0.5H \cdot \frac{(E_{\text{mean}})}{(EI)_{\text{eff}}} \tag{3.90}$$

The maximum stress obtained in Eq. (3.90) should be compared with allowable stresses as applicable by adopted design standard.

3.8 Creep Behavior of CLT in Bending

Creep behavior is critical in CLT design and should be taken into consideration. FPInnovation in Canada proposed two different approaches to account for creep in CLT; refer to the CLT Handbook, Canadian edition, for details on creep behavior of CLT panels in bending [3].

3.9 Vibration of CLT Floors

Vibration behavior of CLT floors is different than conventional wood floors supported by joists and decking. FPInnovation in Canada proposed a design method to control vibration in CLT floors; refer to the CLT Handbook, Canadian edition, for details on the design of CLT floors for vibration [3].

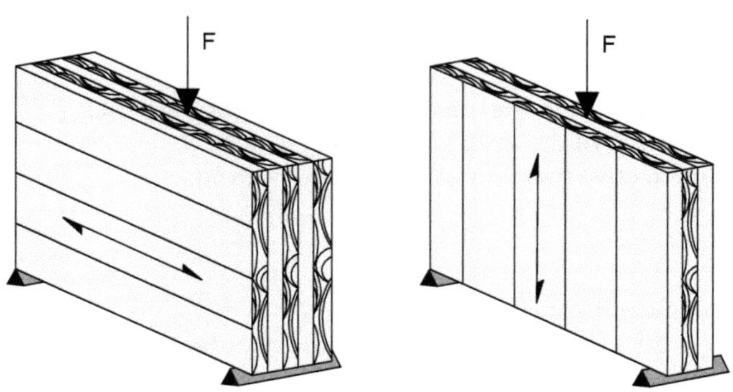

FIGURE 3.23 CLT beams or lintels under axial in-plane loads.

References

[1] Bernasconi A. Il calcolo dell`XLAM Basi, normative, progettazione, applicazione, Corso di approfondimento—Edifici di legno, www.promolegno.com.

[2] European Standard EN 1995-1-1 2004/A1 2008 Eurocode 5 (2004/2008). Design of timber structures—part 401 1-1: general—common rules and rules for buildings/incl Amendment A1. European Committee for 402 Standardization, Bruxelles (Belgium).

[3] Gagnon, S. and M. Popovski. 2011. Structural Design of Cross-Laminated Timber Elements. In: Chapter 3, CLT Handbook (Canadian Edition). FPInnovations. Quebec, Quebec, Canada.

[4] Ross, L. A., S. Gangnon, and E. Keith. 2013. Structural Design of Cross-Laminated Timber Elements. In: Chapter 3, CLT Handbook (U.S. Edition). FPInnovations. Quebec, Quebec, Canada. 2011.

[5] Ceccotti, A. 2003. Composite structures. In Timber Engineering, ed. S. Thelandersson and H. J. Larsen, Chapter 21. West Sussex, England: John Wiley and Sons.

[6] Blass, H. J. and P. Fellmoser. 2004. Design of solid wood panels with cross layers. In Proceedings of the 8th World Conference on Timber Engineering, June 14–17, 2004, Lahti, Finland, 2:543–548.

[7] Kreuzinger, H. 1999. Flächentragwerke—Platten, Scheiben und Schalen—Berechnungsmethoden und Beispiele [Plane loadbearing structures—plates loaded in- and out-of-plane and Eur. J. Wood Prod. (2016) 74:331–351 349 123 shells—calculation methods and examples]. Fachverlag Holz, Dusseldorf (in German).

[8] Mestek, P., H. Kreuzinger, and S. Winter. 2008. Design of cross laminated timber (CLT). Paper presented at the 10th World Conference on Timber Engineering, June 2-5, 2008, Miyazaki, Japan.

[9] Bejtka, I. and F. Lam. 2008. Cross laminated timber as innovative building material. In Proceedings of the CSCE Annual Conference, Québec, QC. Montreal, QC: Canadian Society for Civil Engineering.

[10] Fellmoser, P. and H. J. Blass. 2004. Influence of rolling shear modulus on strength and stiffness of structural bonded timber elements. In Proceedings of CIB-W18 Meeting, Edinburgh, United Kingdom, paper 37-6-5.

[11] Timoshenko, S. and S. Woinowsky-Krieger. 1959. Theory of plates and shells. New York: McGraw-Hill. 580 p.

[12] International Code Council (ICC). 2012. International Building Code. IBC, Washington, DC.

[13] International Code Council (ICC). 2015. International Building Code. IBC, Washington, DC.

[14] International Code Council (ICC). 2018. International Building Code. IBC, Washington, DC.

CHAPTER 4

Structural Design—Connections

Keith Porter

B. Eng, Dalhousie University

Max Closen

Dipl.-Ing (FH); MaSc., MTC Solutions Inc.

4.1 General Overview of Connections

4.1.1 Cross-Laminated Timber and Modern Connection Technology

Mass timber structural systems are attracting the attention of engineers and architects looking for alternatives to concrete and steel. While products like glulam beams and nail-laminated timber panels have been around for decades, the development of cross-laminated timber (CLT) panels starting in the 1990s (Brandner, Flatscher, Ringhofer, Schickhofer, & Thiel, 2016) has inspired considerable interest in timber as a structural material suited to modern needs.

The strength, beauty, and versatility of CLT is in many ways driving new forms of structural design that are exceeding limitations of conventional wood construction. The Wood Innovation Design Centre building in Prince George, BC, topped off at eight stories with 51,000 ft^2 (4700 m^2), making it the tallest mass timber building in North America when it was completed in 2014. Just three years later, the 17-storey gravity frame for the 160,000-ft^2 (15,000 m^2) Brock Commons Tallwood House, composed exclusively of glulam columns and CLT panels, was completed in just 9 weeks.

While it is fair to give credit to the development of CLT, advances in timber connection technology also play a significant role in making mass timber construction a viable in the modern construction landscape. Structural design, after all, consists not only in specifying the appropriate structural elements but also in joining them together to create structural systems.

With the expansion of mass timber, there is a corresponding demand for long, high-strength fasteners that can be site-installed with ease, speed, and precision. New lines of self-tapping structural screws have proven especially suitable for use with CLT for these very reasons. Self-tapping screws were developed over the last few decades largely in response to a demand for threaded fasteners with high axial-loading capacities. Self-tapping screws are widely used throughout Europe, where mass timber construction is more established, and even now most North American producers of CLT specify them in suggested connection details. Self-tapping screws are also highly versatile in application, and, unlike proprietary systems, their design procedures are easily generalized, and so they will form the focus of this chapter.

4.1.2 Overview of Connection Design

CLT and mass timber construction is positioned as an alternative to concrete and steel more so than light frame wood construction. In the years to come, many structural designers will wish to become more familiar with timber design concepts as the demand for mass timber buildings grows. A complete review of the principles of timber connection design is not the focus here, but some key concepts will be reviewed in this section and revisited throughout the chapter where especially relevant.

The engineering design of connections in conventional timber structures focuses largely on connection strength. Connections must be strong enough to provide continuous load paths from applied gravity, wind, and seismic loads on the structure down to the foundation. Under normal service conditions, this means connections must be sufficiently designed to transfer loads from one element to another without yielding or failing. Research programs into fastening technology typically begin with studies of single fasteners loaded to failure. These studies are then expanded to investigate how multiple fasteners perform together as groups. Similarly, procedures for calculating connection strength begin with one fastener and then expand and modify this calculation for connections with multiple fasteners.

In addition to calculating the connection strength, the designer must verify that the wood member or the fastener will not fail before the calculated connection strength is reached. Connection design, then, must satisfy not only connection strength requirements but also timber verification and fastener strength verification checks.

For a given fastener type, connection strength is primarily influenced by fastener size, fastener material characteristics, wood density, and grain direction. Self-tapping screws classify as "dowel type" fasteners (which also includes bolts, nails, and so forth). Larger-diameter self-tapping screws generally provide greater connection strength and fastener strength.

Different wood species possess different densities. Connection strength properties for softwoods, including dowel bearing (embedment) strengths, withdrawal resistance, and head pull-through resistance, tend to increase with increasing wood density. In the National Design Specifications (NDS), the density of a wood species is assigned a specific gravity, "G," which is a unitless measure of density. At present, most CLT is manufactured from either spruce-pine-fir ($G = 0.42$) or Douglas fir-larch (North) trees ($G = 0.49$). Specific gravities are based on mean oven-dry mass and volume, divided by the density of water ($\rho_w = 1000$ kg/m^3). For example, the mean oven-dry density of Douglas fir-larch (North) is 0.49×1000 kg/m$^3 = 490$ kg/m^3.

In addition to sufficient strength, connections must also provide sufficient stiffness to a structural system, primarily for serviceability requirements, such as limiting drift or deflections. Unfortunately, information on the relative stiffness of timber connectors is often difficult to procure from timber design codes (Borg, 2000). The structural requirements of modern mass timber buildings are increasing demand for this type of information.

As the design of modern timber structures trends toward taller buildings, there is also a greater need for ductility in structural timber systems. Designing for ductility requires an understanding of the various types of failure modes encountered in timber connections to avoid brittle failure modes (Smith & Frangi, 2014). Unlike steel, wood is an anisotropic structural material, so it exhibits very different strength characteristics depending on how it is loaded relative to the grain. Timber is generally strongest when loaded parallel to the grain, either in tension or in compression. Of the various failure modes, wood crushing tends to yield large deformations with continuous load-carrying capability—similar to the stress-strain response of metals subject to plastic deformation. Wood loaded perpendicular to the grain in compression shows the greatest "apparent ductility" (Smith, Gong, & Landis, 2003) but also very low strength.

The brittle failure mode of wood loaded in tension parallel to the grain can be accommodated due to its high strength and ability to handle reasonable strains. There are three loading directions, however, possessing both low-strength and brittle failure modes, that require special attention in connection design:

- tension perpendicular to the grain
- shear parallel to the grain
- rolling shear

Although the perpendicular to the grain loading direction can be further differentiated into radial and tangential loading directions, in this chapter "perpendicular to the grain" will imply either one in a general sense. In conventional timber design practice, tension forces perpendicular to the grain are typically avoided wherever possible, as failure modes in this direction also occur at very low strains.

Wood performs optimally as a structural material when loaded in tension or compression parallel to the grain. At the same time, improper connection detailing can activate brittle shear failure modes parallel to the grain. Rolling shear is rarely an issue with conventional timber design, but it warrants attention with CLT due to the crossing layers (see Fig. 4.1). Testing and analysis show that, in addition to shear failure, tensile fracture perpendicular to the grain due to principal stresses acting diagonally frequently initiates brittle failure modes in this loading direction (see Fig. 4.2) (Nie, 2015; Mestek & Dietsch, 2011).

Outside of wood crushing under compression, ductility in timber systems is provided through yielding of mechanical fasteners [this is opposite the convention for steel design, where ductility is assigned to the members (Popovski, 2000)]. Yielding of fasteners

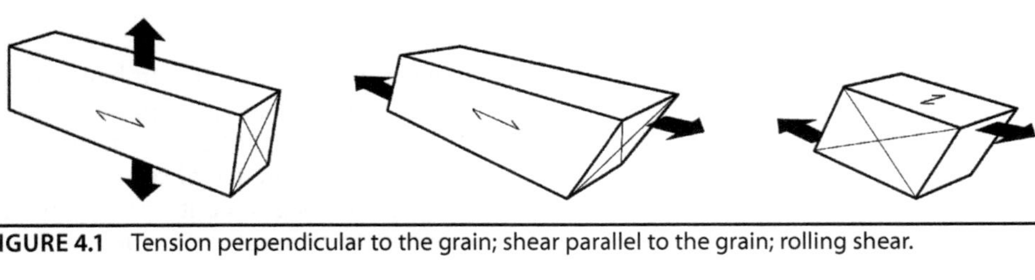

FIGURE 4.1 Tension perpendicular to the grain; shear parallel to the grain; rolling shear.

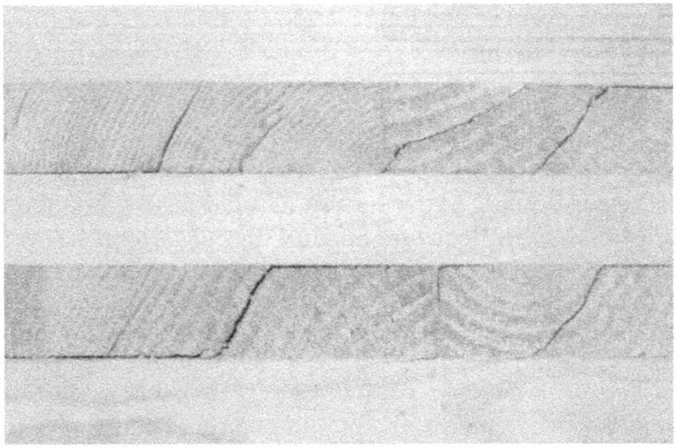

FIGURE 4.2 Typical rolling shear failure of CLT.

in the event of overloading has a crucial role in energy dissipation for the seismic design of timber structures, as explained in Sec. 4.3. Furthermore, in Secs. 4.3 and 4.7, it will be shown how fully threaded self-tapping screws allow designers to reinforce wood members in low-strength loading directions in new ways, which expands the possibilities of wood as a structural material by increasing strength and ductility.

A complete list of strength adjustment concepts unique to timber design can be found in the NDS, but it is worth highlighting two that are most relevant to this chapter. Timber products demonstrate higher-strength capacities for short-term loading and lower-strength capacities for permanent loading conditions. Reference design values in the NDS are calibrated to standard term loading, which is defined as approximately 10 years or less. With short-term loading, such as that caused by wind, designers may increase reference design values by a duration factor of $C_D = 1.6$. For permanent loading, on the other hand, reference design values must be reduced.

Moisture content also has a considerable influence on timber connection design. Wood has the ability to absorb moisture from the atmosphere, eventually reaching an equilibrium moisture content with the surroundings. Moisture content in wood has an influence on connection strength, and penalties are assigned to design values when moisture content is above dry service condition levels. Dry service conditions are satisfied when the average equilibrium moisture content of wood remains below 15 percent over a year, without exceeding 19 percent at any given time. Dimensional changes due to the swelling and shrinking of wood due to moisture changes can also add complexity to connection design. Swelling and shrinking occur primarily in the radial and tangential directions, with little dimensional change along the grain direction. CLT has a unique ability to accommodate reasonable changes in moisture content (see Fig. 4.3), as will be explained toward the end of this section.

One of the major factors influencing choice of connections for timber structures, of course, is price. Perez and Wallwork (2016) estimate that the basic supply cost of connections for a CLT structure may only add up to 5 percent of the total material and construction cost. At the same time, the authors note that the influence of the connection type, especially with regard to installation time, can have a significant impact on a project. With new fastening technologies, it is important to compare costs at the level of the

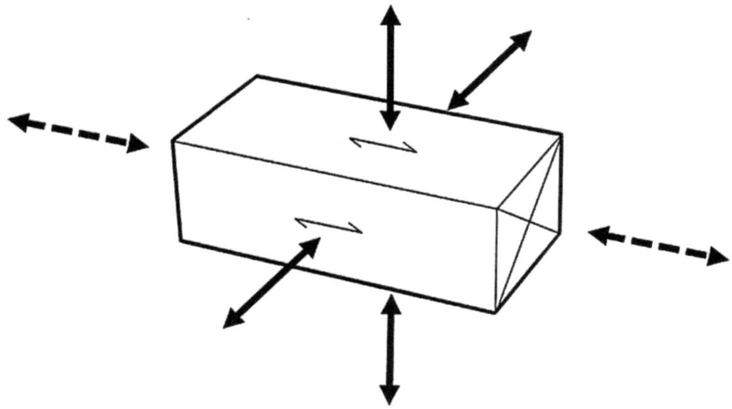

FIGURE 4.3 Radial and tangential dimensional change directions (solid) and longitudinal dimensional change directions (dashed) due to moisture content variation.

complete project. A one-for-one cost comparison between lag screws and self-tapping screws, for instance, ignores logistical and related labor costs, such as the time required for installation, fastener applications, and cost per unit of resistance. For larger projects, an efficient rate of construction has quite a significant influence on the overall construction cost.

Finally, many architects and designers are drawn to incorporate more wood into their projects not only for its sustainability but also for its visual and emotional appeal. The warmth that wood products bring to the built environment should be strengthened by the total appearance, which includes the connections whenever they are visible. As we will see, many modern timber connection systems offer a clean or even concealed architectural appearance that highlights the appeal of modern engineered wood products.

4.1.3 Connection Design for CLT

Designers already familiar with conventional design concepts for mechanical fasteners in wood products will find that design procedures for mechanical fasteners in CLT follow many of the same basic steps and principles. Like general timber connections, mechanical connections in CLT are dependent on the wood species and on the direction of the wood grain mainly. Many design procedures for CLT incorporate some method of accounting for the alternating grain directions between plies for predicting connection strength at service levels. While this means that CLT involves extra steps over homogeneous wood products, simplified procedures have been developed to minimize calculation.

Although CLT is referred to as a "solid wood" product, it is rare for the narrow edges of lamellas to be glued when making the panels. While most lamellas sit directly against one another, gaps between these narrow edges of lamellas are not uncommon (see Fig. 4.4). Additionally, the base material comprising CLT may exhibit wane, which also creates voids (see Fig. 4.5). While gaps between lamellas can affect connection strength, it should not be overemphasized. A sampling of over 900 gaps between adjacent lamellas in European CLT found that gaps tended to be smallest on the outer plies with an average size of under 0.02 in. (0.5 mm). Gaps tended to be larger in the inner layers, with an average size of around 0.04 in. (1.0 mm) (Uibel & Blass, 2007). In Europe, gaps are limited

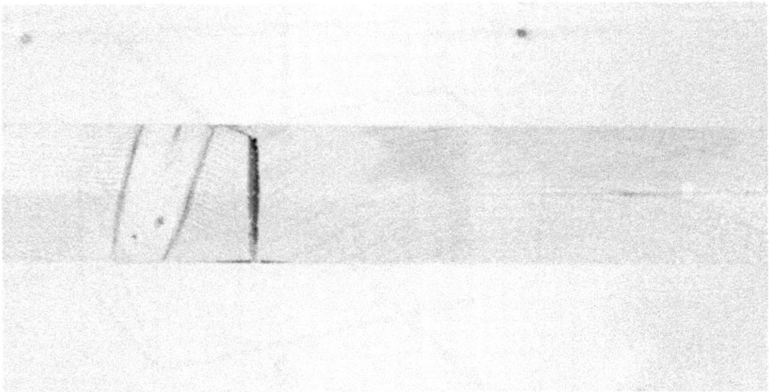

FIGURE 4.4 Gap between inner layer lamellas.

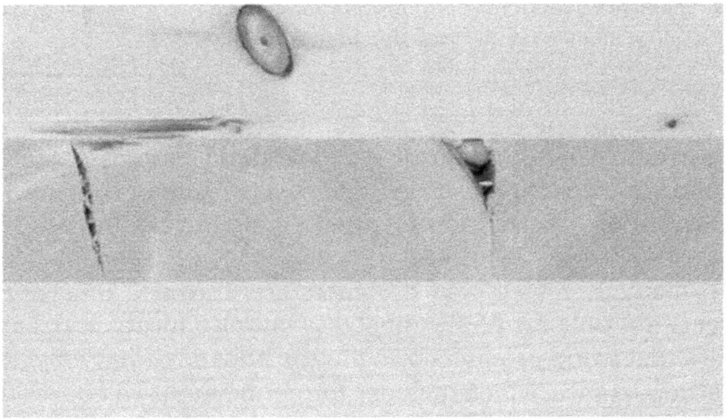

FIGURE 4.5 Void due to wane.

to a maximum width of 6 mm, so designers are thereby advised to use a minimum fastener diameter of 8 mm in the narrow edge of CLT. Furthermore, minimum penetration lengths respective to the wide panel face and narrow panel edge are recommended to account for the presence of gaps. Although maximum gap width specifications are not presently included in APA PRG-320, typical gaps range between 1/16 in. and ⅛ in. (1.5 mm and 3 mm) or less.

Much more pronounced than the influence of gaps (which are irregular) is the presence of end grain in the narrow edge of the panels. Strength reductions associated with end grain application must be considered for fastener application in the narrow face. Geometry requirements also differ with respect to the wide panel face and the narrow panel face.

According to APA PRG 320 (2017), the moisture content of the base lumber at the time of CLT manufacturing is restricted to 12% ± 3%. This range is reflective of the potential equilibrium moisture content for dry service conditions, meaning indoor, heated, and well-ventilated environments. Additionally, in Clause (Cl.) 10.1.5, the NDS (2015) specifies CLT for use in dry service conditions only, unless permitted by the manufacturer. During construction, delivery of CLT should be coordinated so that it arrives as close as possible to the day it will be installed, otherwise staying wrapped to avoid direct wetting (Below, Mohammad, Ni, Malczyk, & Popovski, 2014). One of the advantages of CLT is that it is stabilized for the most part against in-plane dimensional variation due to changes in moisture content. Based on a study by Bradner (2013), in-plane dimensional change for CLT is roughly 0.08 percent per 4 percent change in moisture content, meaning for most applications it can be considered negligible. At the same time, designers should be aware of the potential for CLT to swell in the out-of-plane (perpendicular to the grain) direction due to elevated moisture content, which can compromise connection integrity if not controlled on-site (see Fig. 4.6).

Above all, the most influential parameters when it comes to fastener selection for CLT connection design are the following:

- the thickness of the panels
- the need for high-capacity connections
- the need for speed and simplicity of installation
- cost and environmental impact

CLT panels are typically commercially available in thicknesses ranging from 3 in. to 12⅜ in. (76 mm to 314 mm), although a maximum depth of 20 in. (508 mm) can be produced. Many connections, such as panel to beam or wall to wall connections, require through penetration of deep CLT panels to connect to main members containing the point side of the fastener. Fastener lengths that can exceed the thickness of the CLT panel, therefore, are required. To generalize, large wood members require long fasteners. Furthermore, the toughening aspect of CLT is engaged when fasteners are engaged through cross plies, which also requires fasteners of a suitable length.

There are many examples of older tall wood structures in North America that use large heavy timber members as structural elements. Moffett Air Field Hangers II and III and the Spruce Goose Hanger, all built in the 1940s and still standing today, are some of the most enormous timber structures ever built. Additionally, many cities in North America still contain historic post and beam, truss, and arched industrial buildings built before building codes limited the height of timber buildings. The wood frames of these large structures, however, were designed primarily to carry gravity loads in bending and compression. Stability and resistance to lateral loading for large, industrial-scale wood frame buildings were provided primarily through massive masonry or stone walls

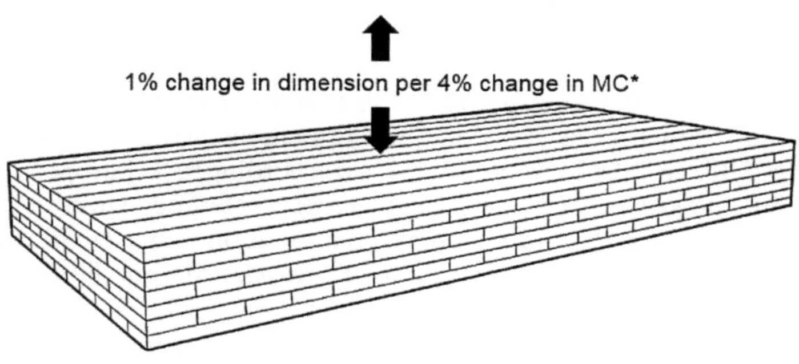

FIGURE 4.6 Dimensional changes in CLT due to MC variation. 1% change in dimension per 4% change in MC*.
*American Wood Council, 2016.

(Mayo, 2015; Smith & Frangi, 2014). Tall modern mass timber buildings, on the other hand, are often stabilized through solid wood shear walls and diaphragms, reinforced-concrete elevator core shafts, and buckling restrained brace systems. Highly stressed connection points involving shear brackets, hold downs, and splice plates require high-capacity connections to transfer these loads effectively in shear and tension.

CLT is a prefabricated structural product. During fabrication, it is common practice for each panel to be made to the correct size with holes for windows, doors, and service conduits cut out. Machining for joinery is also done to the panels edges before any of it reaches the job site, meaning the product is ready to be craned into place in the same day of delivery. CLT connections, therefore, should take advantage of the nature of CLT as a clean, quiet, and above all fast product to install.

CLT is particularly capable of resisting high-magnitude in-plane loading. CLT panels subject to high in-plane loading will typically behave as apparently rigid elements, and respond in the linear elastic range (Sullivan, 2017). The high in-plane strength properties of the panels can only be exploited through successful connection design, as timber connections typically only provide a fraction of the total load-carrying capacity of the structural members they connect. As CLT and mass timber enables taller and taller timber construction, allowances for energy dissipation and ductility must be built into the design through the connections between the rigid elements to respond to overloading through seismic events.

Connection design begins with the consideration of one fastener and then expands to the analysis of groups. A further level of analysis considers the entire structural system. At this level, analysis is directed to predict, for instance, the kinematic behavior of shear walls and to check if sufficient deformations are allowed by the structure in order for the connections to assume their full design strength (Smith & Frangi, 2014). This chapter, however, will focus on the behavior of single fasteners and fastener groups.

Since many new fastening products for CLT continue to be developed in Europe, designers should be aware of the conventions of the European timber design code Eurocode 5 (EC5) and exercise extreme care before adopting European design values for use in North America. A large discrepancy between design values is due to the fact that EC5 is based on the load and resistance factored design (LRFD) approach, with characteristic strength design values generally based on 5th percentile of tested short-term strengths (less than 5 minutes). The NDS, on the other hand, is based on the allowable stress design (ASD) approach. New dowel-type threaded fasteners, such as self-tapping screws, are evaluated according to ASD principles outlined in Acceptance Criteria AC233. Reference withdrawal design values for standard term loading (approximately 10 years) are thereby based on mean ultimate tested values divided by a factor of safety of 5.0. To illustrate the difference between the two approaches, a ½-in. (12 mm) diameter ASSY® self-tapping screw is rated as having a characteristic withdrawal design strength of nearly 700 lb/in. (123 N/mm) in softwood according to its European technical approval (ETA), but the same screw is rated as having a reference withdrawal design value of 188 lb/in. (33 N/mm) according to its ICC Evaluation Service Report (ESR).

Decades of research across the United States, Canada, and Europe have gone into the many accepted design procedures for connections in CLT. Some design procedures have been included in the NDS in 2015 for the first time, which simplifies the design process considerably. With new construction technology, there are many situations which require the designer to search for reliable information. Some procedures are based on new principles that have found their way into common practice which have been verified by testing but have not been included in official codes yet. Some design procedures come from technical approval documents, such as ICC ESRs for self-tapping screws, as well as other proprietary products. There are also times when the designer must make design decisions based on long-established principles in timber engineering and basic mechanics. The increased structural demands of buildings made from CLT and mass timber are prompting a fresh look at the information available in timber design codes, which may inspire further changes intended to assist designers in the years to come. CLT and self-tapping screws are both highly versatile structural components, and connection design for CLT using self-tapping screws will continue to evolve and expand in the years ahead.

4.2 Introduction to Self-Tapping Wood Screws and Types Available

4.2.1 Features and Advantages of Self-Tapping Screws

Self-tapping structural wood screws are highly engineered, high-strength fasteners that are seen by many as the "perfect ally" for CLT construction (Silva, Branco, Ringhofer, Lourenço, & Schickhofer, 2016) (see Fig. 4.7). Long self-tapping screws are increasingly the preferred choice for mass timber construction for three reasons. First, they come in lengths appropriate to the dimensions of CLT and mass timber elements. Second, they offer high-capacity connections required by tall mass timber structures. Third, self-tapping screws offer fast and effective on-site installation, which complements the same-day delivery of premanufactured, ready-to-assemble CLT panels to the construction site.

Self-tapping screws may appear quite similar to conventional threaded fasteners. However, just as CLT is enabling new approaches to wood construction, self-tapping screws have many features that enable them to surpass the limits of conventional threaded fasteners—not just in terms of capacity but also in terms of application. Up until recently, lag screws were the primary resource in the engineer's toolbox for high-performance threaded fasteners. However, as Borg (2000) writes, lag screws are primarily used in place of bolts for metal to wood connections wherever through penetration of the wood member is impractical. Due to their bolt-like design and laborious installation process, the application of lag screws in mass timber construction remains limited. As we will see, self-tapping screws are characterized by unique range of sizes and fastener geometries, high-strength material properties, and simple installation procedures that set them apart from conventional threaded fasteners.

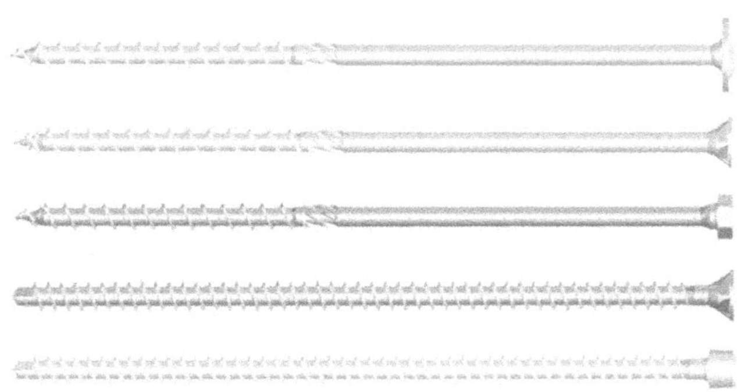

FIGURE 4.7 Typical self-tapping structural wood screws.

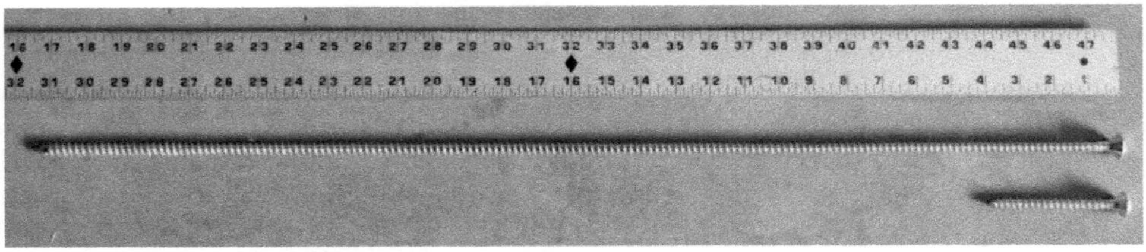

FIGURE 4.8 3/8 in. × 31.5 in. (10 mm × 800 mm) and 3/8 in. × 4 in. (10 mm × 100 mm) ASSY® VG CSK screws.

The main feature that makes self-tapping screws distinct is their size. While lag screws are commonly available in lengths up to 12 in. (305 mm), self-tapping screws come in sizes ranging from 2 in. to 59 in. (51 mm to 1500 mm) long (see Fig. 4.8). A nine-ply CLT panel can be over 12 in. (305 mm) deep, while standard glulam beam depths can reach over 7 ft (2134 mm). Traditional wood fasteners were designed to deal with much smaller dimensioned wood. Quite simply, the member dimensions involved in mass timber construction call for larger fasteners. The range of sizes also allows for appropriate sizes to be selected which achieve high-thread penetration lengths without through penetration, which can be useful in reinforcing CLT connections across alternating plies.

During manufacture, self-tapping screws are double heat-treated to achieve a hard outer layer with some overall ductility. While the process is unique to specific manufacturers, the case-hardening process generally involves heating the screws to temperatures above 1500°F (800°C) before quenching in oil or water. Some manufacturers also apply a process whereby the outer surface of the screws is enriched with carbon while heated, which further increases hardness. The case-hardening procedure ensures that the threads will maintain their integrity under high stresses during installation. The hardening process also significantly increases the torsional strength which is necessary to resist high torque generated during installation. The annealing process that follows restores toughness and ductility to the steel. The double heat-treating process produces a fastener that has increased torsional strength, tensile strength, and bending yield strength. For comparison, ½ in. (12 mm) diameter lag screws are assigned a bending yield strength of 45,000 psi (310 MPa) (according to the NDS), while a similar ½-in. (12 mm) diameter self-tapping screw achieves over 166,000 psi (1147 MPa) (ICC-ESR 3178, ICC-ESR 3179). The high-strength characteristics of self-tapping screws produce a high-performance structural fastener with slender dimensions. It is important to note that differences in base material property and heat treatment procedures between manufacturers make material strength properties unique to specific products.

The threads for self-tapping screws are produced through a rolling process (Dietsch & Brandner, 2015). This rolling process produces large thread wings which exceed the smooth shank diameter of the screws, along with a relatively small root diameter, with the ratio of core (or "root") diameter to outer thread diameter (D_r/D) typically not exceeding 0.65 ratio. Self-tapping screws do not remove wood during installation like a drill—rather, the wood is densified around the threaded portion of the screw. The comparatively small root diameter reduces the wedge effect as the screws are installed, thereby reducing the splitting tendency of the wood. Large thread wings contribute to greater withdrawal capacities, which is why it can be advantageous to load self-tapping screws in the axial direction wherever they are capable of achieving long thread penetration lengths (see Fig. 4.9).

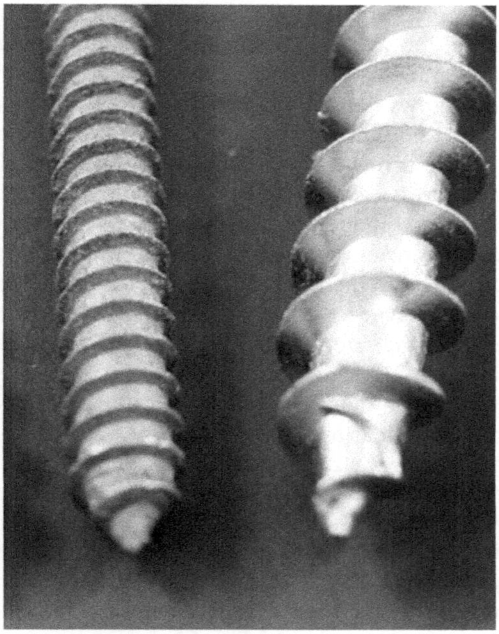

FIGURE 4.9 Lag screw (left) versus self-tapping screw (right).

Self-tapping screws are broadly divided into two categories: partially threaded and fully threaded screws. Conventional wood screws and lag screws are generally classified as partially threaded screws, although, due to their comparatively time-consuming installation process, they have not been used extensively for structural needs in most mass wood construction up to now. Partially threaded and fully threaded screws each serve unique purposes in connection design (see Fig. 4.10).

4.2.2 Partially Threaded Screws

Partially threaded screws have both a threaded portion and a smooth shank portion along the fastener length and are engineered to pull wood members together. Most often,

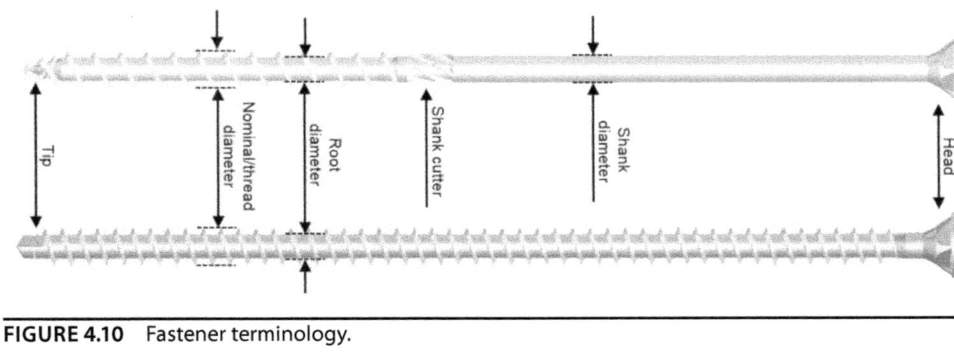

FIGURE 4.10 Fastener terminology.

partially threaded screws are utilized for assembly purposes (holding members together) or to transfer lateral forces with the loading applied perpendicular to the fastener axis (head pull-through can also be a critical design consideration for axial loading). Many partially threaded screws come with a shank cutter between the threaded portion and the shank. The purpose of the shank cutter is to clear a hole slightly larger in diameter than the shank as the screw is installed. This reduces friction during installation and allows the wood to settle freely as required. A typical shear connection using a partially threaded screw has the threaded portion of the screw embedded only in the main member, with the shank intersecting the shear plane. Thus, while the threaded portion of the screw pulls up on the main member, the head of the screw pulls down on the side member, and the two are drawn together. Designers should note that although this pulling action generates a strong friction component between the two members, this is not accounted for in lateral design values calculated according to the NDS. Rather, the importance of pulling the members together in most connections is to eliminate gaps, thereby promoting firm bearing.

Partially threaded screws can be used as "assembly screws" either on their own or in advance of fully threaded screw installation. In the latter case, partially threaded screws ensure that the wood members are held tightly together before the fully threaded screws go in. Otherwise, fully threaded screws could simply "lock in" any existing gaps. Nonstructural partially threaded screws can also be used in conjunction with other materials to help seal the space between elements as a measure to control sound, fire, or smoke transmission, for example (see Fig. 4.11).

4.2.3 Fully Threaded Screws

Fully threaded screws have threads running along the entire length of the shaft and are optimized for axial loading, reinforcing, and holding wood members in place. It is worth mentioning that fully threaded screws can be loaded perpendicular to the axis in shear connections and have even demonstrated enhanced ductility and stiffness for this purpose due to the increased contribution of axial effects. Fully threaded screws, however, are applied more often in lateral connection details that exploit the high withdrawal resistance of fully threaded screws by installing them at angles inclined to the shear plane.

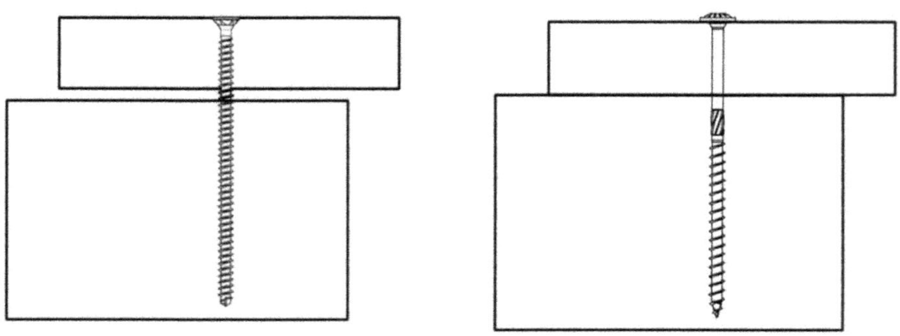

FIGURE 4.11 Non-intentional gap locked in place by fully threaded screw (left); partially threaded screws clamp wood members together across smooth shank (right).

Just as rebar reinforces concrete in tension where it is the weakest, fully threaded self-tapping screws can be used to reinforce wood in low-strength loading conditions such as tension perpendicular to the grain. Reinforcement with fully threaded self-tapping screws can also strengthen wood in shear parallel to the grain, rolling shear, and even compression perpendicular to the grain (Dietsch & Brandner, 2015).

Fully threaded self-tapping screws are used in material handling devices like lifting yokes as well as timber/concrete composite systems, mechanically jointed beams, beam hangers, coupling devices, splice connections, and other proprietary metal-to-wood connectors.

4.2.4 Head Types

Partially threaded and fully threaded self-tapping screws come in a variety of different head shapes for different purposes. Most fastener lines will include some form of a washer head, hex head, cylinder head, and countersunk head.

Washer head screws feature the largest-diameter head amongst self-tapping screws, which increases the effective bearing area for the head (see Fig. 4.12). Washer head screws are optimized for applications which require high-head pull-through capacity and are generally used for wood/wood connections.

Hex head screws are unique amongst self-tapping screws in that while most self-tapping screws are installed using specially designed six-point drives, hex head screws can also be installed using standard socket drives (see Fig. 4.13). Most hex head screws have a reinforced tapered shoulder beneath the head that is the same diameter as the threads. This tapered shoulder makes them especially suitable for steel/wood connections, as the tapered shoulder provides a tight fit in the steel hole as well as a suitable bearing area for the steel plate when loaded laterally.

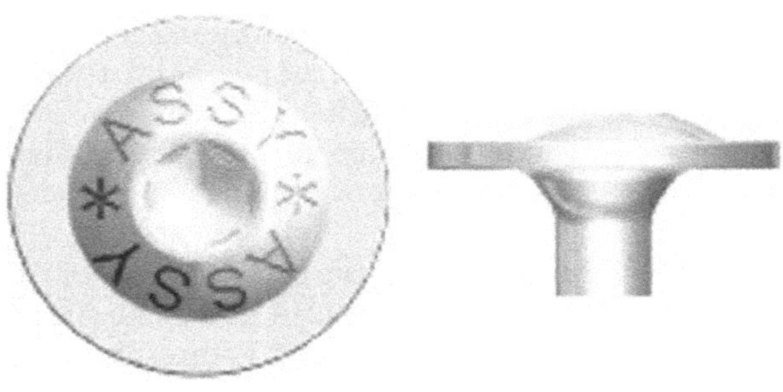

FIGURE 4.12 Washer head.

FIGURE 4.13 Hex head.

FIGURE 4.14 Cylinder head.

Cylinder head screws feature a head engineered to the smallest size possible that can still transmit sufficient torque during installation (see Fig. 4.14). The small head size allows cylinder head screws to be driven below the surface of the wood where they remain inconspicuous, leaving the wood surface flush and neat in appearance. Cylinder head screws are most often used for connections with inclined fastener angles such as panel-to-panel shear connections as well as for various reinforcement applications where a clean architectural appearance is required.

Finally, countersunk head screws come in both partially threaded and fully threaded versions and serve multiple purposes in mass timber construction. Countersunk head screws offer medium head pull through resistance by virtue of the head diameter, but unlike the washer head screws, these can be driven flush with the wood member. Countersunk screws can easily be fitted with a 90° washer to achieve greater bearing area or a 45° washer for inclined screw applications where the side member is a steel plate (see Fig. 4.15). Like cylinder head screws, fully threaded countersunk screws can be used for all reinforcement applications and are especially suitable where the screw is installed at 90° to the wood bearing surface to provide compression reinforcement perpendicular to the grain.

4.2.5 Installation, Safety, and Reliability

Traditional large diameter fasteners require predrilling before installation to avoid wood splitting due to the "wedge effect." For a lag screw, at least two different diameter holes are required for each screw: one for the threaded portion of the fastener and one for the shank (an additional hole with a third diameter is required if the screw is to be sunk flush with the wood surface). Self-tapping screws, on the other hand, require no predrilling in most cases.

Self-tapping screws may feature either a counterthread or ringthread tip for partially threaded screws or a drill tip for fully threaded screws. Regardless, the tip is designed to bite into the wood during installation. Once the threads engage, the screw pulls itself into the wood, tapping its own hole as it progresses. The large thread wings on self-tapping screws provide greater mechanical interaction with the wood, increasing withdrawal resistance, while the smaller core diameter reduces the splitting tendency due to the wedge effect. Unlike a drill, which removes material, self-tapping screws displace material.

FIGURE 4.15 Countersunk head, 90° washer, and 45° washer.

Therefore, the wood is densified around the screw. In order to avoid splitting, self-tapping screws are assigned their own unique spacing, edge, and end distance requirements which are distinct from lag screws. It is important to note that geometry requirements are also unique to specific products (varying between even ETAs) and therefore should not be generalized for all self-tapping screws. The following example geometry requirements are differentiated between partially threaded and fully threaded screws, with the subscript "L" indicating a loaded edge or end (see Figs. 4.16 and 4.17 and Tables 4.1 and 4.2):

The self-tapping feature has a few important benefits. Without predrilling, installation time is greatly reduced. This is a tremendous advantage when considering continuous line

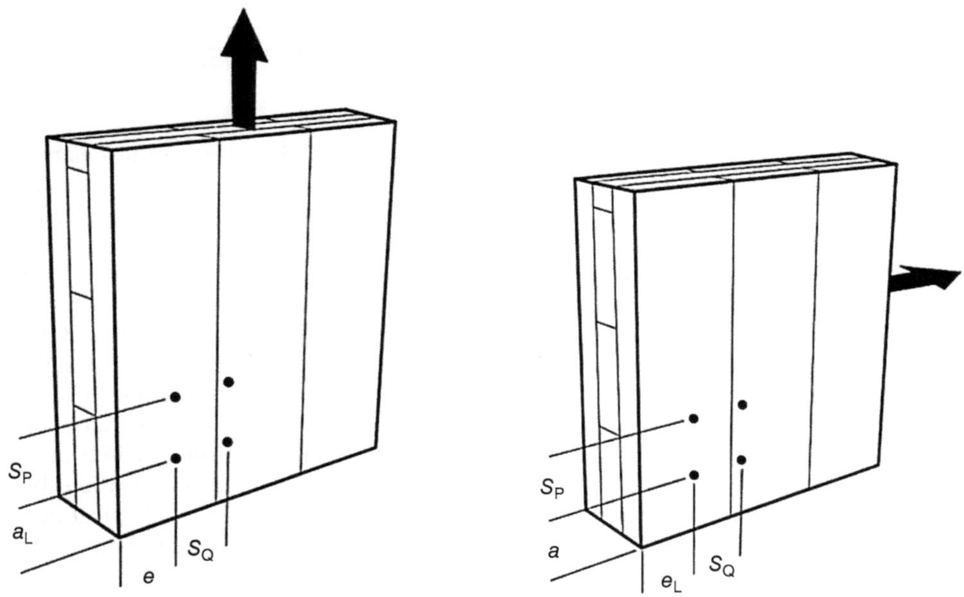

FIGURE 4.16 Geometry requirements for ASSY® self-tapping screws in the wide panel face of CLT (subscript "*L*" denotes loaded end or edge).

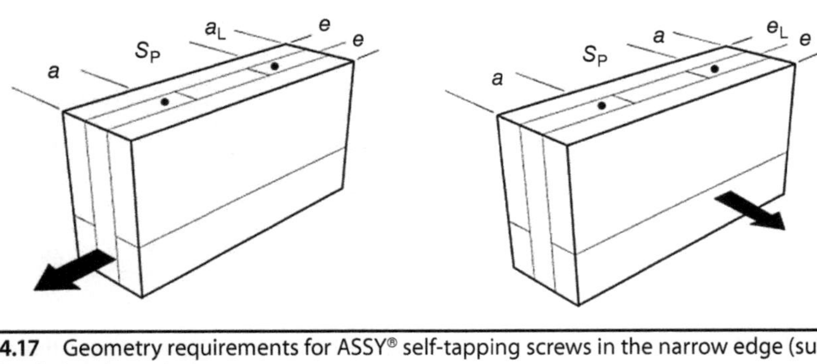

FIGURE 4.17 Geometry requirements for ASSY® self-tapping screws in the narrow edge (subscript "*L*" denotes loaded end or edge).

End distance		Edge distance		Spacing	
a_L	a	e_L	e	S_P	S_Q
6*D*	6*D*	6*D*	2.5*D*	4*D*	2.5*D*

TABLE 4.1 Geometry Requirements for ASSY® Self-Tapping Screws in the Wide Panel Face of CLT

End distance		Edge distance		Spacing	
a_L	a	e_L	e	S_P	S_Q
12*D*	7*D*	6*D*	3*D*	10*D*	4*D*

TABLE 4.2 Geometry Requirements for ASSY® Self-Tapping Screws in the Narrow Edge of CLT

joints in panel to panel connections, for instance. An efficient rate of installation means that CLT and mass timber elements can be assembled almost as fast as they can be lifted into place. Keeping this in mind, specifying designers are advised to consider the entire cost of connections, including additional labor and crane time when pricing fasteners.

Another advantage of the self-tapping feature concerns reliability. Since lag screws generally require at least two different diameter predrilled holes, the performance of each screw is highly dependent on the workmanship involved. For instance, if the hole to receive the threaded portion is drilled too wide for the threads to bite into the wood or if the larger diameter hole for the shank extends into the area of the threaded portion of the screw, the withdrawal capacity will be reduced. Similarly, bolts require that holes drilled in advance, and oversized holes can contribute to uneven load distribution between fasteners acting as a group (Popovski, 2000). Self-tapping screws, on the other hand, "tap" their own hole as they are installed, ensuring both a tight mechanical fit between the threaded portion and the surrounding wood and a precise alignment between the hole and the fastener (see Fig. 4.18).

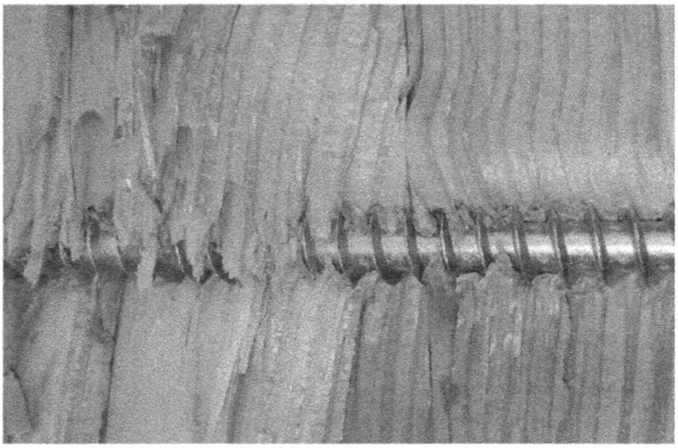

FIGURE 4.18 Opened specimen showing tight mechanical interaction between threads and surrounding wood (densification of wood can be observed around screw).

Self-tapping screws typically utilize some type of six-point drive designed for a tight, stable fit to reduce cam-out, facilitating easy installation with a single tool. Low-rpm high-torque electric drills can be fitted with wide-handle grips to counteract the torque from the larger fasteners. Lag screws are often installed using soap as a lubricant to reduce torque. Self-tapping screws, on the other hand, are sometimes factory coated with a lubricating finish, so no additional lubricant is necessary.

Predrilling is not a requirement for self-tapping screws in wood species not prone to splitting, but in certain cases it may be beneficial to use pilot holes. For example, pilot holes are recommended for any situations calling for a high degree of precision regarding the point of installation, such as screws installed close to the ends and edges of wood members, or cases where the screw is inclined to the surface of the wood (at low angles, the drill tip of fully threaded screws may travel across the wood rather than biting into it). Pilot holes for inclined screws are created using a jig and are only required to extend a few inches at most (see Fig. 4.19).

Many installers face difficult angles and challenging heights when erecting structures, so the ease of application for self-tapping screws can be a welcome advancement. Furthermore, without all the predrilling and on-site fabrication, many contractors appreciate the cleanliness of the work areas since so much sawdust is no longer coating every surface.

Among the many advantages to using mass timber is enhanced fire safety—not just throughout the service life of a building but also during construction before fire suppression systems have been installed. CLT and mass timber structural elements arrive to the job site ready to be lifted into place, with no additional cutting or sawing required. Since the wood does not require additional working, the presence of sawdust is greatly reduced, thereby reducing the risk of fire. Structural self-tapping screws, therefore, complement the nature of mass timber construction not just in terms of installation speed but also by keeping the sawdust level on the job site down to a minimum.

Architects often find the new opportunities for clean, neat finishes on structural connections to be greatly appealing. Structural engineers, who must specify the fastener type, can benefit from an expanded range of high-performance connection details and reinforcing applications backed by proprietary ICC ESRs with the added reliability and consistency that comes along with the self-tapping installation procedure.

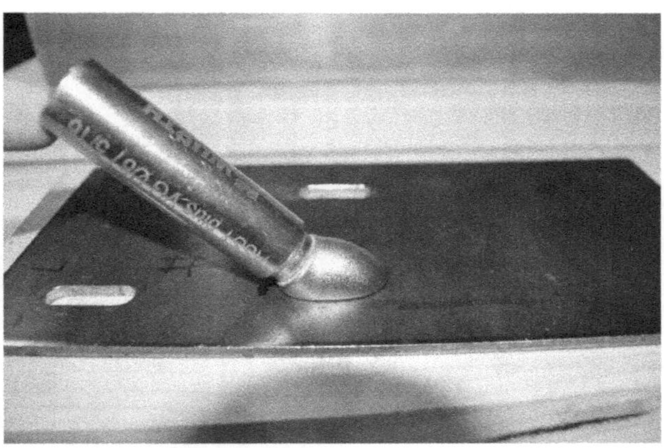

FIGURE 4.19 45° jig for steel side plate using 45° washers.

4.3 Review of Connection Concepts, Failure Modes and Testing

4.3.1 Axial Loading

Axial loading of screws concerns head pull-through resistance and withdrawal resistance of the connection as well as the tensile strength of the screw (see Fig. 4.20).

Withdrawal resistance is a function of the wood density, the angle of the fastener axis relative to the grain, the fastener diameter, and the thread penetration length. The lower of the withdrawal resistance and the tensile resistance will govern the design value for the threaded portion of the screw. Withdrawal resistance is considered equal to "push-in" resistance in the opposite direction along the fastener axis, so resistance due to thread engagement with the wood is the same whether the screw is "pushed" or "pulled." For screws loaded directly in compression, an additional buckling check for the fastener must also be performed.

Head pull-through resistance is influenced by wood density and the size and shape of the screw head (Blass, 2003). Large round washer heads offer the greatest head pull-through resistance for partially threaded self-tapping screws. For partially threaded screws loaded axially, head pull-through resistance generally governs the design value for the side member, and withdrawal resistance from any threaded portion in the side member is disregarded. On the other hand, head pull-through resistance is generally not considered when using fully threaded screws since the thread penetration lengths in side members will demonstrate higher stiffness and strength in withdrawal (exceptions may apply to thin side members like plywood splines). Since head pull-through and withdrawal resistance are each associated with different connection stiffness, they should not be summed in a single loading direction. At ultimate loads, head pull-through can demonstrate apparently ductile behavior through wood crushing, for instance, if the wood is loaded perpendicular to the grain and prevented from splitting (see Fig. 4.21).

A research project at the University of British Columbia (Lam, Zhang, & Lee, 2013) consisting of withdrawal tests of thousands of screws covering four different wood species (plus engineered wood), three installation angles, and four thread penetration

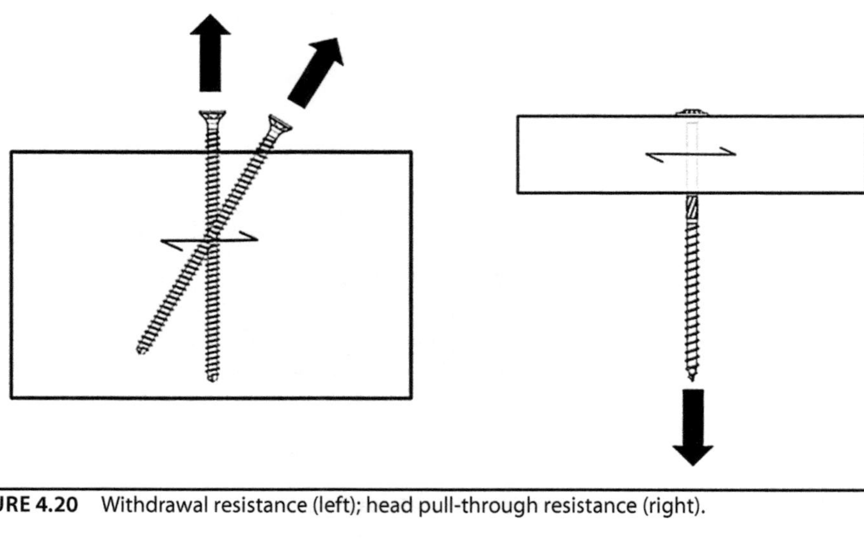

FIGURE 4.20 Withdrawal resistance (left); head pull-through resistance (right).

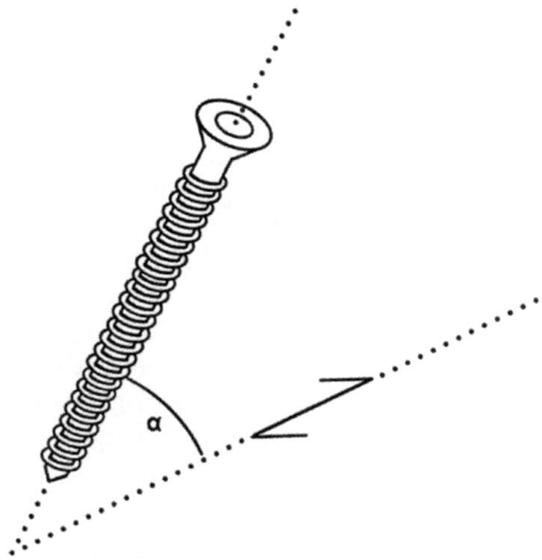

FIGURE 4.21 Angle, α, between fastener axis and wood grain.

lengths was used to establish the first reference withdrawal design values for self-tapping screws to be recognized by ICC Evaluation Service Reports. Maximum withdrawal resistance per unit thread length is achieved when the screw axis is oriented 90° relative to the grain direction and decreases with lowered angles. Screws installed 45° to the grain showed approximately 85 percent of the withdrawal strength of a screw oriented 90° to the grain, while screws installed 30° to the grain demonstrated approximately 80 percent of the withdrawal strength of a screw oriented 90° to the grain. (In this chapter, a factor of 0.85 is applied to 90° reference withdrawal design values to estimate reference withdrawal design values where the fastener is installed 45° to the grain). Reference withdrawal design values from ICC ESR-3178 and ESR-3179 are reproduced in Table 4.3.

	Fully threaded screws		Partially threaded screws	
D	*G* = 0.42	*G* = 0.49	*G* = 0.42	*G* = 0.49
¼"	126 [22]	166 [29]	126 [22]	166 [29]
5/16"	177 [31]	183 [32]	177 [31]	183 [32]
3/8"	188 [33]	194 [34]	188 [33]	194 [34]
½"	188 [33]	211 [37]	188 [33]	211 [37]

See ICC ESR-3178 and ICC 3179 for up-to-date design values (values expected to increase).

Reference withdrawal design values valid for effective thread penetration length ≥ 8*D*.

TABLE 4.3 Reference Withdrawal Design Values (W90) for ASSY® Self-Tapping Screws (α = 90°) lb/in. [N/mm]

Subsequent proprietary withdrawal testing at the University of British Columbia (Hossain, 2016a) of 180 screws in samples of Canadian-made spruce pine fir CLT at 90° and 45° and two penetration lengths while avoiding obvious gaps demonstrated the suitability of using the withdrawal design values previously established for homogeneous wood products in CLT under similar applications, as the approach generally mimicked how large numbers of screws are actually installed into CLT.

Many studies have focused on how the presence of gaps between adjacent lamellas affects withdrawal resistance of self-tapping screws in CLT. One study found that screws installed directly between adjacent lamellas where the gap width = 0 mm show no decrease in withdrawal capacity (Silva, Branco, Ringhofer, Lourenço, & Schickhofer, 2016). The worst-cast scenario for screws installed in the wide face of a CLT panel would be to have the screws lined up directly with large gaps in all penetrated layers. As Uibel and Blass (2007) point out, the probability of having gaps line up for a screw in all plies is highly unlikely. Furthermore, it is even more improbable that any such gaps would be very wide. Silva, Branco, Ringhofer, Lourenço, & Schickhofer, (2016) investigated the effect of various combinations of gaps between plies in CLT. In their worst-case scenario, three 1/6-in. (4 mm) wide gaps lined up along the axis of 5/16-in. (8 mm) fully threaded screws. At 12 percent moisture content, the mean withdrawal resistance was still over 70 percent of the reference test values representing a screw installed through a specimen with no gaps. Uibel and Blass (2007) performed over 900 withdrawal tests on various gap configurations in CLT and found that, while mean withdrawal values compared favorably to homogeneous wood samples, characteristic 5th percentile values were found to be lower for CLT on account of the gaps, with small ¼-in. (6 mm) screws affected the most. Aside from the fact that the NDS considers mean, and not 5th percentile characteristic values, self-tapping screws are currently evaluated under AC233 in the United States, which assigns a factor of safety of 5 to average ultimate tested values, which seems sufficient to account for the influence of gaps in the wide face of the panel if in compliance to CLT manufacturing standards. Finally, the NDS does not assign a reduction factor to the withdrawal resistance of lag screws installed in the wide face of CLT. Accordingly, the same approach is taken with self-tapping screws. As a general rule, Uibel and Blass (2007) recommend that screws should penetrate at least three layers when loaded axially in the wide panel surface to compensate for gaps.

When self-tapping screws are installed into the wide face of a CLT panel perpendicular to the surface, the grain direction of both major and minor direction lamellas will engage the screws at 90°, so withdrawal design values do not need to account for alternating layers. Interestingly, Ringhofer, Brandner, and Schickhofer (2013) found that withdrawal strength for self-tapping screws installed 90° in the wide panel face tends to increase not just with the effective length but also with the number of layers penetrated, as the layer with the highest density controls the fastener performance. If the screws penetrate through multiple plies at an inclined angle, then the screw will penetrate alternating plies at different angles to the grain, as shown in Fig. 4.22. Design procedures calculating the withdrawal strength for self-tapping screws installed through alternating grain directions are not established at this point. In this chapter, the lowest angle will be assumed for conservatism.

For application into the narrow edge of a CLT panel, the NDS assigns an end grain factor to lag screws of $C_{eg} = 0.75$, regardless of the direction of the penetrated lamella, as per Cl. 12.2.1.5 (exceptions to the rule can be found in Cl. 12.2.1.5). The end grain factor assumes an angle to grain of 0°, so it is intended to be applied to the reference withdrawal design values of screws installed perpendicular to the grain. This factor can

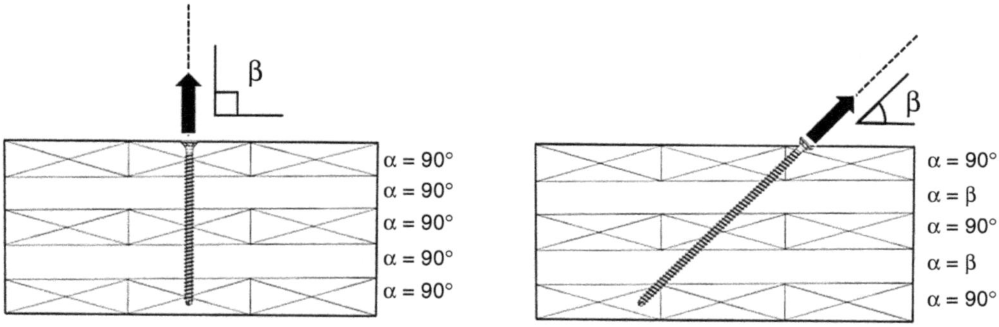

FIGURE 4.22 Angle to grain, α.

be applied to self-tapping screws, but reduction factors derived for self-tapping screws (mentioned previously) can be applied that already consider the angle to grain relationship. Designers are discouraged at present from loading fasteners in withdrawal installed below 30° relative to the grain, especially beyond short-term durations. Screws should not be installed in gaps in the narrow edge, as they can run along the fastener axis, weakening the connection. The influence of gaps can be minimized by installing fasteners at small angles (≥15°) which allow the fasteners to cross gaps whenever they are encountered. Long screws can also help to compensate for strength reductions due to the presence of end grain. Uibel and Blass (2007) recommend a minimum penetration depth of 10*D* for screws installed perpendicular to the narrow edge surface CLT. Self-tapping screws with diameters 5/16 in. (8 mm) and above are also recommended to compensate for gaps and voids in the narrow panel edge.

When withdrawal resistance of a screw is exceeded, the wood usually fails in shear around the outside of the threads (Blass, 2003), with a sudden drop in load-carrying capacity. Testing of single screws shows that failure in withdrawal should be considered a brittle failure mode for design purposes (see Fig. 4.23).

Axial loading of the fastener must be checked against the allowable screw tensile strength to ensure it is not exceeded. Self-tapping screws come in lengths capable of generating enough withdrawal resistance to exceed the tensile strength of the screw. Tensile fracture of the screw (screw breakage) is also considered a brittle failure mode (however, at a very predictable point with small variation). Once sufficient thread penetration length has been achieved such that the withdrawal resistance matches the tensile strength of the screw, further threat penetration length will not increase the strength of connection any further. Allowable tensile strengths from ICC ESR-3178 and ESR-3179 are reproduced in Table 4.4.

It is worth mentioning that there may be reason to achieve greater thread penetration lengths—reinforcing the wood against parallel to grain shear failure, for instance. Design convention assigns the minimum of the allowable tensile strength and the adjusted withdrawal design value to the connection strength. At the same time, it is helpful to know when screw breakage is the more likely failure mode rather than withdrawal. Unfortunately for the designer, knowing when this point is reached can be difficult to predict based on design values alone. One of the reasons is that the allowable tensile strength of the screw is based on a factor of safety of 3.0, while the reference withdrawal design value is based on a factor of safety of 5.0, as per AC233. To assist the designer, Table 4.5 contains a summary of the thread penetration lengths whereby tensile fracture of the screw becomes highly probable at F_u.

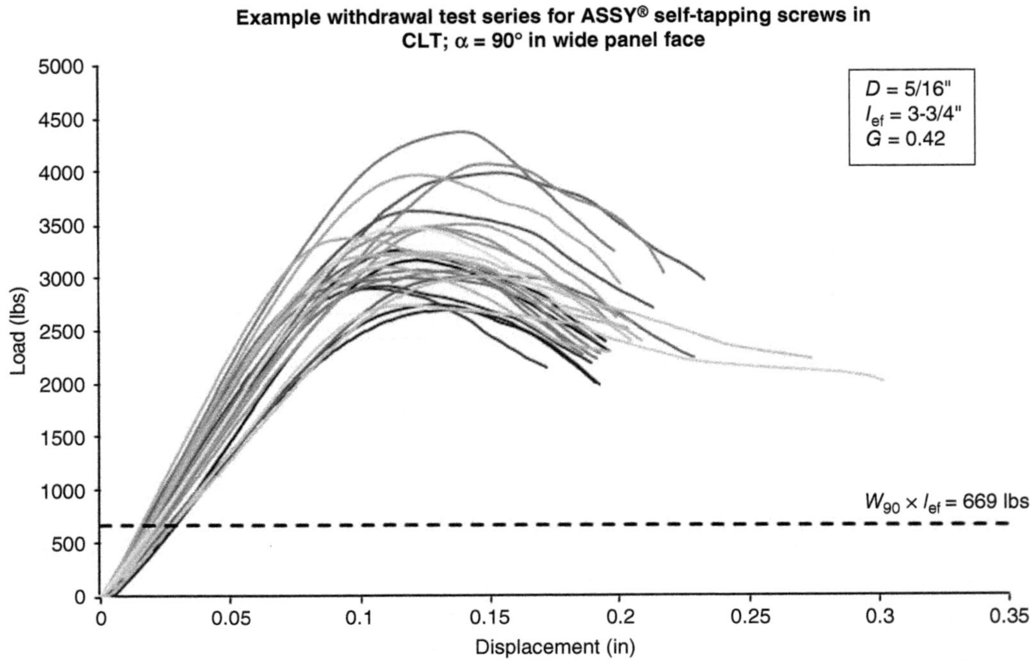

FIGURE 4.23 Example load displacement curve for withdrawal testing in CLT, with reference withdrawal value, $W_{90} \times l_{ef}$, shown with dashed line. [*Source: Hossain (2016a).*]

D	Partially threaded	Fully threaded
¼"	1150 [5070]	1165 [5120]
5/16"	1950 [8590]	1775 [7820]
3/8"	2780 [12,240]	2550 [11,220]
½"	3070 [13,510]	3470 [3470]

See ICC ESR-3178 and ICC ESR-3179 for up-to-date design values.

TABLE 4.4 Allowable Tensile Strength for ASSY® Self-Tapping Screws, lb [N]

4.3.2 Lateral Loading

Lateral loading describes forces acting perpendicular to the fastener axis. Screws loaded laterally are commonly called "shear screws." Lateral connections are generally designed using partially threaded screws, as they are a more economical fastener choice for this connection and also because they pull the members together, closing gaps. In order to

	D	SPF ($G = 0.42$)	DF ($G = 0.49$)
in. [mm]	¼ [6]	5 [125]	4 [100]
in. [mm]	5/16 [8]	6½ [165]	5 [125]
in. [mm]	3/8 [10]	8 [200]	6 [160]

TABLE 4.5 Estimated Thread Penetration Lengths with High Probability of Tensile Strength Governing Ultimate Connection Strength

exploit the capacity of partially threaded screws to pull members together, fastener size should be selected so that the threaded portion completely penetrates the main (or point side) member with the smooth shank intersecting the shear plane (see Fig. 4.24). If threads extend across both sides of the shear plane in a wood-to-wood connection, there is a risk of "locking in" gaps between both members, just like fully threaded screws, as discussed in Sec. 4.2. Fully threaded screws are optimized for axial loading, so they are not often used for lateral connections where the screw is loaded perpendicular to the axis, although they too can be highly effective shear screws for reasons having to do with the "rope effect," explained later in this chapter.

Lateral design values for timber connections using connectors like screws are predicted using a set of equations known as the "European Yield Model." This model was presented in 1949 by Johansen and adopted and modified by timber building codes in the decades to follow. The European Yield Model was incorporated in the NDS in 1991, adjusted to ASD design methods. Although the design procedure appears cumbersome without a spreadsheet or software, the model has proven to be a reliable method and a powerful tool for engineers designing with new fasteners. Without the Johansen model, countless experiments must be performed on all varieties of fasteners, wood geometries, and loading directions with the design values subsequently derived and tabulated. The European Yield Model, on the other hand, predicts the strength of a shear connection

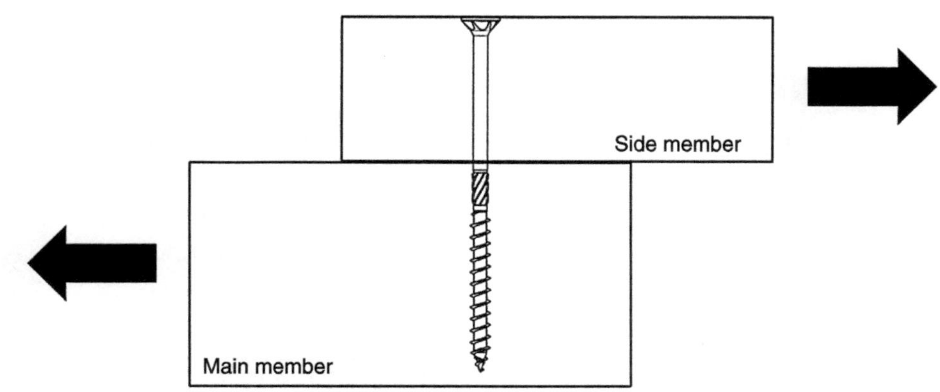

FIGURE 4.24 Lateral loading; side member (s), main member (m).

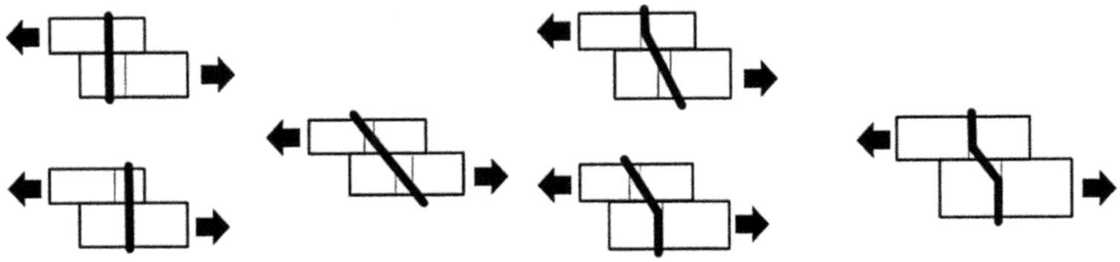

FIGURE 4.25 Four failure modes, as per NDS Cl. 12.3.

using the embedment strength of the wood (referred to as "dowel bearing strength" in the NDS), the bending yield strength of the fastener, and the geometry of the connection.

For single shear connections, the predicted strength of four possible failure modes are calculated, with the lowest strength governing the design of the connection (see Fig. 4.25).

Failure modes include wood crushing in the main or side member (*I*), fastener rotation without bending (II), fastener yielding with wood crushing forming one plastic hinge (modes III), and fastener yielding with wood crushing forming two plastic hinges (mode IV) (see Fig. 4.26).

The dowel-bearing strength (F_e) of wood is based on compressive tests using cylindrical bearing surfaces, which distinguishes it from compressive strength. F_e can be calculated from formulas in the NDS, based on the wood species (*G*), fastener diameter (*D*), and grain direction (parallel or perpendicular to grain loading). The bending yield strength for new fasteners is determined through testing according to AC233 and reported in technical approvals such as International Code Council Evaluation Service Reports (ICC ESR) (see Table 4.6).

A number of simplifying assumptions are made in the European Yield Model: Dowels (screws) are assumed to be loaded evenly along their embedded length, the elements are assumed to be in static equilibrium, and material behavior (both timber and fastener) is idealized as rigid-plastic. The European Yield Model, as presented in the NDS, assumes that screws are simply held in place and that there is no normal force acting between the main and side members. Strength contributions stemming from end fixity, friction, and axial resistance are not included.

The convention to ignore the effects end fixity, friction, and axial resistance is justified on the assumption that the effects are hard to quantify and generalize for all types of

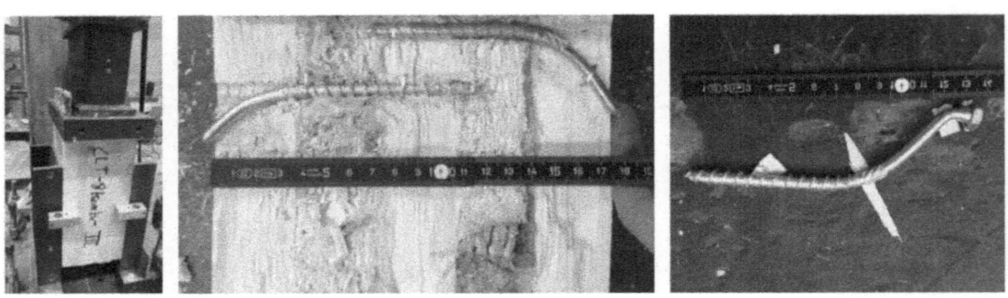

FIGURE 4.26 5/16 in. × 5½ in. (8 mm × 140 mm) ASSY® Kombi hex head screws tested to failure in five-ply CLT with steel side plate; one plastic hinge formed within first CLT layer, causing wood crushing; second hinge formed at steel/wood interface. (*Courtesy of Jan Niederwestberg.*)

D in. [mm]	Partially threaded psi [MPa]	Fully threaded psi [MPa]
¼ [6]	169,500 [1169]	129,200 [891]
5/16 [8]	150,200 [1036]	132,500 [914]
3/8 [10]	136,600 [942]	136,600 [942]
½ [12]	166,300 [1147]	166,300 [1147]

See ICC ESR-3178 and ICC ESR-3179 for up-to-date design values.
Due to differences in fastener geometries and material properties, bending yield strengths in Table 4.6 are valid only for ASSY® screws.

TABLE 4.6 Bending Yield Strength for ASSY® Self-Tapping Screws

connections and all types of fasteners. Furthermore, resistance due to friction may not be reliable over the service life of a structure, as the wood may shrink, generating a gap between the members. One of the consequences of ignoring these effects is that the design strength of lateral connections using self-tapping screws using the NDS Cl. 12.3 is assumed to be the same as connections made with smooth dowels, despite testing which shows considerable increases in design strength, which are simply neglected. Ignoring these effects, therefore, can lead to less economical connection design and higher overall hardware cost. Future research should address these issues in an attempt to advance to a more economical design procedures.

The European Yield Model can be applied to CLT using new fasteners with only slight modifications to the procedure for homogeneous wood. When following Cl. 12.3 in the NDS, one must decide which screw diameter to use as parameter "D" in the yield equations: outside thread diameter, D; smooth shank diameter, D_s; or root diameter, D_r. In the NDS, tabulated reference lateral design values are made using the root diameter, D_r, as this models the most conservative case with a reduced dowel bearing area in the main and side members and fastener bending taking place within the threaded portion of the screw. Technical Report-12 by the American Wood Council (2012) allow the shank diameter, D_s, to be used for certain cases where the threaded portion of the screw is sufficiently far away from the shear plane. Until self-tapping screws are directly addressed by the NDS, it is worth mentioning that there is perhaps a case to be made for using D_s with self-tapping screws—the European approach recognizes the additional strength provided from the material in the threads, assigning an effective diameter, $D_{ef} = 1.1 \cdot D_r$, which approximates the shank diameter. In this chapter, D_r will be used for lateral design purposes (see Fig. 4.27).

For connections in the wide face of CLT, F_e is assigned based on the grain direction of the surface plane with respect to the direction of the loading, as per NDS Cl. 12.3.3.5. For self-tapping screws with $D_r \geq$ ¼ in. (½-in. or 12-mm screws), the dowel bearing strength is differentiated for parallel or perpendicular to grain loading. For self-tapping screws with $D_r <$ ¼ in., the dowel bearing strength is considered equal in both directions.

FIGURE 4.27 Cross section of threaded portion showing contribution of thread wing in addition to the root diameter.

For the ½-in. (12 mm) screw (D_r > ¼ in.), the strength contributions of the alternating plies must be accounted for through an "effective" dowel bearing length. For cases where loading is parallel to the grain in the shear plane, the length of the lower strength cross-wise plies is reduced by a factor $F_{e\perp}/F_{e//}$. For cases where loading is perpendicular to the grain direction of the surface lamination, the thickness of the cross-wise plies can remain unadjusted, or, according to Cl. 12.3.5.2, the thickness of the cross-wise plies can be increased by a factor of $F_{e//}/F_{e\perp}$. While not an analytically precise application of the European Yield Model, this approach is practical, and, as Uibel and Blass (2007) have shown, a simplified approach can yield reliable results which deviate only slightly from a more precise application of the (highly idealized) European Yield Model. Similar procedures have been used in Europe and Canada to approximate dowel bearing strength in CLT without overcomplicating the design procedure. Designers should keep in mind that the actual dowel-bearing length must be used to check minimum penetration requirements (see Fig. 4.28).

For lateral loading in the narrow edge of panels, designers should be mindful of the influence of end grain on connection strength. For fasteners with $D_r \geq$ ¼ in., the use of an end grain factor of $C_{eg} = 0.67$ must be applied to lateral design values to account for the presence of end grain in the narrow panel edge. For smaller-diameter fasteners, the end grain factor must also be applied unless the fastener is installed into the side grain of a lamella.

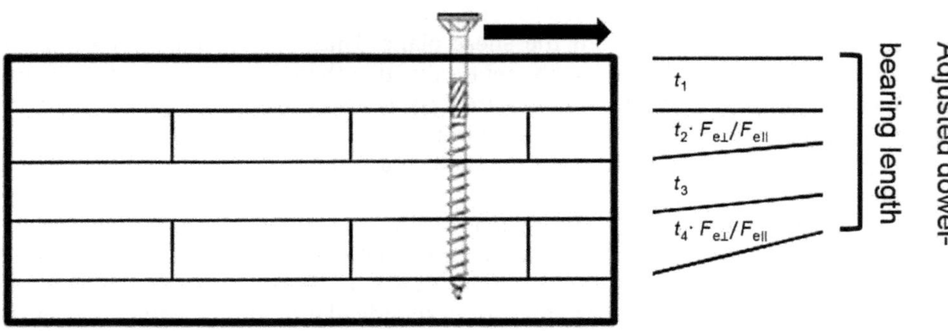

FIGURE 4.28 Example of adjusted dowel-bearing length with lateral force in major strength direction.

All four failure modes in the European Yield Model are considered ductile to varying degrees. As mentioned in Sec. 4.1, even wood crushing is considered as "apparently ductile" (Smith, Gong, & Landis, 2003), as the material maintains load-carrying capability even with significant deformation. Predictions made using the European Yield Model, however, are only valid as long as premature brittle failure modes are avoided. Unique spacing, edge, and end distances are specified to avoid splitting of the wood and to provide sufficient material for load transmission (see Fig. 4.29).

When multiple fasteners are placed together, wood member limit states, such as net tension strength and group tear-out, need to be checked against the connection strength. For the most part, laterally loaded self-tapping screws in the wide panel face of CLT tend to show high ductility and resistance to premature brittle failure modes, in part due to the fact that the crossing layers in CLT act as reinforcement against tensile and shear failures (Sadeghi & Smith, 2014). To exploit the toughening aspect of CLT for lateral connection design, fasteners should be firmly secured into at least one cross lamination (see Fig. 4.30).

Group tear-out tends to be a failure mode limited to tightly spaced groups of fasteners with steel side plates located near the end of members. Design procedures for checking

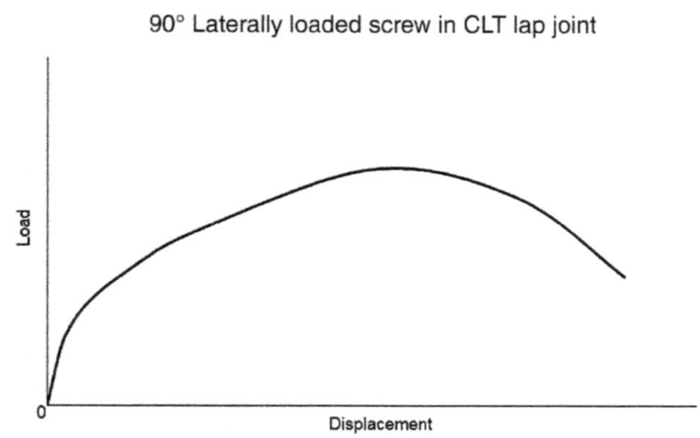

FIGURE 4.29 Representation of load displacement curve for laterally loaded screws in CLT.

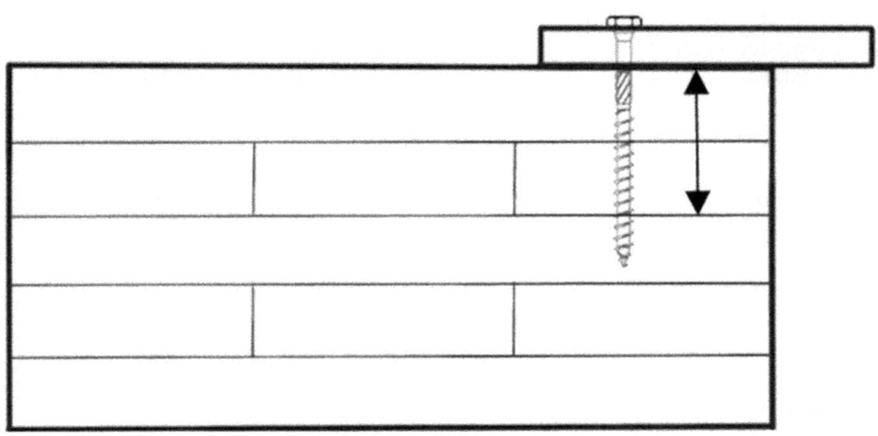

FIGURE 4.30 Shear screw secured through >2 laminations including one cross lamination.

this failure mode are not firmly established for CLT as of yet. Research by Uibel and Blass in 2007 found that splitting and plug shear failure modes tended to be limited to wood in the outer laminations at the shear plane. The authors note that even when these failure modes did occur, they did not lead to premature brittle failure of the connection, as the load-carrying capacity remained constant or fell only slightly (Uibel & Blass, 2006; Blass & Schädle, 2011). Conventional methods of checking group tear-out are based on checking the strength of the assumed shear and tension failure planes. CLT, however, is generally not edge glued, meaning shear transfer between adjacent parallel lamellas cannot be assumed. Tension and shear failures are brittle and associated with different stiffness, so summing their strengths can also be problematic. A stiffness-based approach has been proposed by Zarnani and Quenneville (2014) which models the reinforcing effect of the cross layers while neglecting any contribution from the lateral shear planes. By relating strength to stiffness, the model allows the designer to predict the proportion of the load carried through tensile stress at the head of the connection, shear stress at the bottom of the connection, and shear stress at the interface between the crossing layers in order to identify the critical failure plane. According to the model in their work, crossing layers reinforce the connection by distributing stresses outside the connection area (which then get distributed to parallel planks outside the connection area).

In general, highly stressed steel plate connections are made more robust in CLT by spreading out the fasteners over a wider area and by anchoring the fasteners through as much of the panel depth as possible.

With any lateral connection where fasteners are placed in a row, uneven load sharing between fasteners (due to unequal wood deformation around the fasteners or perhaps due to construction tolerances) can also cause some of the fasteners in a group to be loaded to failure before others. Connection strength calculations determined with the European Yield Model are adjusted to account for this through group effect modifiers, which reduce total strength of a connection group.

Group effects are accounted for in North American timber design codes through strength reduction factors formulated as a function of the cross area of the wood member, which makes them difficult to apply to massive panels like CLT. The approach in EC5 is to use an effective number of fasteners, n_{ef}, such that $n_{ef} \leq n$, which is formulated as a function of the number of fasteners (n) acting together and/or fastener spacing. While the purpose is the same, the European approach is recommended since it is easier to apply to CLT, it is worth noting that European research into group effects tends to evaluate the ultimate load achievable, and not just the load at the elastic limit of the outermost fasteners. Interestingly, since self-tapping screws installed in the wide face of CLT tend to exhibit ductile failure modes and group effects tend not to produce premature splitting, many have taken the position that group effect modifiers are not necessary for such applications. This is true in general for any group of laterally loaded fasteners where plastic deformations are possible such that load-carrying capacity of n fasteners is not limited by premature splitting of the wood (Ehlbseck & Werner, 1995). Uibel and Blass (2006), for instance, recommend $n_{ef} = n$ for joints with multiple fasteners in the wide panel face of CLT but recommend an effective number of fasteners for self-tapping screws applied in the narrow edge (Uibel & Blass, 2007b). Initial stiffnesses and ultimate loading capacities, however, should consider n_{ef}. In general, n_{ef} for self-tapping screws can be calculated using formulas from EC5 for nails and bolts (Ehlbseck & Ehrhardt, 1995).

For ¼-in. (6 mm) self-tapping screws,

$$n_{ef} = n^{k_{ef}}$$

where n = the number of screws acting together in a row parallel to the grain

$$k_{ef} = 1.0 \quad \text{for } S_p \geq 14D$$
$$k_{ef} = 0.85 \quad \text{for } S_p = 10D$$
$$k_{ef} = 0.7 \quad \text{for } S_p = 7D$$
$$k_{ef} = 0.5 \quad \text{for } S_p = 4D$$

where S_p is defined in Figs. 4.16 and 4.17
D = outside thread diameter (nominal thread diameter)

For 5/16-in. (8 mm), ⅜-in. (10 mm), and ½-in. (12 mm) self-tapping screws,

$$n_{ef} = \min: \begin{bmatrix} n \\ n^{0.9} \cdot (S_p/13D)^{0.25} \end{bmatrix}$$

Most applications of narrow edge loading will be in the in-plane loading direction. While unusual for a primary loading direction, there may nonetheless be instances when panels will be loaded out of plane as well. In these cases, designers should detail connections to minimize and tendency toward prying behavior of the larger-diameter fasteners (Smith & Frangi, 2014) and to avoid overstressing the wood in tension perpendicular to the grain (see Figs. 4.31 and 4.32).

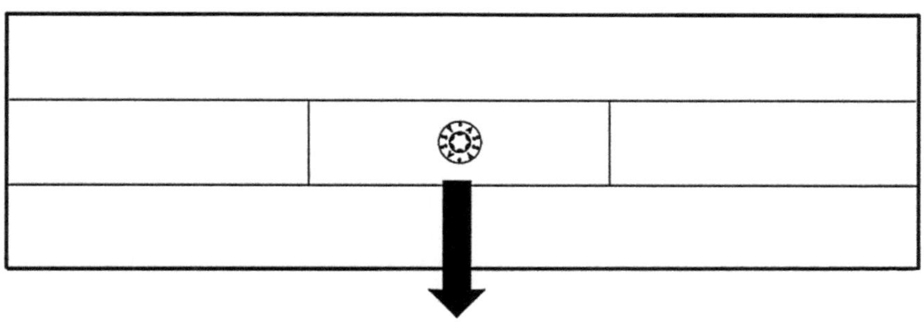

FIGURE 4.31 Out-of-plane loading in the narrow panel edge.

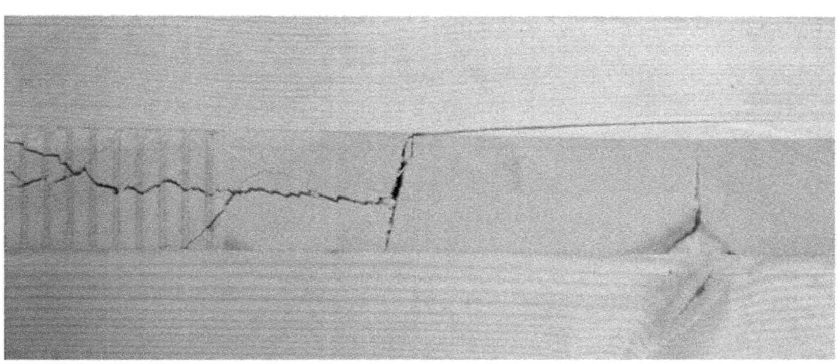

FIGURE 4.32 Splitting due to excessive tensile stresses perpendicular to the grain caused by fastener loaded out of plane in the narrow panel edge.

4.3.3 Inclined Screws

The European Yield Model predicts the strength of connections made with screws by modeling them as smooth dowels simply held in place to carry load in shear. Fully threaded screws, however, are optimized for axial loading. By inclining fully threaded screws toward the direction of the load, the axial strength and stiffness of the fastener is engaged by loading it also in tension. As the angle between screw axis and the direction of the load is lowered, the connection is characterized less by the embedment strength of the wood and the bending strength of the fastener and more by the withdrawal and tensile strength of the screw. Inclined screw connections made with fully threaded self-tapping screws are a stronger, stiffer alternative to conventional 90° connections, and a powerful design tool for use with mass timber and CLT (see Fig. 4.33).

Design models for inclined fully threaded screws began to emerge in the early 2000s. In 2002, Bejtka and Blass published an extension of the European Yield Model for inclined screws so that connection strength would be characterized by the embedment strength of the wood, the bending strength of the fastener, the withdrawal strength of the screw, and the friction along the shear plane. In the same year, Kevarinmäki (2002) published a simplified method for calculating the strength of inclined screw connections installed at an angle of 45° by only considering the axial and friction components in the loading direction. The simplified method considers the connection as a truss, with the screw loaded in tension. The applied shear force is considered as the cosine component of the tensile force in the screw. As shown in Figs. 4.34 and 4.35, once the connection is loaded, a normal force is generated between the members.

Friction is typically neglected when considering 90° laterally loaded screws because, as mentioned earlier, it is not always a reliable component of the connection. With inclined screw connections, however, friction forces can be assumed to be active once the connection is engaged (Krenn & Schickhofer, 2009) since the normal force is a result of the screw being loaded in tension. The inclusion of the friction force should be left to the discretion of the designer, with the condition being met that the screws are to be installed correctly using the right installation torque, with sufficient bearing area in the connection and no gaps between the members.

The term "inclined screws" can refer to screws loaded primarily either in tension or compression. The term "inclined shear-tension screws" will hereby be used whenever

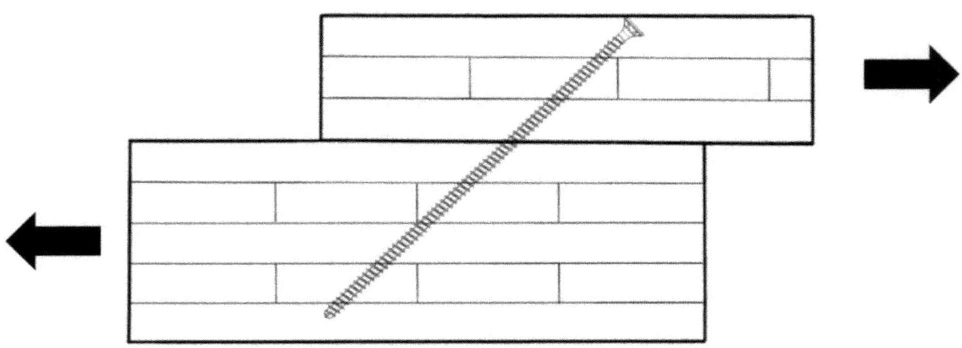

FIGURE 4.33 Inclined shear-tension screws.

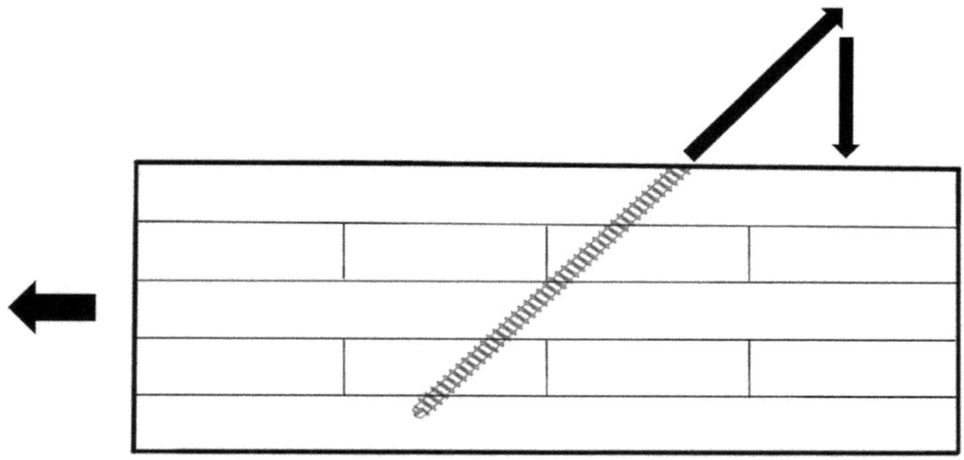

FIGURE 4.34 Truss model of an inclined screw connection.

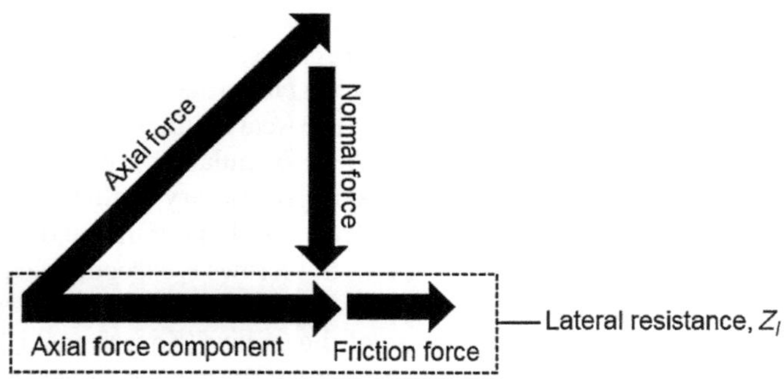

FIGURE 4.35 Truss model force diagram.

necessary to distinguish them from "inclined shear-compression screws," which will be introduced later.

The general approach to calculating the resistance of inclined shear-tension screws using the simplified method is as follows:

$$Z'_l = n_{ef} \cdot R'_T \cdot (\cos \beta + \mu \cdot \sin \beta)$$

where Z'_l = adjusted lateral design value for inclined shear-tension screw lb (N)

R'_T = adjusted axial resistance of shear-tension screw

$$R'_T = \min: \begin{bmatrix} T_a \\ W_\alpha \cdot l_{ef} \cdot C \end{bmatrix}$$

T_a = allowable tensile strength of the screw, lb (N)

W_α = reference withdrawal design value for the screw at angle, α, to the grain, lb/in. (N/mm)

$$l_{ef} = \min: \begin{bmatrix} l_{t,s} \text{ (side member thread penetration length), in. (mm)} \\ l_{t,m} \text{ (main member thread penetration length, not including tip) in. (mm)} \end{bmatrix}$$

α = angle between screw axis and grain direction ($90° \geq \alpha \geq 30°$)
β = angle between screw axis and line of force ($45° \geq \beta \geq 30°$)
C = all adjustment factors pertaining to wood properties (e.g., load duration factors)

$$n_{ef} = \max: \begin{bmatrix} 0.9 \cdot n \\ n^{0.9} \end{bmatrix}$$

n = number of screws acting together in a connection
μ = kinetic friction coefficient (recommended: 0.25)

For application in the wide face of CLT,

$$R'_T = \min: \begin{bmatrix} T_a \\ W_{\alpha,ef} \cdot l_{ef} \cdot C \end{bmatrix}$$

where $W_{\alpha,ef}$ = effective reference withdrawal design value lb/in. (N/mm)

The formula for calculating the effective number of screws, $n_{ef} = n^{0.9}$, is an empirical formula taken from EC5 for axially loaded screws. Since inclined screw connections can sometimes have large numbers of screws acting together, this formula can quickly become overly conservative (Pirinen, 2014). Various research (including Krenn and Schickhofer, 2009) led to the proposal for the $n_{ef} = 0.9 \cdot n$ formula for inclined shear-tension screws, which has since been recognized by various proprietary evaluation reports or approvals for inclined screws in general. Testing of small numbers of inclined screws acting together shows that once the withdrawal strength of the screws is exceeded, load carrying drops off, usually in the order of 50 percent or less of F_{max}. Subsequently, any residual load-carrying capacity is taken up by the dowel action of the screws (Kevarinmäki, 2002; Closen, 2012). Failure modes characterized by tensile fracture of the screws, however, prevent any further dowel effects from acting in the connection. In either case, group action should be accounted for in connections with multiple inclined screws acting together.

In North America, the simplified method is recommended for angles of $45° \geq \beta \geq 30°$. It is worth mentioning that some European standards accept $60° \geq \beta \geq 30°$ for strength verification, but since most of the testing in North America has been performed using $45° \geq \beta \geq 30°$, this narrower range is recommended. Lower angles engage the axial strength of screw more than higher angles, creating generally higher-capacity connections, but angles below 30° are not recommended since end grain effects increasingly characterize the connection. Furthermore, it can become difficult to install screws with flat installation angles using large power drills. It should be emphasized that installation angles of 45° are by far most common, as they balance high performance with ease of installation.

Various studies since have investigated the inclined screw model. Two important studies worth mentioning that account for forces acting perpendicular to the fastener (dowel effects) include Tomasi, Crosatti, and Piazza (2010) and Jockwer, Stieger, and Frangi (2014). For most practical purposes, the simplified method accounting for axial effects has been shown to be a reliable and easy-to-use model (Bejtka & Blass, 2002) with only modest reduction in design strength. The simplified method will be used in this chapter.

To illustrate the potential of connections using inclined fully threaded screws, consider the following example (see Figs. 4.36 and 4.37):

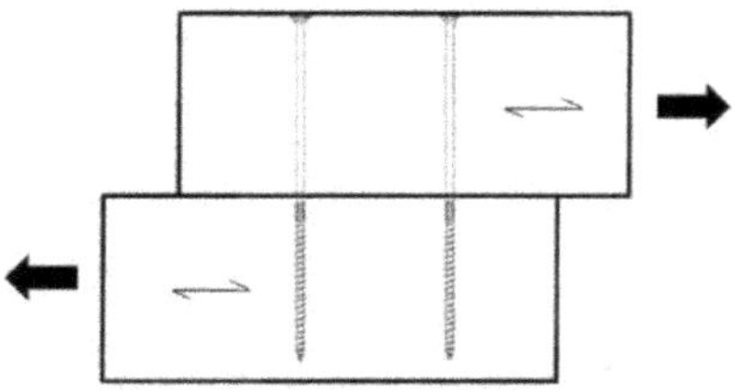

FIGURE 4.36 90° shear screws.

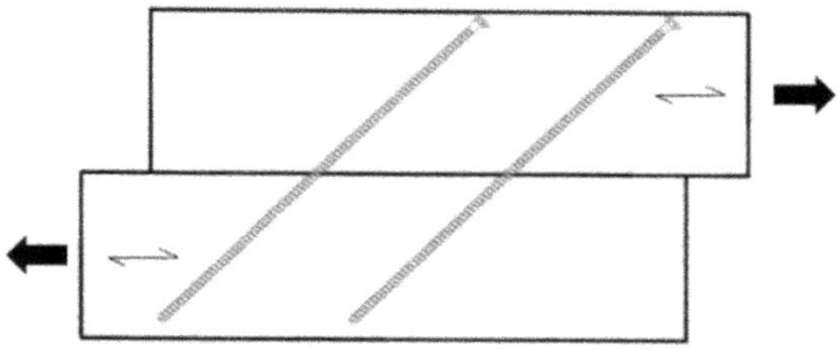

FIGURE 4.37 45° inclined shear-tension screws.

90° shear screws	45° inclined shear-tension screws
Screws:	Screws:
3/8 in. × 9½ in. (10 mm × 240 mm)	3/8 in. × 13-3/8 in. (10 mm × 340 mm)
ASSY® ECOFAST	ASSY® VG CSK
(Partially threaded)	(Fully threaded)
$G = 0.42$	$G = 0.42$
$l_s = 5$ in. (125 mm)	$l_{ef,min} = 13\text{-}3/8$ in. − 5/(cos 45) in. − 3/8 in.
$l_m = 9½$ in. − 5 in. − 3/8 in.	= 5.92 in. (150 mm)
= 4-1/8 in. (105 mm)	
$C_D = 1.0$	$C_D = 1.0$
	$R_t' = 0.85 \cdot 188$ lb/in. · 5.92 in. · 1.0
	= 946 lb (< T_a) (4,208 N)
$Z' =$ **255 lb (1134 N)** (per screw)	$Z_l = 946$ lb · (cos 45 + 0.25 · sin 45)
	= **836 lb (3718 N)** (per screw)

This example clearly shows that inclined screws have tremendous load-carrying capacity over laterally loaded screws. As a general rule, inclined screw connections achieve at least 1.5 times the design strength of their counterpart 90° screw connections if the screw length is kept constant between both arrangements (this was the basis of analysis taken by Bejtka & Blass, 2002). The example above shows that there is even greater potential by increasing the length of the screws, as accommodated by the lower angle of installation. Since inclined screws create stronger connections, they can be the ideal choice for high-intensity loading applications such as hold-downs, drag straps, tension splices in trusses or long-span beams, and also in moment connections. Since there is more carrying capacity per screw, there is an opportunity for more economical connections as well.

In addition to greater strength, inclined screws create much stiffer connections. Testing shows that inclined screw connections can yield at least 8 to 12 times the stiffness of a 90° connection with the same screws (Bejtka & Blass, 2002; Tomasi, Crosatti, & Piazza, 2010), with displacements at F_y typically in the range of 0.02 in. to 0.08 in. (0.5 mm to 2.0 mm) (Closen, 2012) and 0.04 in. to 0.16 in. (1.0 mm to 4.0 mm) at F_{max} (Kevarinmäki, 2002; Krenn & Schickhofer, 2009).

The stiffness of an inclined shear-tension screw can be estimated considering axial effects only in a serial-spring model using the following formula, put forth by Kevarinmäki (2002):

$$K_v = 1/[(1/K_{\parallel,s}) + (1/K_{\parallel,m})]$$

where K_v = initial stiffness along the shear plane ($45° \geq \beta \geq 30°$)

$K_{\parallel,i}$ = initial axial stiffness in member i

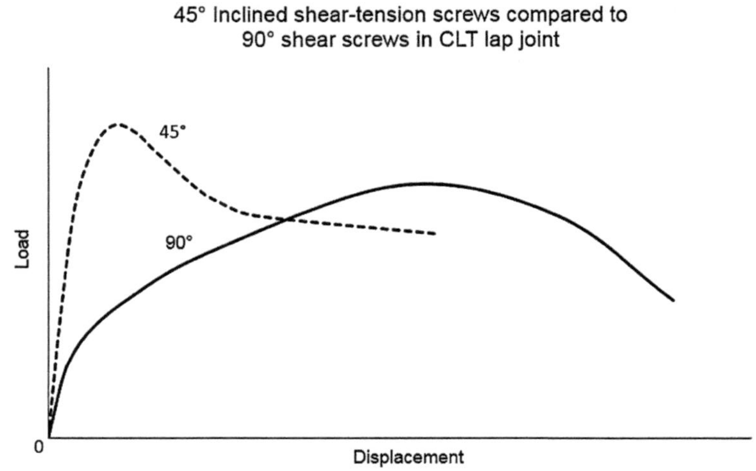

FIGURE 4.38 Representation of load displacement curve for 45° inclined shear-tension screws and 90° shear screws.

Axial stiffnesses, $K_{\parallel}$, are determined through testing and often contained in product approval reports.

Inclined screw connections are characterized by the strength, stiffness, and failure modes of screws loaded axially, which makes the load-displacement behavior of inclined screws distinct with respect to 90° laterally loaded screws, as shown in Fig. 4.38.

Examination of test specimens loaded to failure typically show limited screw bending, with wood crushing contained to small area around the shear plane (see Figs. 4.39–4.41).

The procedure discussed in this section applies to inclined shear-tension screw connections only, where loading is limited to one direction along the shear plane (no reverse loading—see Fig. 4.42). To promote even load sharing, screws of equal size and with equal thread penetration within either member are to be used within groups. Screws

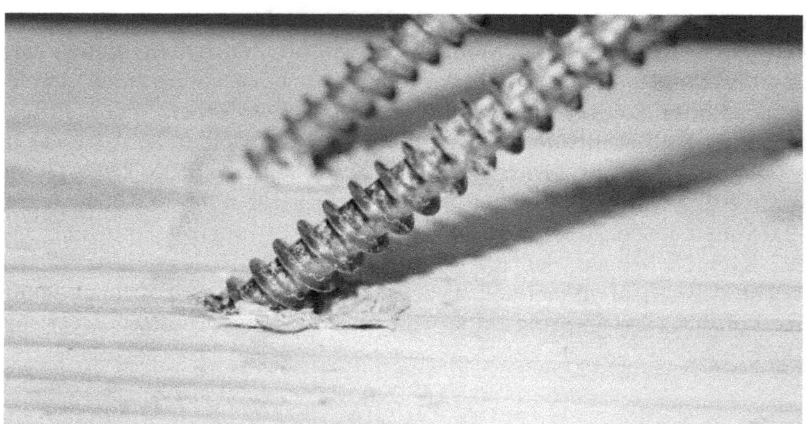

FIGURE 4.39 Minimal wood crushing after withdrawal failure.

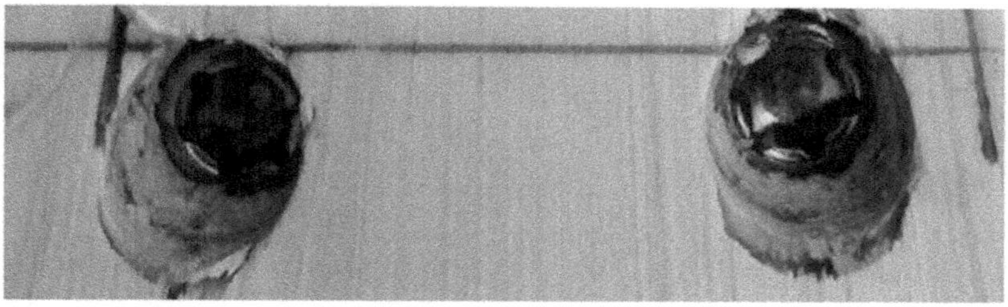

FIGURE 4.40 Withdrawal failure (pull-in) with shear failure of wood surrounding threads.

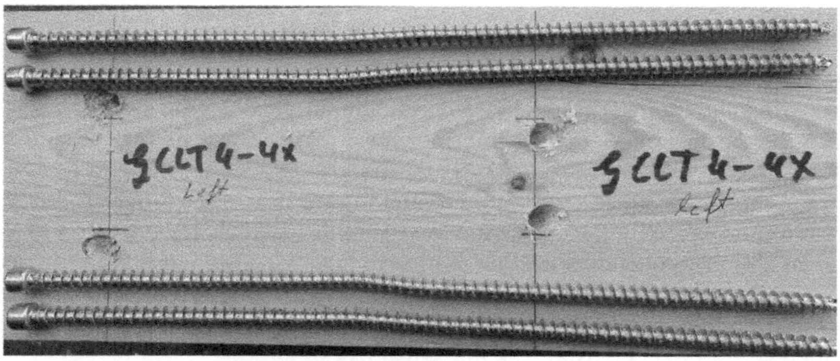

FIGURE 4.41 Limited screw bending at ultimate loading.

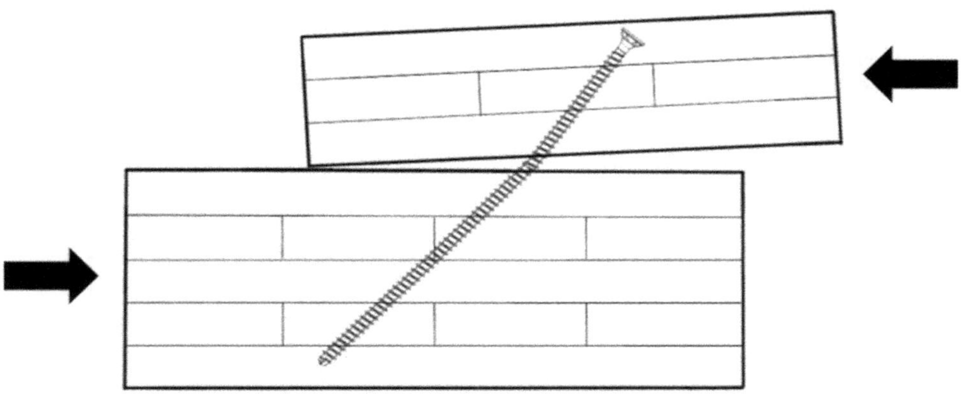

FIGURE 4.42 Reverse loading leading to panel separation and screw bending.

within groups must also be installed mutually parallel (in other words, at the same installation angle). For steel-to-wood connections, power tools with torque clutch mechanisms should be used to install inclined screws to encourage even pretensioning among the screws, which also promotes even load sharing.

Since inclined screws introduce a force component perpendicular to the direction of the force, care should be taken to avoid detailing that risks splitting in the wood for

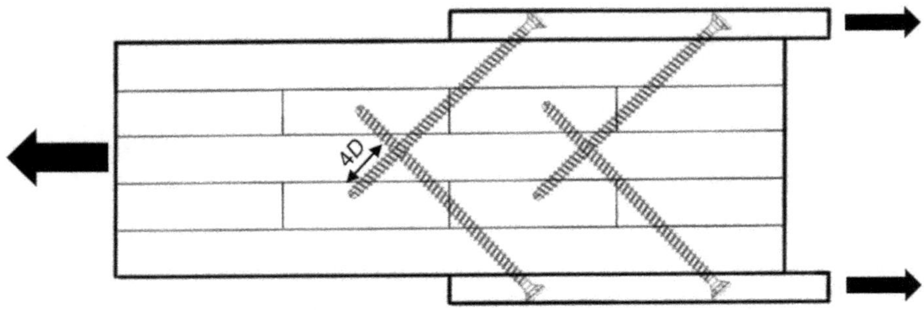

FIGURE 4.43 Symmetrical joint with screws overlapping 4*D* as reinforcement (not including tip).

symmetrical, three-member joints (for instance, for steel to panel connections). Following the instruction of Krenn and Schickhofer (2009), symmetrical joints should have the screws overlap by $\geq 4D$, measured along the axis (not including the tip), as shown in Fig. 4.43.

Furthermore, designers should keep in mind that since tensile force transfer cannot be assumed across adjacent cross-wise lamellas (which have unglued edges), inclined screws should typically penetrate sufficient layers loaded parallel to the grain.

Geometry requirements for inclined screws are measured from the center of gravity of the embedded thread section in each member. Geometry requirements for axially loaded screws may be used (see Fig. 4.44).

Since inclined screw connections utilize fully threaded screws, it is necessary to ensure that both members are pressed tightly against one another during installation. If the self-weight of the panel does not supply sufficient pressure, then partially threaded assembly screws or clamps can be used for this purpose. Assembly screws may be spaced at approximately 2 ft to 4 ft (600 mm to 1200 mm) on center, depending on the CLT panel thickness.

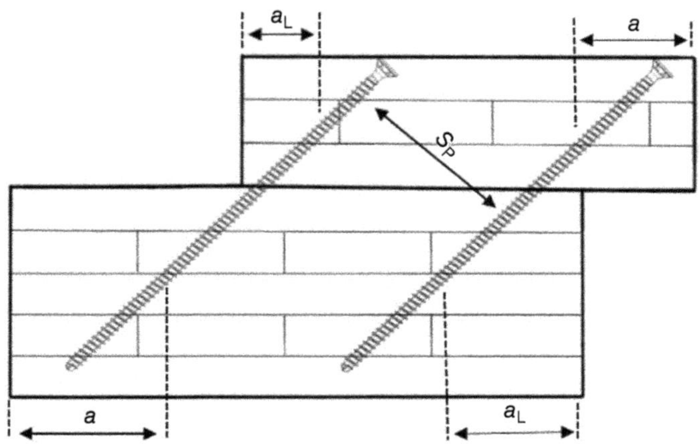

FIGURE 4.44 Geometry requirements for edge and end distances for inclined screws reference the center of gravity in the threaded portion of the screw for each member. Spacing between screws in a row is measured perpendicular to the screw axis.

Inclined screws can be installed "freehand" by engaging the wood straight on (just enough for the threads to hold the screw in place) and then establishing the correct angle manually before installing the remainder of the screw. Pilot holes are often used to ensure precise placement of the screws while the panel is still on the ground before being lifted into place by crane. With CLT, it can be especially important to establish precision when using inclined screws since many connections are located near the edges of the wide face of the panels. Angles that are too steep may also penetrate through the panels when not intended. For wood-to-wood connections, jigs made from simple plywood, framing squares, or more advanced aluminum jigs can be made to accept a drill bit at the appropriate angle. Pilot holes are generally no more than a few inches deep and serve two purposes:

1. Without pilot holes, self-tapping screws may travel across the surface rather than engaging the fiber. Pilot holes initiate the path of installation with precise placement on the surface of the wood. This can be especially important for long, smaller-diameter screws in slender beams or into the narrow panel edge.
2. Pilot holes establish the angle of installation. Once the screw is engaged in the pilot hole, it will self-tap the remainder of its hole along the desired path.

Inclined screws can be exploited for a variety of applications in CLT where strength and stiffness is sought over yielding, such as reinforcement applications, as well as mechanically jointed bending systems, such as timber concrete composite slabs.

4.3.4 Screw Crosses

When loading is anticipated in only one direction, screws are oriented mutually parallel to resist the load in tension. When designing for reverse loading, inclined screws can be installed as symmetrical screw crosses in both directions. While self-tapping screws can be produced with asymmetrical threads, it is commonly accepted that fully threaded screws generally exhibit the same withdrawal resistance whether they are pulled or pushed. When screws are arranged as crosses, one screw will be loaded axially in tension, while the other one will be loaded axially in compression (see Fig. 4.45).

Again, considering the truss analogy (Fig. 4.46), it can be seen that one of the consequences of adding shear-compression screws alongside shear-tension screws is that the normal force between the elements is canceled out, which means friction cannot be added to connection strength. According to Kevarinmäki (2002), the absence of friction explains why screw crosses can show 25 to 30 percent less capacity over equal numbers of

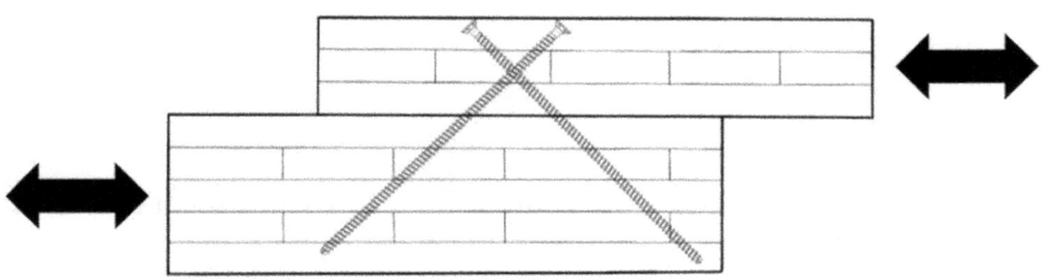

FIGURE 4.45 Symmetrical screw cross.

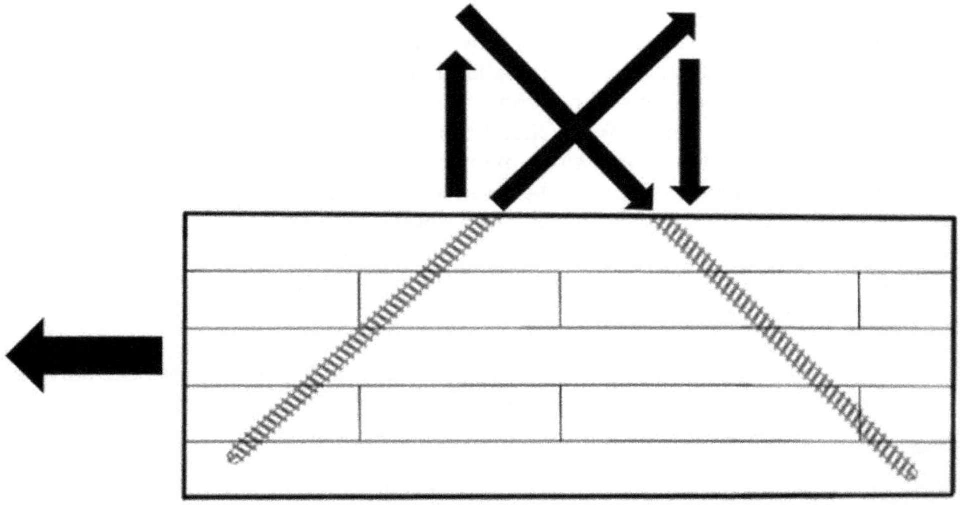

FIGURE 4.46 Truss model for symmetrical screw crosses; resultant normal force from shear-tension screw canceled out by shear-compression screw.

inclined shear-tension screws at 45°. Since screw pairs require only a small offset, however, larger numbers of screws can be fit into a small area to increase connection strength. There are also some instances where the lack of a normal force can be exploited, such as preventing crushing of soft insulating interlayers in timber concrete composite systems (see Fig. 4.47) (Gerber, 2016).

The simplified method for calculating inclined screws is useful for screw crosses as well, although it is formulated instead for design strength *per screw cross*:

$$Z'_X = n_{x,\text{ef}} \cdot (R'_C + R'_T) \cdot \cos\beta$$

where Z'_X = adjusted lateral design value for screw cross connection lb (N)

n_x = number of screw crosses acting together in a connection

$$n_{x,\text{ef}} = \max: \begin{bmatrix} 0.9 \cdot n_x \\ n_x^{0.9} \end{bmatrix}$$

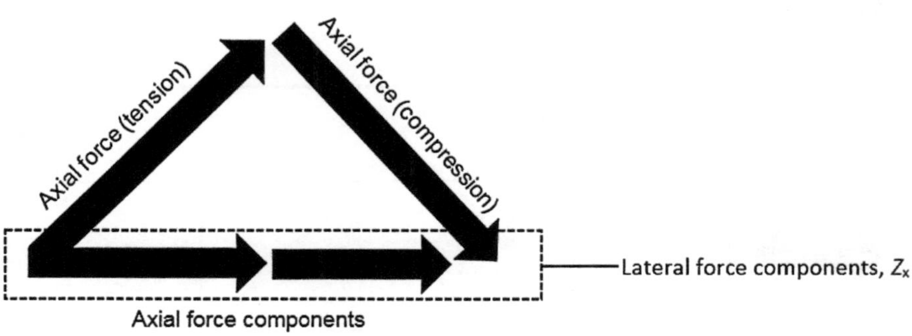

FIGURE 4.47 Axial force diagram for symmetrical screw crosses.

R'_C = adjusted axial resistance of shear-compression screw
R'_T = adjusted axial resistance of shear-tension screw

$$R'_C = \min: \begin{bmatrix} 0.8 \cdot T_a \\ W_\alpha \cdot l_{ef} \cdot C \end{bmatrix}$$

$$R'_T = \min: \begin{bmatrix} T_a \\ W_\alpha \cdot l_{ef} \cdot C \end{bmatrix}$$

T_a = allowable tensile strength of the screw, lb (N)
W_α = adjusted withdrawal design value for the screw at angle α to the grain lb/in. (N)

$$l_{ef} = \min: \begin{bmatrix} l_{t,s} \text{ (side member thread penetration length), in. (mm)} \\ l_{t,m} \text{ (main member thread penetration length, not including tip) in. (mm)} \end{bmatrix}$$

α = angle between screw axis and grain direction ($90° \geq \alpha \geq 30°$)
β = angle between screw axis and line of force ($45° \geq \beta \geq 30°$)

For application in the wide face of CLT,

$$R'_C = \min: \begin{bmatrix} 0.8 \cdot T_a \\ W_\alpha \cdot l_{ef} \cdot C \end{bmatrix}$$

$$R'_T = \min: \begin{bmatrix} T_a \\ W_\alpha \cdot l_{ef} \cdot C \end{bmatrix}$$

The method above is another example of the simplified approach to calculating the design strength of inclined screws proposed by Kevarinmäki (2002). The 0.8 factor assigns a lower fastener strength for screws loaded in compression. A research project of screw crosses at the University of McGill (Prat-Vincent, 2011), however, applied the above approach and showed that, with sufficient withdrawal and push-out resistance, failure tended to be controlled by tensile fracture of the shear-tension screws.

Screws acting as a pair should be installed at the same angle, opposite to one another, with the same thread penetration lengths within either member. Screws in a pair need to be offset from one another only 1.5*D* to accommodate tolerances on-site. It is important to note that screws are not required to cross in the shear plane. In fact, it is generally better to spread the entry points for connections using small numbers of crosses to stabilize the connection.

The stiffness of a screw cross can be estimated using the following equation:

$$K_x = 2/[(1/K_{\parallel,s}) + (1/K_{\parallel,m})]$$

where K_x = initial stiffness along the shear plane
$K_{\parallel,i}$ = initial axial stiffness in member i

An extension of the screw cross concept that is especially applicable to CLT involves installing fully threaded screws at double (or "compound") angles across butt joints to connect panel edges to one another. Testing carried out in at the University of British Columbia on screws installed 45° horizontally away from the line of the joint and approximately 30° vertically away from the panel surface confirmed that double inclined screw connections behave primarily in the same manner as inclined screws. Furthermore, the testing confirmed that the simplified design process considering only axial effects was suitable for this arrangement (Hossain, Danzig, & Tannert, 2016). With regard to the design angles, the 45° horizontal angle ensures that, regardless of the vertical angle, α will remain equal for both grain directions in the CLT panel. The vertical angle should be kept low (approximately 30°) so that the screw remains loaded primarily in the axial direction. The angle to the grain, α, and the angle to the line of the force, β, are measured diagonally from the axis of the screw. In this case, a 45° horizontal angle and a 30° vertical angle produce a compound angle of approximately 52°, which is reasonably close to the recommended boundary condition of 45° for inclined screws. Again, screws are assumed to carry load primarily in the axial direction, with alternating shear-tension and shear-compression screws arranged evenly along line joints (see Fig. 4.48).

Like inclined shear-tension screws, screw crosses exhibit high initial stiffness, leading to high maximum loading capabilities at low displacements. Likewise, screw cross connections are characterized by a steep drop in capacity beyond F_{max} once axial strength properties are exceeded, with dowel effects providing continuing load-carrying capabilities at further displacements of approximately 50 percent of F_{max}. Failure modes for screw crosses are characterized by screw fracture, compression screw push-out, or tension screw pull-in. Depending on the screw pattern, failure modes for screw crosses can include some kind of panel separation and/or rotation if the tension screws fail first in withdrawal and the compression screws push the panel out (see Fig. 4.49).

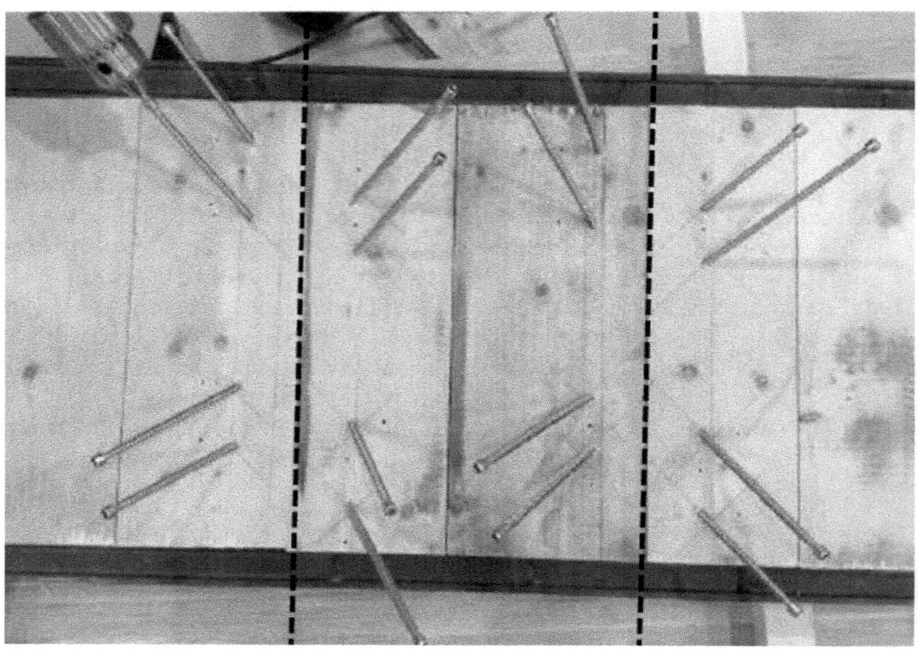

FIGURE 4.48 Double angle screw configuration for a three-member test specimen; joints are indicated with dashed lines.

FIGURE 4.49 Shear-compression screw push-out.

Inclined screws and screw crosses can be exploited for mechanically jointed beams (also known as "screw laminated beams") and composite elements. The efficiency of the composite action between the two elements is determined by the rigidity of the shear connection between them. Mechanical connections will always exhibit some degree of slip (relative displacement), which inhibits full composite action. Inclined screws and screw crosses, however, are very stiff, which means they are favorable for creating highly efficient connections for composite bending elements. Spanning capability for CLT is often limited by service limit states, like vibration or deflection. Mechanically jointed solutions can offer stiffer floor elements and therefore enable longer spans. CLT and glulam, for instance, can be joined to create hollow box beams or rib-stiffened panels to add structural depth and therefore strength and stiffness to a CLT floor panel (Montgomery, 2014). According to Dietsch and Brandner (2015), it is possible to achieve a connection efficiency factor γ of around 0.8 using inclined fully threaded self-tapping screws, as calculated according to the "gamma method" for mechanically jointed timber elements.

Timber concrete composites (TCC) system can also produce panels with greater bending stiffness, with the concrete loaded in compression and the CLT in tension. While performing structural work, TCC systems can also enhance fire safety, acoustic performance, serviceability performance, thermal mass, and weight for an otherwise light timber building if desired. Small-scale testing of shear joints can be an effective method of measuring the stiffness of these connections. For instance, shear stiffness values for various configurations of self-tapping screws in CLT timber concrete composite systems can be found in a comparative study performed at the University of British Columbia by Gerber (2016) (see Fig. 4.50).

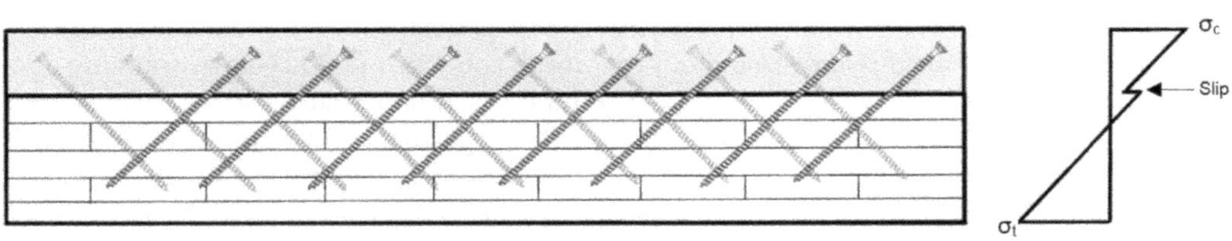

FIGURE 4.50 Timber concrete composite slab element and example bending stress diagram.

For mechanically jointed elements, the effective number screws, $n_{ef} = n$ (or $n_{x,ef} = n_x$ in the case of screw crosses). Design procedures can be found in Annex B of EC5.

4.3.5 Ductile Connections in CLT

Mechanical connections play a crucial role in supplying ductility to CLT and mass timber structures. As mentioned in Sec. 4.1, wood is limited in its ability to provide ductility, as tension and shear failures (including tension failures due to bending) are brittle. Additionally, since CLT is rigid when loaded in plane, yielding behavior must come from the connections between panels.

Ductility describes the ability to carry load under plastic deformations. Modern CLT buildings are often taller and more slender than historic tall wood buildings (Smith & Frangi, 2014). As these buildings become larger, load distribution among the elements can become more difficult to predict with precision. Ductility in CLT structures is necessary to provide visible warnings that signal unexpected overloading. Ductility also plays a role in enhancing structural robustness. The ability for fasteners to yield under high loading promotes stress redistribution to surrounding fasteners, which allows for the formation of alternative load paths. This not only helps to prevent localized overloading but it also helps to prevent progressive structural failure mechanisms (Jorissen & Fragiacomo, 2011).

Compared to steel and concrete, there is less information available concerning the ductility in timber structures. Furthermore, there is no established method of quantifying ductility. According to Munoz, Mohammad, Salenikovich, and Quenneville (2008), there are six available methods of measuring ductility in timber structures. Three commonly cited methods include the following:

$$\mu_1 = \Delta_{max}/\Delta_y \tag{1}$$

$$\mu_2 = \Delta_u/\Delta_y \tag{2}$$

$$\mu_3 = \Delta_u - \Delta_y \tag{3}$$

Method (1) compares the displacement at F_{max} (Δ_{max}) to the displacement at F_y (Δ_y). This method is helpful in measuring the reserve strength in a system under increasing loading. The second method compares the ultimate displacement at $F_u(\Delta_u)$ to the displacement at (Δ_y), which is helpful in predicting deformation ability but not reserve strength, for instance, in the case of inclined screws where load-carrying capacity drops after F_{max}. Jorissen and Fragiacomo (2011) make the case for an absolute definition in the form of total displacement rather than a ratio, as this method is helpful in predicting capacity for, among other things, stress redistribution. The designer must decide which method is most useful for the intended purpose and also be aware of how others measure ductility, as the different methods can yield very different results.

Further complicating things is the fact that the yield point for timber connections is not as sharply defined in load-displacement curves as it is for steel. Various standards exist for defining the yield point throughout the world, and, like methods of determining ductility, these standards yield very different results. The dowel-bearing strength of wood is determined for the NDS using the 5 percent diameter offset, as outlined in ASTM D5764, while the equivalent energy elastic-plastic method (EEEP), as outlined in ASTM E2126, can be used for wall and floor assemblies (Below, Mohammad, Ni, Malczyk, & Popovski, 2014).

Classification	Average ductility ratio, $\mu = \Delta_{max}/\Delta_y$
Brittle	$\mu \leq 2$
Low ductility	$2 < \mu \leq 4$
Moderate ductility	$4 < \mu \leq 6$
High ductility	$\mu > 6$

TABLE 4.7 Ductility Classification System Proposed by Smith et al. (2006)

Smith, Asiz, Snow, and Chui (2006) propose a classification system for determining ductility based on method (1), using an offset method for obtaining the yield point (see Table 4.7).

Earlier in this chapter, it was shown that screws loaded laterally tend to have highly ductile capacities, while screws loaded in withdrawal are characterized by a steep drop in capacity at low displacements after reaching F_{max}. Ductile connections using screws in CLT, therefore, will generally involve screws loaded laterally. CLT has been shown to have greater capacity for ductile behavior in lateral connections over homogeneous wood of comparable grade since the cross laminations tend to reinforce the panels against splitting (Pirinen, 2014; Smith & Frangi, 2014). The authors also make note of the fact that slender self-tapping screws tend to exhibit high ductility for transferring shear forces between CLT panels. Lateral design values based on the European Yield Model are considered as ductile for all four single-shear failure modes. Failure types I and II are characterized by wood crushing, while failure types III and IV include fastener yielding. To achieve maximum ductility, designers should target failure types III and IV (see Fig. 4.51).

FIGURE 4.51 Opened *D*-fir glulam connection utilizing ½-in. (12 mm) partially threaded screw, exhibiting type IV failure with two plastic hinges (gap visible in picture opens up only after the load is released).

Designing using the European Yield Model can be an iterative procedure, but a general way to predict the type of failure mode is to consider fastener slenderness. Fastener slenderness (λ) is expressed as a ratio:

$$\lambda = t/D$$

where t = panel thickness
D = fastener diameter

Slenderness ratios are unique to connection material as well as dimensional properties, as they are influenced by fastener bending strength and the dowel-bearing strength of the wood (Pirinen, 2014). As shown in Fig. 4.51, increasing fastener slenderness pushes the failure type into fastener yielding, with type IV fully achieved through a slenderness ratio of 7 for proprietary self-tapping screws in CLT ($G = 0.42$) (see Fig. 4.52).

Using self-tapping screws in CLT, type III or IV failure modes are easily achievable, even despite the high bending yield strength of hardened self-tapping screws. Even plywood splines ¾ in. (19 mm) thick can produce type III failure using ¼-in. (6 mm) screws (Sadeghi, 2015; Hossain, 2016b).

At the end of this section, it will be explained why fastener yielding (type III or IV), is desirable for seismic design. Types II, III, or IV are also more advantageous over type I because they offer increased ultimate capacity and ductility.

The European Yield Model, as it is presented in the NDS, does not consider the effects of end fixity, friction, or axial resistance. Failure types II, III, and IV, however, are characterized by some form of inclination of the fastener. End fixity (for example, through a large washer bearing on the surface) can resist rotation, which contributes to connection strength. In a series of tests on panel to panel joints in CLT, Sadeghi (2015) has shown that adding a washer to self-tapping screw can significantly increase the connection capacity at both yielding and maximum levels (see Table 4.8). Additionally, Sadeghi

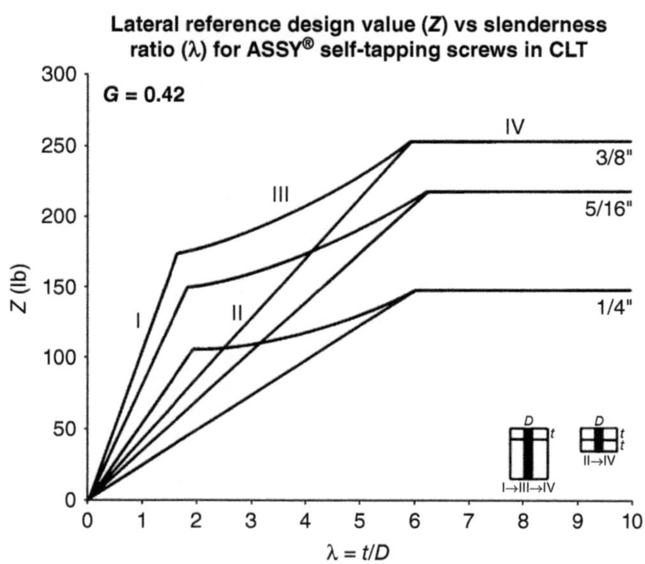

FIGURE 4.52 Increased slenderness leading to higher failure mode types.

D in. [mm]	No washer F_y lb [N]	No washer F_{max} lb [N]	With washer F_y lb [N]	With washer F_{max} lb [N]
¼ [6]	423 1,880	991 4,410	798 3,550	1,965 8,740
5/16 [8]	946 4,210	2,221 9,880	1,288 5,730	3,138 13,960

TABLE 4.8 Effect of Washer on Surface Spline Connection Capacities (Results Reprinted from Sadeghi, 2015)

showed that the addition of a washer pushed the failure of the screw from mode III to mode IV through the production of a second plastic hinge on account of increased end fixity.

The axial resistance of the screw can also have considerable effects. For a screw loaded in shear, only dowel effects are engaged. However, as the screw becomes inclined to the force, either through rotation (II) or through the formation of one or two plastic hinges (III, IV), axial resistances take effect. For partially threaded screws, this includes head pull-through in the side member and withdrawal resistance in the main member. For fully threaded screws, the withdrawal strength in both members is engaged.

In addition to dowel effects, screws subjected to inclination generate additional resistance through an axial force component along the shear plane. Additionally, as with inclined screws, axial forces in the screw produce a normal force between the members. This normal force generates friction, which further increases connection strength as displacement increases. These phenomena are referred to as the "rope effect" (see Fig. 4.53).

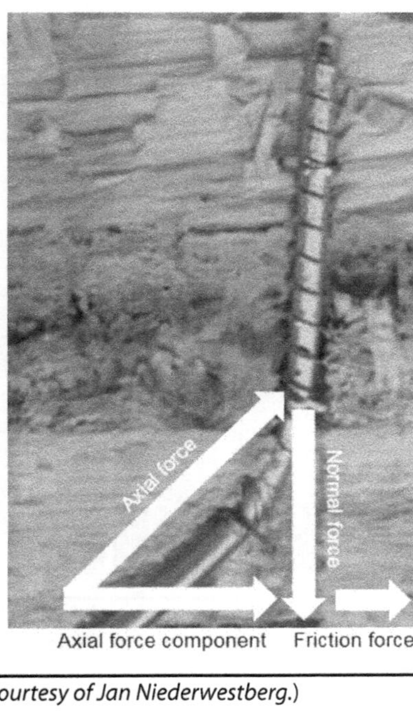

FIGURE 4.53 Rope effect. (*Courtesy of Jan Niederwestberg.*)

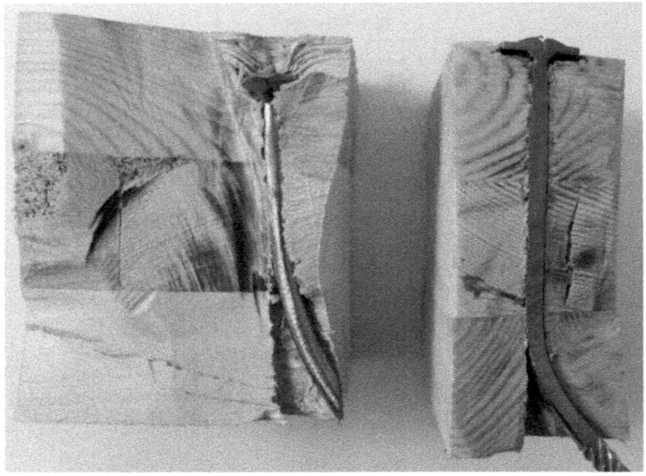

FIGURE 4.54 The larger washer head (right) contributes more resistance due to the rope effect than the countersunk head (left), which was pulled in to the CLT.

The rope effect can have considerable influence on the ductility of the connection, and, since fully threaded screws are optimized for axial loading, large ductility ratios can be achieved using fully threaded screws in shear (see Figs. 4.54 and 4.55). Tomassi, Piazza, Angeli, and Mores (2006) achieved ductility ratios $\mu_1 \approx 35$, $\mu_2 = 58$, and $\mu_3 = 57.5$ using fully threaded screws in monotonic shear loading.

Shear screws and inclined screws can be combined in one connection to combine the advantages of both—producing a connection that is initially very stiff but also

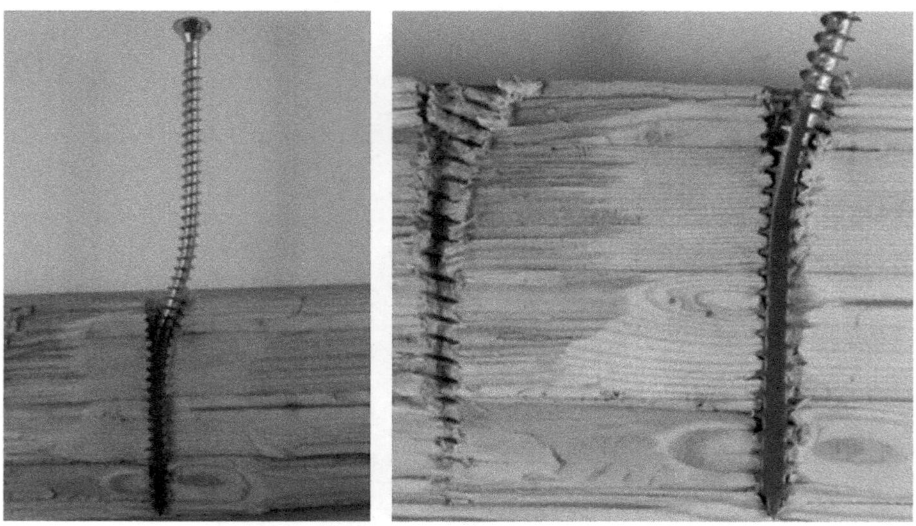

FIGURE 4.55 Threads remain engaged despite screw yielding, showing that significant axial forces can be generated in threaded fasteners subject to lateral loading.

highly ductile. According to Tomassi, Piazza, Angeli, and Mores (2006), the performance of connections composed of screws at different inclinations can be modeled as the superposition of both types in terms of their load-displacement behavior. It should be emphasized that the design strength of both types cannot simply be summed as a connection strength, as each is associated with a different stiffness. A simplified approach is to assign the connection design strength to the inclined screws, with the 90° screws contributing ductility. A study conducted at University of British Columbia on CLT panel-to-panel connections confirmed that by combining both screw types, the stiffness of screw crosses could be supplemented by the ductility of 90° shear screws, effectively producing connections that are strong, stiff, and also highly ductile (Hossain, 2016b). The performance of specific arrangements and screw lengths should be determined through testing (see Fig. 4.56).

One final topic to mention on ductility concerns the prevention of premature brittle failure modes through capacity-based design. Capacity-based design is based on the idea of designing brittle elements with sufficient overstrength to assign probable failure to ductile elements. Overstrength factors (OSF) are properties of connections that relate the 95th percentile connection strength ($R95$) to the design strength (R_d):

$$\mathrm{OSF} = R^{95}/R_d$$

OSF are used to assign failure to ductile connection elements so that they may act as "fuses" in the event of overloading. Brittle elements, meanwhile, maintain integrity through appropriate oversizing. In order to assign failure to ductile connections with sufficient probability, CLT wood member limit states in the CLT panels should be designed for the 95th percentile connection strength, $\mathrm{OSF} \cdot R_d$. Steel side plates also display low ductility, for instance, when failure is characterized by tensile yielding or punch-through of anchor bolts (Fragiacomo, 2014). Accordingly, plastic behavior is assigned to the screws unless steel plates are designed specifically for ductile behavior. Since overstrength ratios are a function of design values, overstrength ratios determined for European markets should not be applied to ASD. In the absence of generalized OSF, capacity-based design should rely on test data.

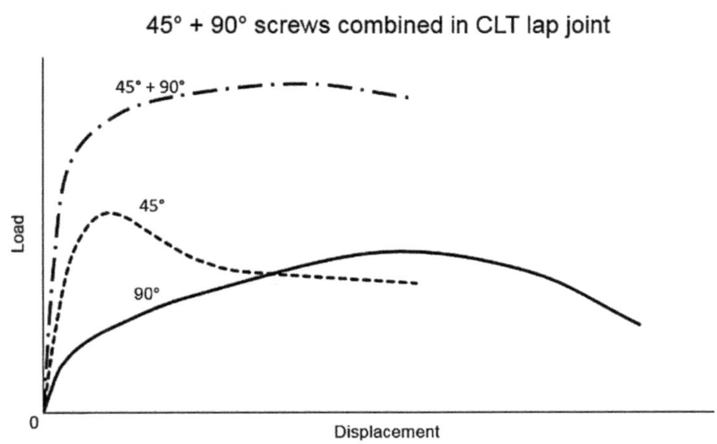

FIGURE 4.56 Representation of connection behavior of inclined screws and shear screws.

4.3.6 Reinforcement of CLT

The prevention of premature brittle failure modes can also be prevented by reinforcing CLT in weak loading directions. Just as concrete is reinforced with rebar to overcome its weakness in tension, CLT can be reinforced using fully threaded screws to boost its tensile, shear, and compressive capacities.

Wood is weakest when loaded in tension perpendicular to the grain. Tensile failure modes in this loading direction are brittle and occur at very low strains. CLT is largely protected against this type of failure when loaded from the wide face of the panel. Connections in the narrow edge, however, present risk of splitting when panels are loaded out of plane. Fully threaded screws installed across the assumed crack position can be designed to transmit perpendicular-to-the-grain forces to the unloaded side of the panel. Alternatively, fully threaded screws can be installed at an angle to cross multiple plies, locking them together (see Fig. 4.57).

Similarly, self-tapping screws are often used to reinforce glulam bending members notched at the tension side of the supports (see, for example, Dietsch & Brandner, 2015). This design method, however, should not be used as a design verification method for CLT. The model assumes a parabolic shear stress curve based on classic beam theory, which does not apply to CLT. Additionally, little research has been carried out to understand how notching affects CLT in the transverse layers. As Jockwer (2014) points out, the standard design model for reinforcing notches accounts for only tensile, not shear, forces, which can lead to underdesigned reinforcement when shear effects are prominent. Perhaps most importantly, gaps in unglued edges of CLT in the crossing layers can cause cracks to deviate from the normally assumed crack plane, as shown in Fig. 4.58.

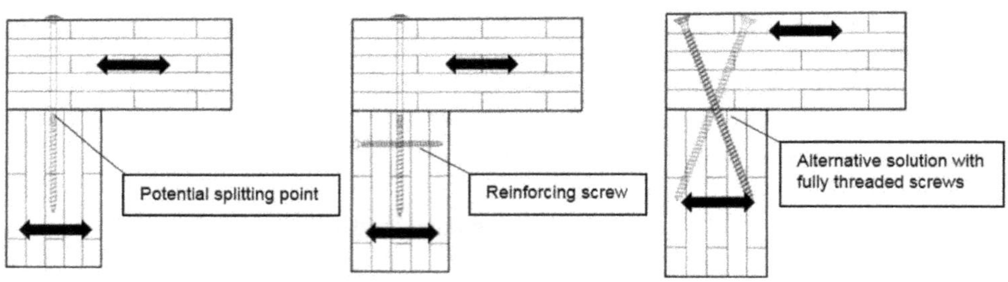

FIGURE 4.57 Out-of-plane loading in the narrow edge.

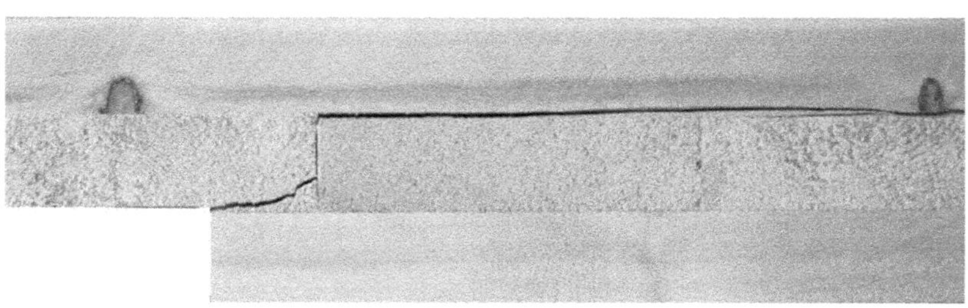

FIGURE 4.58 Crack propagation along unglued gap.

One of the weak mechanical properties of CLT when used as floor panels is the rolling shear strength of the cross-wise laminations. Shear stresses resulting from point loading produce in principle tensile stresses in the 45° direction, which effectively pull the cross-wise layers in tension. Using the analogy of a truss model, Mestek and Dietsch (2011) have proposed a method of reinforcing the cross-wise layers with inclined fully threaded screws to restore the shear capacity of the panel around locations of high-intensity loading (see Figs. 4.59 and 4.60).

Although wood failures due to crushing are considered ductile, designers must still design against wood crushing due to high-compressive stresses. Compression reinforcement can be especially useful for CLT structures where localized bearing loads exceed the bearing capacity of the wood. Examples include line loading from walls bearing on panels and point loads from columns. To reinforce panels against crushing, fully threaded screws are installed flush with the bearing surface, which is loaded using a steel plate. The design process checks for the minimum of three failure modes: (1) wood crushing at the entire surface under the bearing plate with screw push-in, (2) wood crushing at the entire surface under the bearing plate with screw buckling, (3) wood crushing at the screw tips, where the effective bearing area is increased due to stress dispersion. Since the model was designed for beam supports, extrapolation is necessary to adapt the principles for panel reinforcement. Taking cues from Brandner and Schickhofer (2012), the designer may assume a 45° stress distribution angle in the

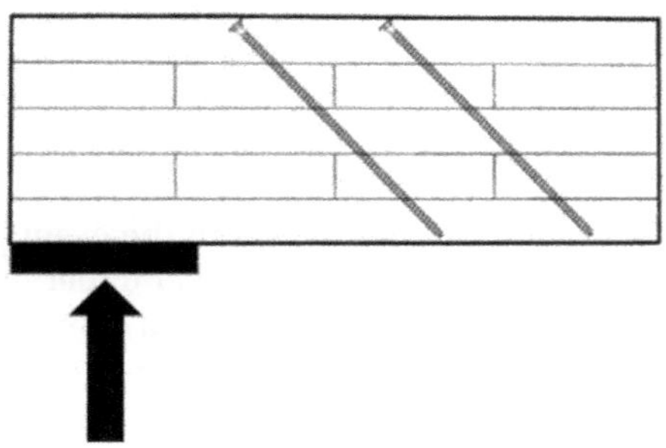

FIGURE 4.59 Point loading with fully threaded self-tapping screws for shear reinforcement, as proposed by Mestek and Dietsch (2011).

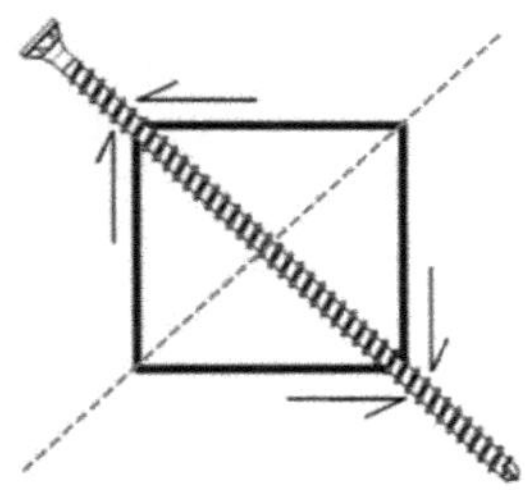

FIGURE 4.60 Fully threaded self-tapping screw in line with principle tensile stress direction (diagonal crack position indicated with dashed line).

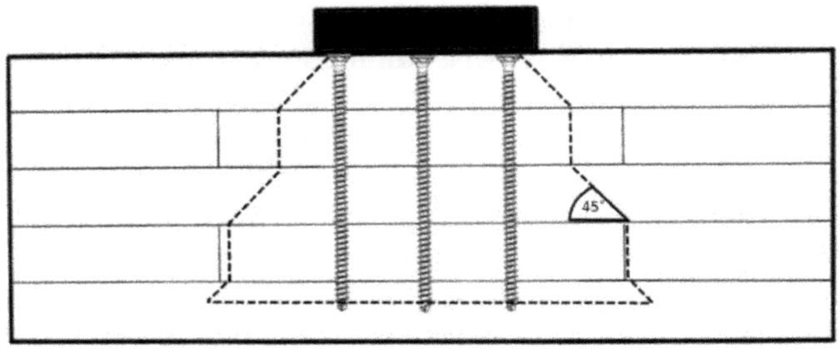

FIGURE 4.61 Estimated stress distribution for directly loaded CLT, assuming 45° distribution angle in the longitudinal direction and 0° in the transverse direction, measured from screw heads.

longitudinal direction and 15° in the transverse direction, although 0° is recommended in this chapter for conservatism due to the presence of unglued gaps in this direction. Stress distribution is measured from the screw heads, and the effective number of fasteners for compression reinforcement, $n_{ef} = n$. Allowable screw buckling strength (B) is estimated for ASD methods by adjusting European procedures and assuming a factor of safety of 3.0 (see Fig. 4.61).

By reinforcing CLT, strength and ductility can be increased, and the ability to use CLT as a structural element is further expanded. Reinforcing can also play an important role in seismic design with CLT, as discussed in Sec. 4.3.7.

4.3.7 Dynamic Performance of CLT Connections for Seismic Design

The ability for mechanical connections to yield under extreme loading is a critical feature of CLT structures designed for high-seismic zones. Under normal service conditions, the design is kept in the elastic range. In the event of high-seismic loading, however, if the connections can be designed to yield under cyclic loading, then substantial energy can be absorbed, which reduces the lateral demands on the structure.

Monotonic load-displacement curves acquire information on connection strength, stiffness, yield point, and ductility (Karacebeyli & Popovski, 2003). To understand how a connection may perform in the event of an earthquake, test specimens are subject to cyclic (reversed) loading as well. The cyclic load-displacement curves produced through cyclic loading are called hysteresis curves (or loops). By comparing the monotonic load curve with the hysteresis curve, information is gathered on how the performance of the connection degrades over repeated loading cycles as well as how effective the connection is at dissipating energy (see Fig. 4.62).

The deviation between the monotonic (or "virgin") curve and the backbone curve connecting the peaks of the hysteresis loops shows the degradation of connection strength. The "pinched" shape, typical of hysteresis loops for timber structures, shows the degradation of connection stiffness at the beginning of each loading cycle.

Connections in specified energy dissipative zones should be designed for fastener yielding (type III or IV). Type IV is preferred due to its appreciated ability to dissipate energy through two plastic hinges. As the screws yield and incline at the shear plane, localized wood crushing displaces wood around the fastener, resulting in a loss of initial stiffness.

Energy dissipation potential is indicated by the total area contained within the hysteresis loops. Hysteresis loops that are more pinched (due to wood crushing) show

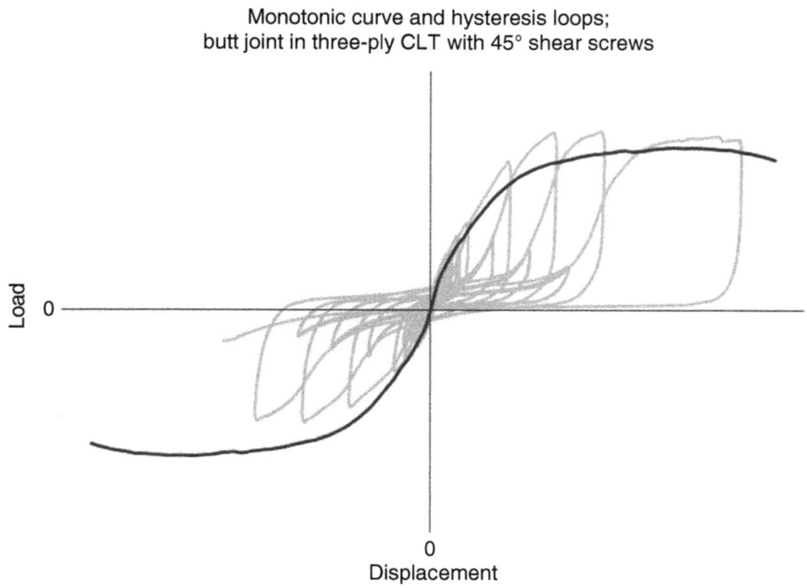

FIGURE 4.62 Monotonic curve (black) and hysteresis loops (gray); from research data gathered by Hossain (2016b).

depreciated energy potential, as the area contained within them is reduced. Wood crushing dissipates energy, but only through the initial crushing cycle. This explains why slender fasteners, which can dissipate energy through repeated cyclic yielding with minimal wood crushing, are preferred.

Connections outside of dissipative zones should be designed to maintain their integrity during a seismic event. Under cyclic loading, CLT may be subject to weaker loading directions not normally assumed under normal conditions. Reinforcing critical areas with fully threaded screws can prevent panel failure in such cases.

4.4 Review of Connection Types

4.4.1 Lap Joints

Lap joints are a common panel-to-panel connection due to their ease of assembly. Equal potential dowel-bearing length on either side of the joint means that higher strength and stiffness can be obtained from lap joints using 90° shear screws when compared to spline joints (Sadeghi, 2015; Hossain, 2016b). Partially threaded screws can be employed to transfer in-plane forces and to assist in sealing gaps between members, but inclined screws can be used as well for stiff, high-strength connections. Lap joints are generally designed to transfer shear along the joint, but they can also be designed as horizontal ties to resist forces pulling the panels apart (Sadeghi, 2015; Perez & Wallwork, 2016). The lap joint design should consider cost and machining considerations, as each lap takes away area that a CLT panel can cover. Designing the lap joint to the minimum required to allow fasteners to develop their full capacity through maintaining proper edge and end distances is essential. Furthermore, the lap joint width is important for efficient

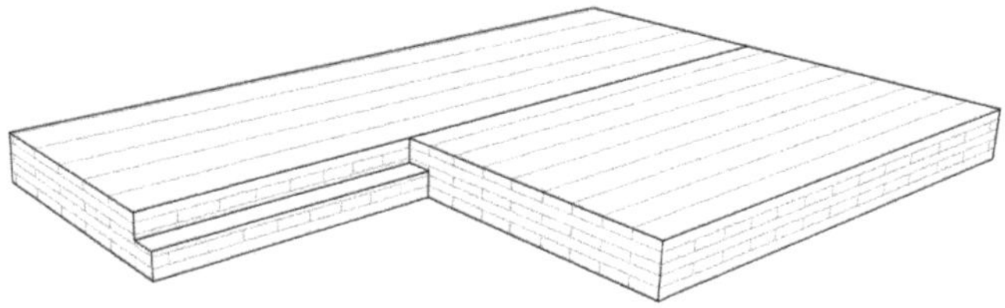

FIGURE 4.63 Lap joint.

machining so that modern computer numerical control (CNC) machines can machine the required lap joint without the need of rotating or turning over the panel. If the CLT panel is cut on a CNC machine, the underside cut depth will be limited to approximately 3½ in. (90 mm) due to the dimension of the cutterhead and tooling units. If the lap joint exceeds 3½ in. (90 mm), the panel needs to be flipped, which doubles the machine time and increases tolerances from around 0.08 in. to 0.4 in. (2 mm to 10 mm). Designers should be mindful that lap joints are essentially notched at half the panel thickness, so the location of lap joints in the floor area of a building should minimize vertical load transfer across the joint and bending in the minor strength direction. Otherwise, reinforcement measures with fully threaded screws can be considered (see Fig. 4.63).

4.4.2 Surface Spline Joints

Surface splines are a common choice to connect floor panels due to their economy and ease of assembly. Although less stiff and less strong compared to lap joints, they involve less material wastage due to machining and are easily installed using slender fasteners. The depth of surface splines does not provide the minimum penetration depths for lateral design (6*D*, for example). Testing by Hossain (2016b) and Sullivan (2017), however, shows that plywood splines are highly resistant to splitting due to their cross laminations (see Fig. 4.64).

4.4.3 Butt Joints

Butt joints do not require any special machining and therefore require no material waste. Butt joints are typically joined with shear screws driven in on an angle to the panel suface; 45° shear screws can be employed for ductility, while double angle screws that capitalize on the high withdrawal strength of the screws are especially suitable for high-magnitude loading. Designers should consider methods of pulling the panels together

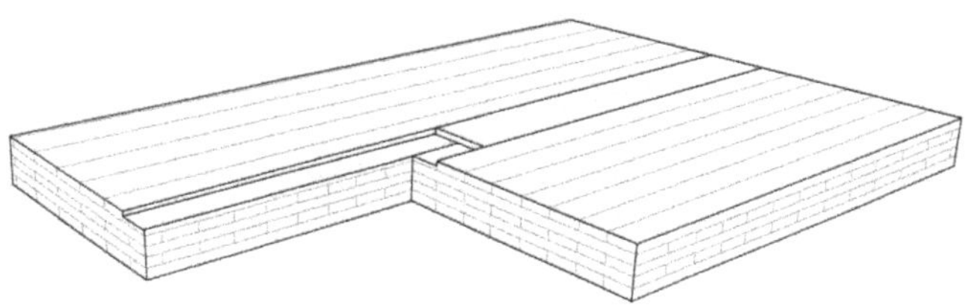

FIGURE 4.64 Surface spline.

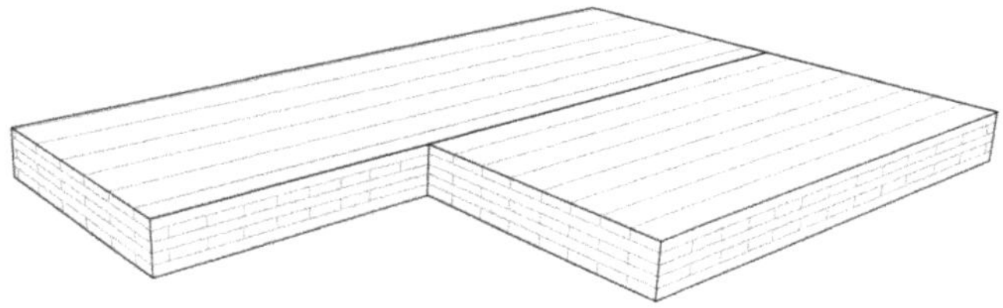

FIGURE 4.65 Butt joint.

before the screws are installed, such as panel ratchets, so that fully threaded screws do not lock in unwanted gaps between the panels (see Fig. 4.65).

4.4.4 Corner Joints

Floor-to-wall and wall-to-wall junctions are connected with panels butting against one another at right angles. A practical design assumption is that end grain will be engaged when installing screws into the narrow panel edges for most panel thickness and layups. Designers are reminded to avoid detailing that would produce prying effects with large fasteners and excessive tension stresses perpendicular to the grain. In some cases, reinforcement measures discussed in Sec. 4.3 may be necessary to transfer tensile stresses across the grain. The simplest way to connect panels at corners is with screws driven straight into the narrow edge of the main member. Partially threaded washer head screws are a common choice, as they pull the panels tightly together. Countersunk head screws can also be driven in at an angle to avoid strength reductions associated with end grain applications or to strengthen the connection against splitting forces perpendicular to the grain. Screws may also be installed at an angle, for instance, where a 90° angle is impractical from a constructability standpoint (see Figs. 4.66 and 4.67).

4.4.5 Panel-to-Beam Connections

In post and beam structures, diaphragm forces are transferred to collector elements, which are then channeled to the lateral load-resisting system. Screw crosses are an ideal

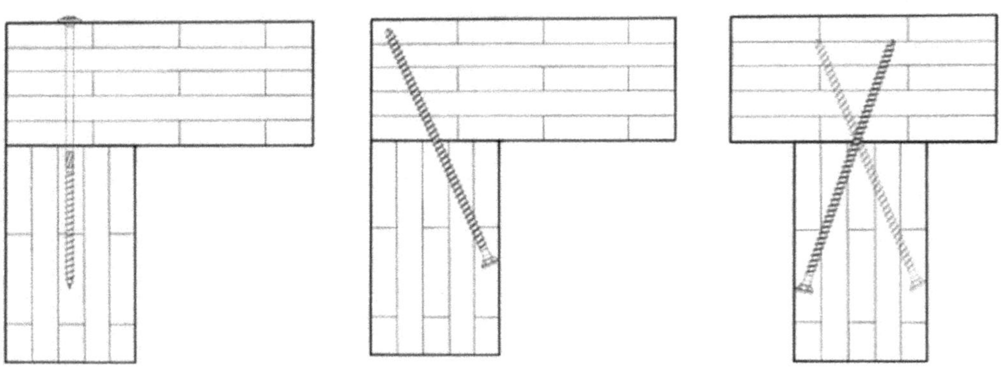

FIGURE 4.66 Examples of wall to wall corner joints.

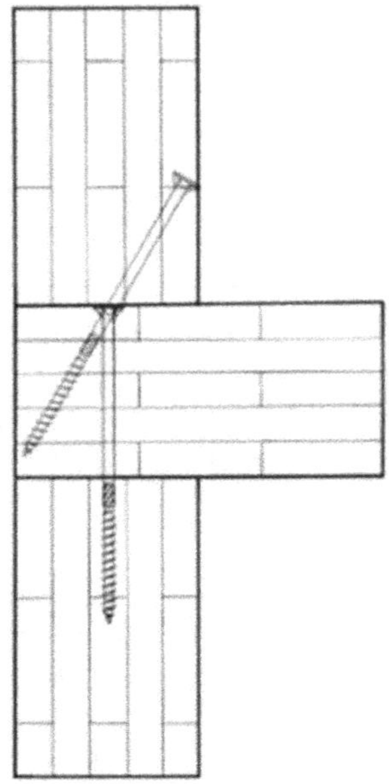

FIGURE 4.67 Example of wall-to-floor/floor-to-wall corner joint.

choice, as their stiffness and strength are especially suited to high-intensity loading. Panel-to-beam connections can also be exploited for composite action, again using screw crosses, to increase the effective bending stiffness of the elements working together (see Fig. 4.68).

4.4.6 Ledger Connections

Ledger connections are primarily employed in balloon frame construction to provide a bearing surface for CLT floor panels, but they can also be employed in platform structures to support roof panels, allowing for a parapet (Perez & Wallwork, 2016). Inclined shear-tension screws can be designed to resist the vertical force, while 90° shear screws act as assembly screws, resist pull-out effects due to the moment, and provide ductility. Toe screws installed from the floor down hold the floor panel in place and resist uplift (see Fig. 4.69).

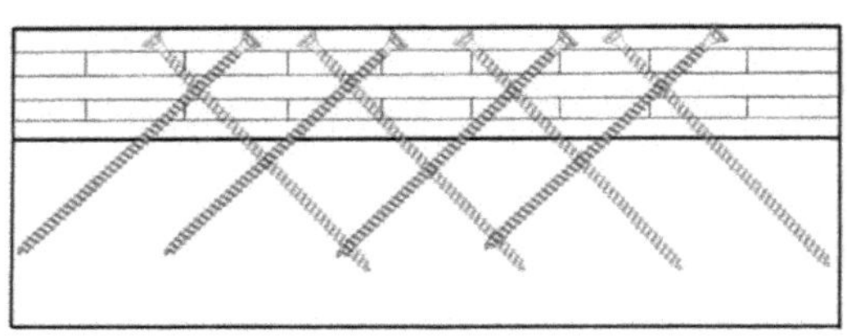

FIGURE 4.68 High-capacity panel-to-beam shear connection with symmetrical screw crosses.

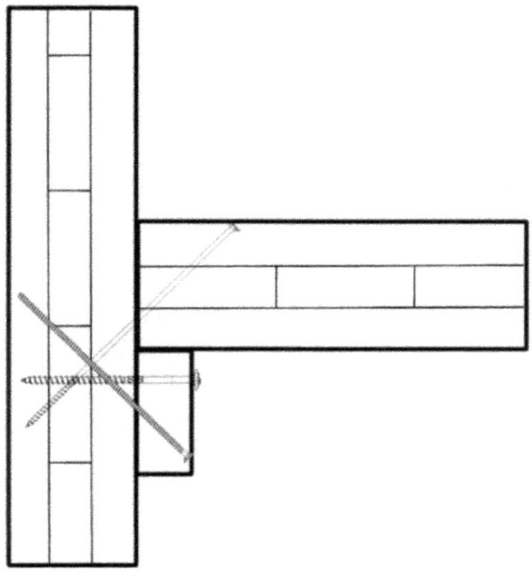

FIGURE 4.69 Ledger connection.

4.4.7 Pre-Engineered Connections

Many proprietary pre-engineered connection systems are now available for the mass timber building market. Pre-engineered connection systems simplify design by offering off-the-shelf solutions to connection needs. Most pre-engineered systems rely on self-tapping screws to transfer design forces to the wood members, so an understanding of the basic principles of screw design also helps when specifying pre-engineered connection systems. Many pre-engineered connections have the option of being fully concealed in routed sections, which enhances fire protection of the steel elements and offers a clean architectural finish (see Fig. 4.70).

FIGURE 4.70 On-site assembly using pre-engineered connection system.

4.5 Connection Design Examples

Example 1: Surface Spline with 90° Shear Screws in Three-Ply CLT (see Figs. 4.71–4.74)

Screws:

5/16 in. × 3⅛ in. (8 mm × 80 mm) ASSY® ECOFAST
$D = 5/16$ (8 mm)
$D_r = 0.209$ in. (5.3 mm) (<¼ in.)
$l = 3⅛$ in.
$F_{yb} = 150{,}200$ psi (1036 MPa) [ICC ESR-3179]
$n = 8$

Plywood:

$G = 0.49$
$t = ¾$ in. (19 mm)
$F_e = 16{,}600\ G^{1.84} = 4468$ psi (30.8 MPa)
$l_s = ¾$ in. (19 mm)

CLT:

$G = 0.42$
$t_1 = 1.3$ in. (33 mm)
$t_2 = 1.37$ in. (35 mm)
$t_3 = 1.3$ in. (33 mm)
$F_e = 16{,}600\ G^{1.84} = 3364$ psi (23.2 MPa)
$l_m = 3⅛$ in. – ¾ in. – 5/16 in. = 2.09 in. (53 mm)
$\Theta = 90°$ (suggested)

Z (NDS Cl. 12.3.1):

$I_m = 454$ lb (2019 N)
$l_s = 216$ lb (960 N)
$II = 159$ lb (707 N)
$III_m = 192$ lb (854 N)
$III_s =$ **137 lb (609 N)** $= Z$
$IV = 187$ lb (832 N)
$n_{ef} = n$
$C_D = 1.6$ (short-term loading)
$Z' = n_{ef} \cdot Z \cdot C_D = 8 \cdot 137$ lbs$\cdot 1.6 =$ **1754 lb (7802 N)**

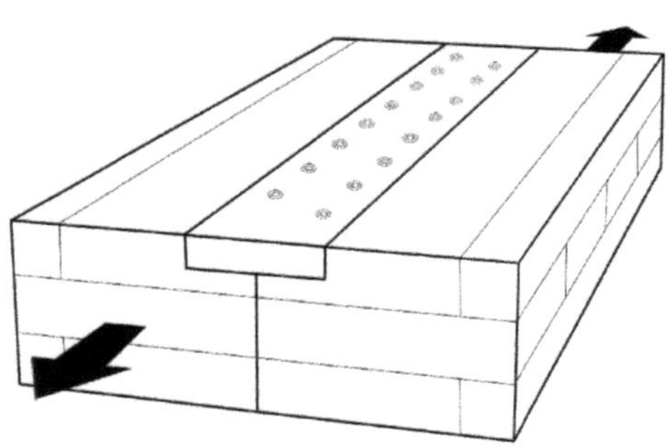

FIGURE 4.71 Surface spline with 90° shear screws in three-ply CLT, iso view.

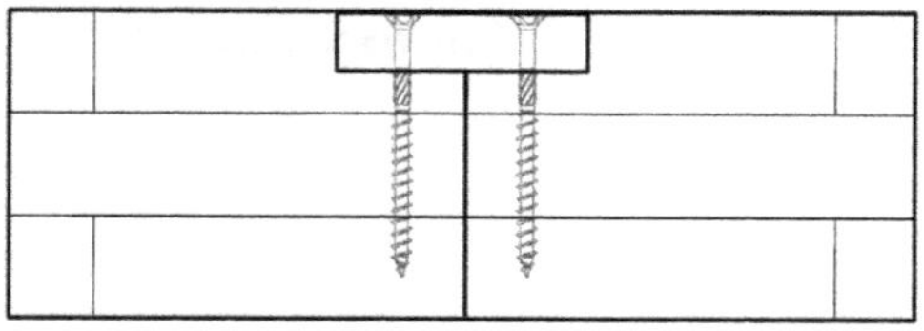

FIGURE 4.72 Surface spline with 90° shear screws in three-ply CLT, front view.

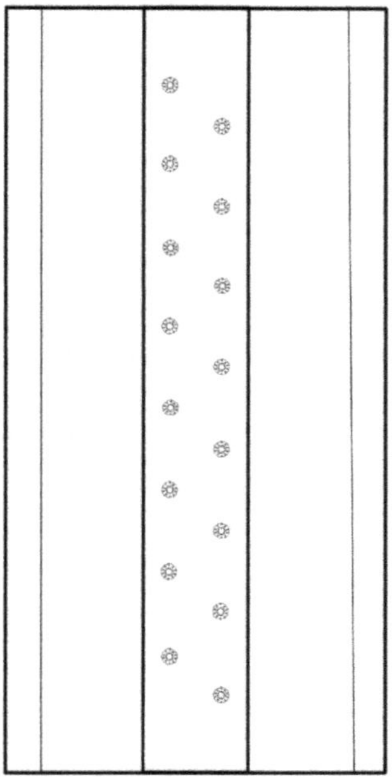

FIGURE 4.73 Surface spline with 90° shear screws in three-ply CLT, top view.

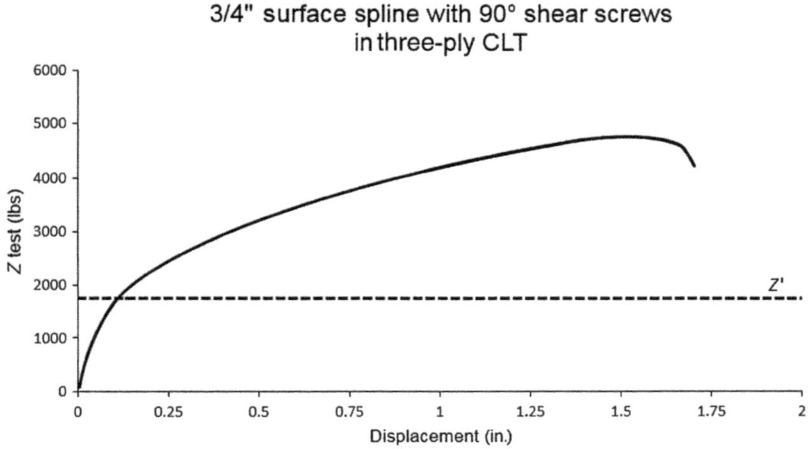

FIGURE 4.74 Monotonic testing curve of ¾-in. (19 mm) surface spline in three-ply CLT with adjusted lateral design value, Z', indicated.

Example 2: Lap Joint with 90° Shear Screws in Three-Ply CLT (see Figs. 4.75–4.78)

Screws:

5/16 in. × 3⅛ in. (8 mm × 80 mm) ASSY® ECOFAST
$D = 5/16$ (8 mm)
$D_r = 0.209$ in. (5.3 mm) (<¼ in.)
$l = 3⅛$ in. (80 mm)
$F_{yb} = 150{,}200$ psi (1036 MPa) [ICC ESR-3179]
$n = 8$

CLT:

$G = 0.42$
$t_1 = 1.3$ in. (33 mm)
$t_2 = 1.37$ in. (35 mm)
$t_3 = 1.3$ in. (33 mm)
$F_e = 16{,}600\ G^{1.84} = 3364$ psi (23.2 MPa)
$l_s = 2$ in. (51 mm)
$l_m = 3½$ in. – 2 in. – 5/16 in. = 1-3/16 in. (30 mm)*
(*6D penetration is required to fufill specifications in ICC ESR-3179. This example, however, references acutal testing geometry and data and is used for illustrative purposes.)
$\Theta = 90°$

Z (Cl. 12.3):

$I_m = 258$ lb (1148 N)
$I_s = 434$ lb (1931 N)
$II = 151$ lb (672 N)
$III_m =$ **138 lb (614 N) = Z**
$III_s = 178$ lb (792 N)
$IV = 175$ lbs (778 N)
$n_{ef} = n$
$C_D = 1.6$ (short-term loading)
$Z' = n_{ef} \cdot Z \cdot C_D = 8 \cdot 138\ \text{lbs} \cdot 1.6 =$ **1765 lb (7851 N)**

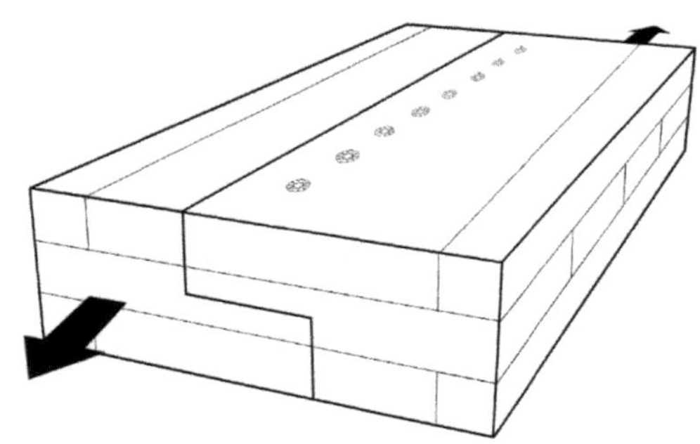

FIGURE 4.75 Lap joint with 90° shear screws in three-ply CLT, iso view.

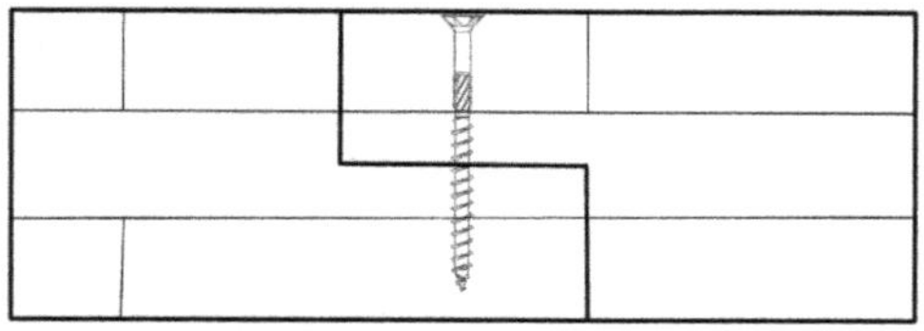

FIGURE 4.76 Lap joint with 90° shear screws in three-ply CLT, front view.

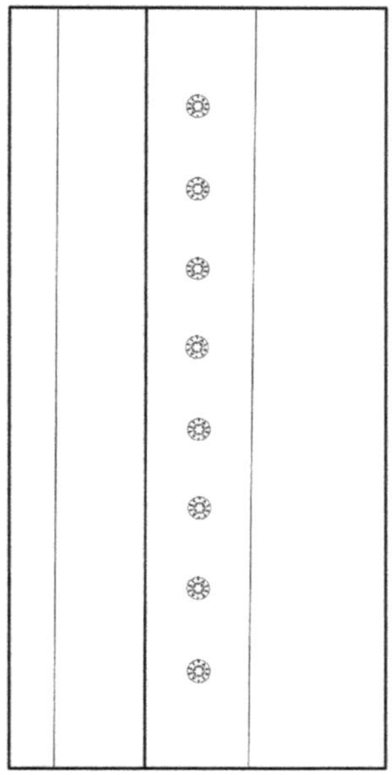

FIGURE 4.77 Lap joint with 90° shear screws in three-ply CLT, top view.

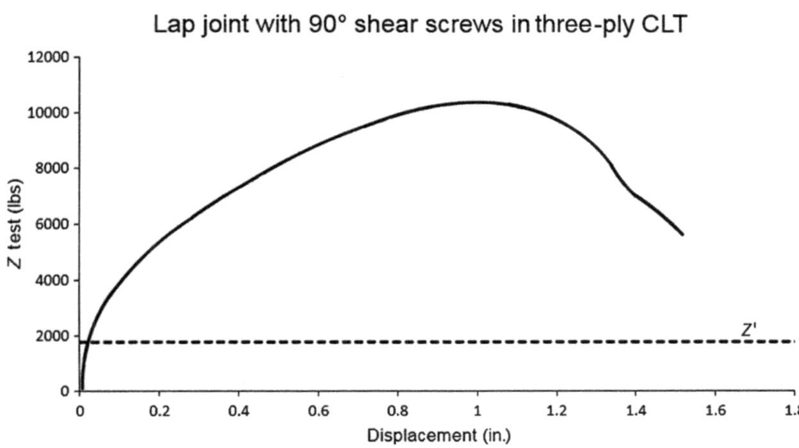

FIGURE 4.78 Monotonic testing curve of lap joint with 90° shear screws in three-ply CLT with adjusted lateral design value, *Z′*, indicated.

Example 3: Butt Joint with 45° Shear Screws in Three-Ply CLT (see Figs. 4.79–4.82)

Screws:

5/16 in. × 5½ in. (8 mm × 140 mm) ASSY® VG CSK

$D = 5/16$ in. (8 mm)

$D_r = 0.196$ in. (5.0 mm) (<¼ in.)

$l = 3½$ in.

$F_{yb} = 132{,}500$ psi (914 MPa) [ICC 3178]

$n = 8$

CLT:

$G = 0.42$

$t_1 = 1.3$ in. (33 mm)

$t_2 = 1.37$ in. (35 mm)

$t_3 = 1.3$ in. (33 mm)

$F_e = 16{,}600\ G^{1.84} = 3364$ psi (23.2 MPa)

$l_s = 2¾$ in. (70 mm)

$l_m = 5½$ in. − 2¾ in. − 5/16 in. = 2.44 in. (62 mm)

$\Theta = 90°$

Z (NDS Cl. 12.3.1):

$I_m = 523$ lb (2326 N)

$l_s = 590$ lb (2,624 N)

$II = 231$ lb (1028 N)

$III_m = 196$ lb (872 N)

$III_s = 216$ lb (961 N)

$IV =$ **152 lb (676 N) =** $\boldsymbol{Z}$

$n_{ef} = n$ ($n_{ef} < n$ for fasteners loaded parallel to the grain)

$C_{eg} = 1.0$ ($\alpha \neq 0°$)

$C_D = 1.6$ (short-term loading)

$Z' = n_{ef} \cdot Z \cdot C_D = 8 \cdot 152\ \text{lb} \cdot 1.6 =$ **1946 lb (8656 N)**

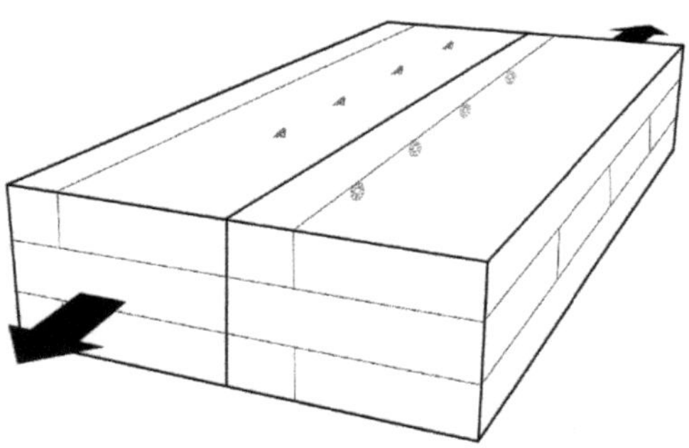

FIGURE 4.79 Butt joint with 45° shear screws in three-ply CLT, iso view.

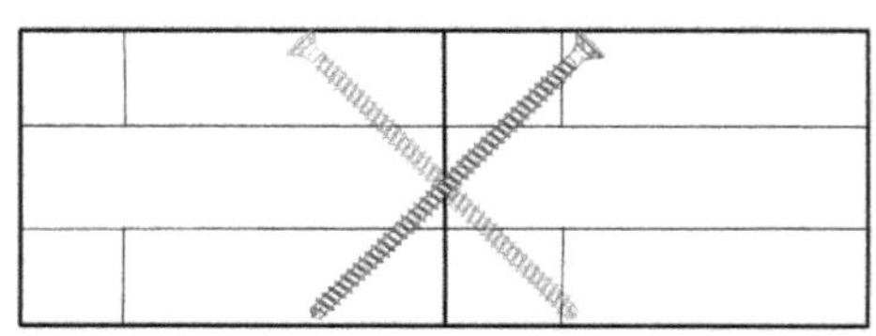

FIGURE 4.80 Butt joint with 45° shear screws in three-ply CLT, front view.

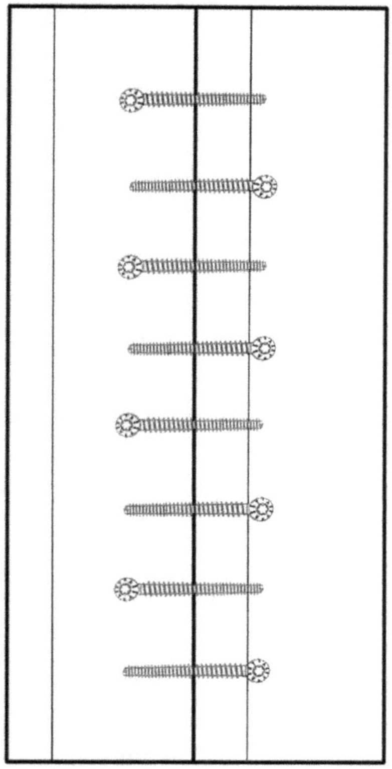

FIGURE 4.81 Butt joint with 45° shear screws in three-ply CLT, top view.

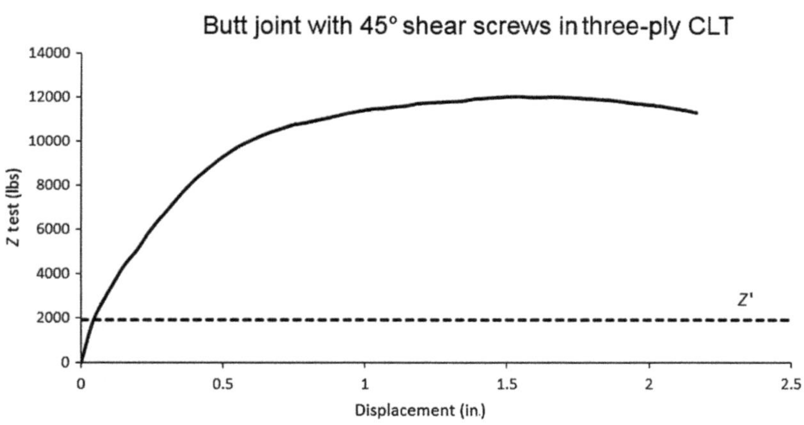

FIGURE 4.82 Monotonic testing curve of butt joint with 45° shear screws in three-ply CLT with adjusted lateral design value, Z', indicated.

Example 4: Butt Joint with Double Angle Screw Crosses (see Figs. 4.83–4.86)

Screws:

5/16 in. × 7⅛ in. (8 mm × 180 mm) ASSY® VG CYL

$D = 5/16$ in.

$D_r = 0.196$ in. (5.0 mm) (<¼ in.)

$L_{head} = 0.295$ in. (7.5 mm)

$l = 7⅛$ in. (180 mm)

$T = 1775$ lb (7820 N) [ICC ESR-3178]

$n_x = 4$

In this example, the screw angle is 45° to the line of the joint and 33° to the surface plane, making a compound angle (α) of approximately 53° between the axis and the grain in both the major strength and the minor strength direction. The screw angle (β) is also 53° to the line of the force.

CLT:

$G = 0.42$

$t_1 = 1.3$ in. (33 mm)

$t_2 = 1.37$ in. (35 mm)

$t_3 = 1.3$ in. (33 mm)

$W_{90} = 177$ lb/in. (31 N/mm) [ICC ESR-3178]

$W_{53} \approx W_{45} = 0.85 \cdot 177 = 150.5$ lb/in. (26 N/mm)

$l_s = 3.563$ in. − 0.295 in. = 3.3 in. (84 mm)

$l_m = 7⅛$ in. − 3.543 in. − 5/16 in. = 3.3 in. (84 mm)

$C_D = 1.6$ (short-term loading)

$R'_C = 150.5$ lbs/in. $\cdot$ 3.2 in. $\cdot$ 1.6 = 770.6 lb (3428 N) ($< 0.8 \cdot T_a$)

$R'_T = 770.6$ lb (3428 N) (T_a)

$n_{x,ef} = \max: [n_x \cdot 0.9; n_x^{0.9}]$

$C_{eg} = 1.0$ ($\alpha \neq 0°$)

$Z'_x = n_{x,ef} \cdot (R'_C + R'_T) \cdot \cos 53$

$= 0.9 \cdot 4 \cdot (770.6 + 770.6) \cdot \cos 53 =$ **3339 lb (14,853 N)**

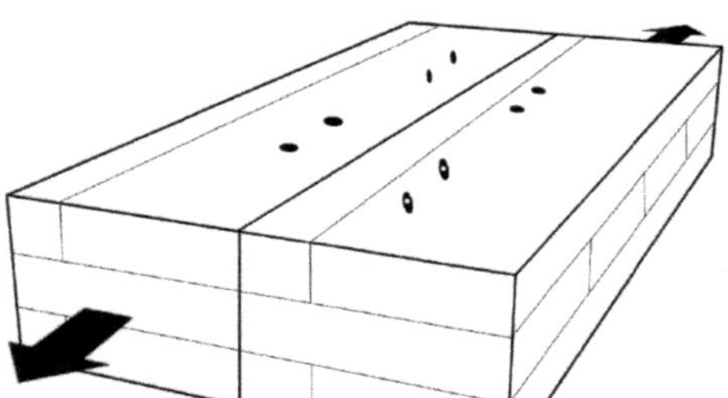

FIGURE 4.83 Butt joint with double angle screw crosses, iso view.

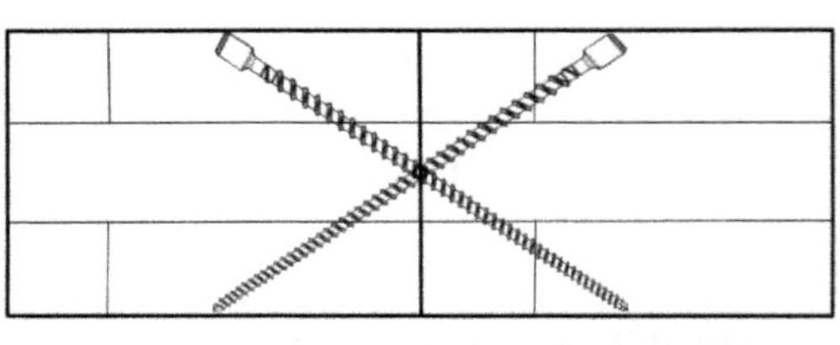

FIGURE 4.84 Butt joint with double angle screw crosses, front view.

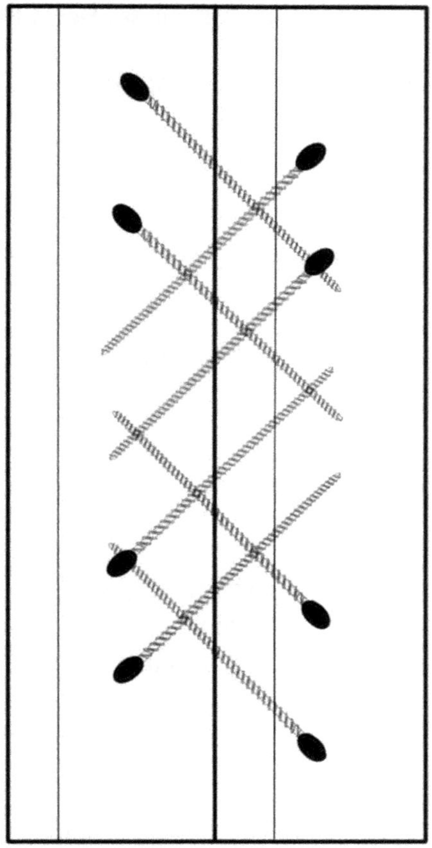

FIGURE 4.85 Butt joint with double angle screw crosses, top view.

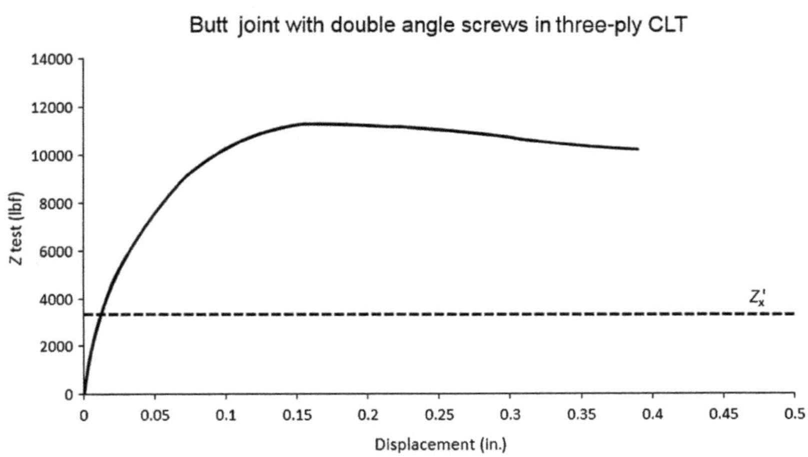

FIGURE 4.86 Monotonic testing curve of butt joint with double angle screws in three-ply CLT with adjusted lateral design value, Z_x', indicated.

Example 5: Screw Cross; CLT to Glulam Beam (see Figs. 4.87–4.89)

Screws:

5/16 in. × 11⅞ in. (8 mm × 300 mm) ASSY® VG CSK

$D = 5/16$ in. (8 mm)

$D_r = 0.196$ in.(5.0 mm) (<¼ in.)

$L_{\text{head}} = 0.181$ in. (4.6 mm)

$l = 7⅛$ in. (300 mm)

$T = 1775$ lb (7820 N) [ICC ESR-3178]

$n_p = 4$

Screw angle (β) is 45° to the line of the force.

CLT:

$G = 0.42$

$t_1 = 1.3$ in. (33 mm)

$t_2 = 1.37$ in. (35 mm)

$t_3 = 1.3$ in. (33 mm)

$W_{90} = 177$ lb/in. (31 N/mm) [ICC ESR-3178]

$W_{45} \approx 0.85 \cdot 177 = 150.5$ lb/in. (26 N/mm)

$l_{t,s} = 5.6$ in. $-$ 0.181 in. = 5.4 in. (137 mm)

Glulam:

$G = 0.49$

$W_{90} = 183$ lb/in. (32 N/mm) [ICC ESR-3178]

$W_{45} \approx 0.85 \cdot 183$ lbsf/in = 155.6 lb/in. (27 N/mm)

$l_{t,m} = 11⅞$ in. $-$ 5.6 in. $-$ 5/16 in. = 6 in. (152 mm)

$C_D = 1.6$

$R'_C = 150.5$ lb/in. · 5.4 in. · 1.6 = 1300 lb (<0.8 · T)

$R'_T = 1300$ lb (<T)

$n_{x,\text{ef}} = \max\text{: } [0.9 \cdot n_x;\ n_x^{0.9}]$

$Z'_x = n_{x,\text{ef}} \cdot (R'_C + R'_T) \cdot \cos 45$

$= 0.9 \cdot 4 \cdot (1300 + 1300) \cdot \cos 45 =$ **6620 lb (29,443 N)**

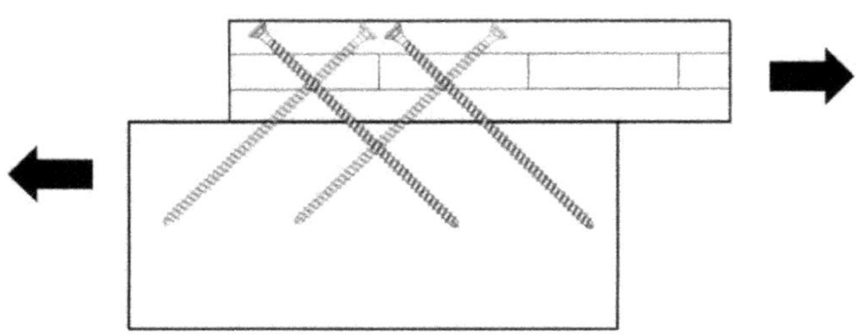

FIGURE 4.87 CLT to glulam beam connection using screw crosses, side view.

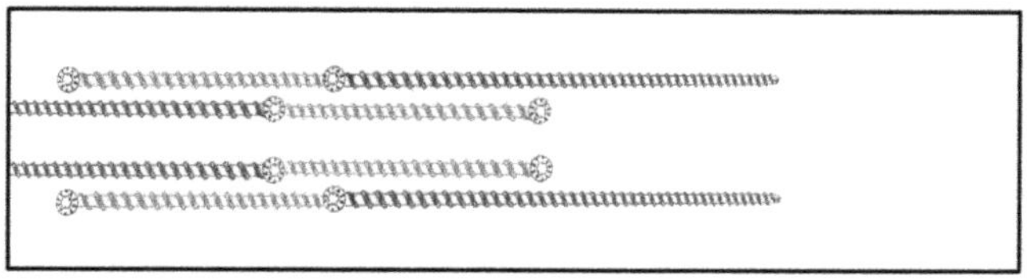

FIGURE 4.88 CLT to glulam beam connection using screw crosses, top view.

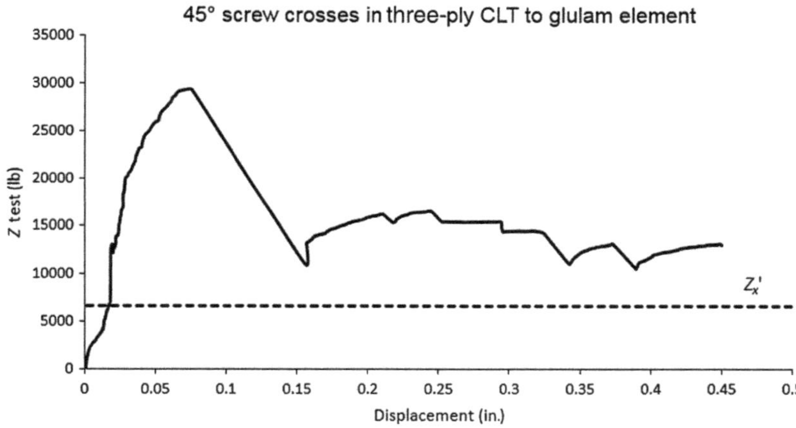

FIGURE 4.89 Monotonic testing curve of 45° screw crosses in three-ply CLT to glulam element, with adjusted lateral design value, Z_x', indicated.

4.6 Brittle and Ductile Connection (45° + 90° Screws) (see Figs. 4.90 to 4.97)

4.6.1 45° Screws (see Figs. 4.90 to 4.92)

Screws:

5/16 in. × 8⅝ in. (8 mm × 220 mm) ASSY® VG CSK

$D = 5/16$ in. (8 mm)

$D_r = 0.196$ in. (5.0 mm) (<¼")

$l_{head} = 0.181$ in. (4.6 mm)

$l = 8⅝$ in. (220 mm)

$T = 1775$ lb (7820 N) [ICC ESR-3178]

$n_x = 2$

Screw angle (β) is 45° to the line of the force.

CLT:

$G = 0.42$

$t_1 = 1.25$ in. (31.8 mm)

$t_2 = 1.4$ in. (35.6 mm)
$t_3 = 1.4$ in. (35.6 mm)
$t_4 = 1.4$ in. (35.6 mm)
$t_5 = 1.25$ in. (31.8 mm)

$W_{90} = 177$ lb/in. (31 N/mm) [ICC ESR-3178]
$W_{45} \approx 0.85 \cdot 177 = 150.5$ lb/in. (26 N/mm)
$l_{t,s} = 4.79$ in. $- 0.181$ in. $= 4.6$ in. (117 mm)
$l_{t,m} = 8.625$ in. $- 4.79$ in. $- 5/16$ in. $= 3.5$ in. (89 mm)

$C_D = 1.6$

$R'_C = 150.5$ lb/in. $\cdot 4.6$ in. $\cdot 1.6 = 1108$ lb (4929 N) $(<0.8 \cdot T_a)$
$R'_T = 1108$ lb (4929 N) $(<T_a)$

$n_{x,\text{ef}} = \max: [n_x \cdot 0.9; n_x^{0.9}]$

$$Z'_x = n_{\text{ef}} \cdot (R'_C + R'_T) \cdot \cos 45$$
$$= 2^{0.9} \cdot (1108 + 1108) \cdot \cos 45 = \mathbf{2924\ lb\ (13{,}007\ N)}$$

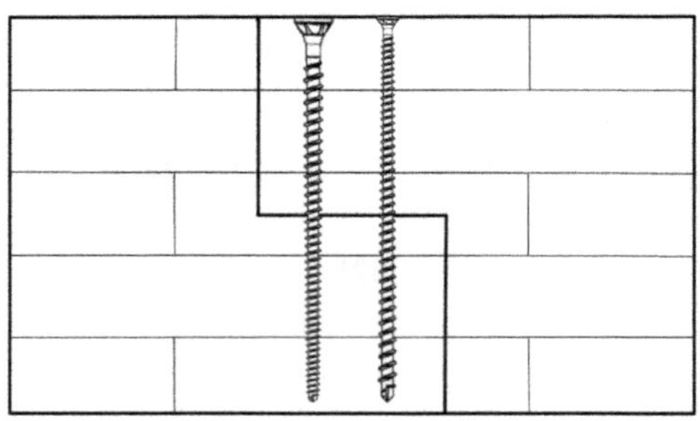

FIGURE 4.90 45° screw crosses in five-ply CLT panel lap joint, front view.

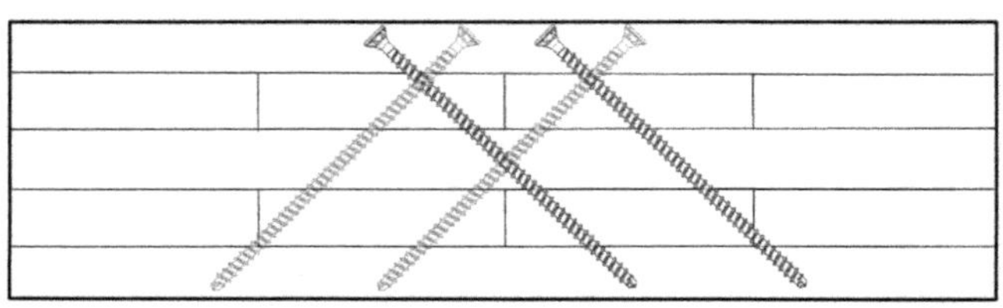

FIGURE 4.91 45° screw crosses in five-ply CLT panel lap joint, side view.

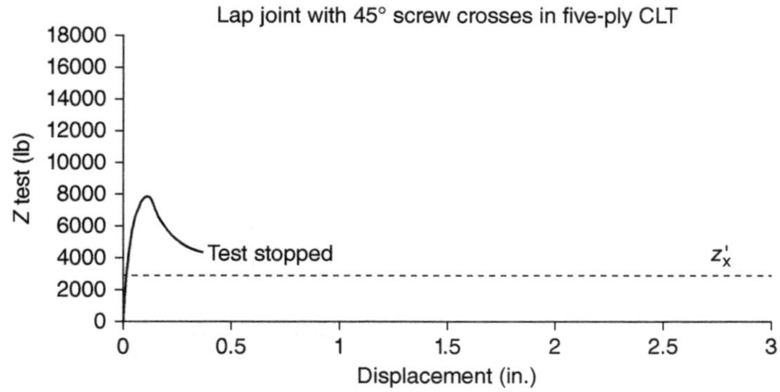

FIGURE 4.92 Monotonic testing curve of 45° screw crosses in five-ply CLT panel lap joint.

4.6.2 90° Screws (see Figs. 4.93 to 4.95)

Screws:

5/16 in. × 6¼ in. (8 mm × 160 mm) ASSY® ECOFAST
$D = 5/16$ in. (8 mm)
$D_r = 0.209$ in. (5.3 mm) (<¼")
$l = 6¼$ in. (160 mm)
$F_{yb} = 150{,}200$ psi (1036 MPa) [ICC ESR-3179]
$n = 4$

CLT:

$G = 0.42$
$t_1 = 1.25$ in. (31.8 mm)
$t_2 = 1.4$ in. (35.6 mm)
$t_3 = 1.4$ in. (35.6 mm)
$t_4 = 1.4$ in. (35.6 mm)
$t_5 = 1.25$ in. (31.8 mm)
$F_e = 16{,}600\, G^{1.84} = 3364$ psi (23.2 MPa)
$l_s = 3.39$ in. (86 mm)
$l_m = 6.25$ in. $-$ 3.39 in. $-$ 5/16 in. = 2.55 in. (65 mm)
$\Theta = 0°$
$n_{ef} = n$
$C_D = 1.6$

Z (NDS Cl. 12.3.1):

$I_m = 692$ lb (3078 N)
$I_s = 920$ lb (4092 N)
$II = 340$ lb (1512 N)
$III_m = 264$ lb (1174 N)
$III_s = 332$ lb (1477 N)
$IV =$ **219 lb (974 N)** $= Z$
$Z' = n_{ef} \cdot Z \cdot C_D$
$Z' = 4 \cdot 219 \cdot 1.6 = 1401.6$ lb (6236 N)
$Z'_x = n_{ef} \cdot (R'_C + R'_T) \cdot \cos 45$
$= 2^{0.9} \cdot (1108 + 1108) \cdot \cos 45 =$ **2924 lb (13,007 N)**

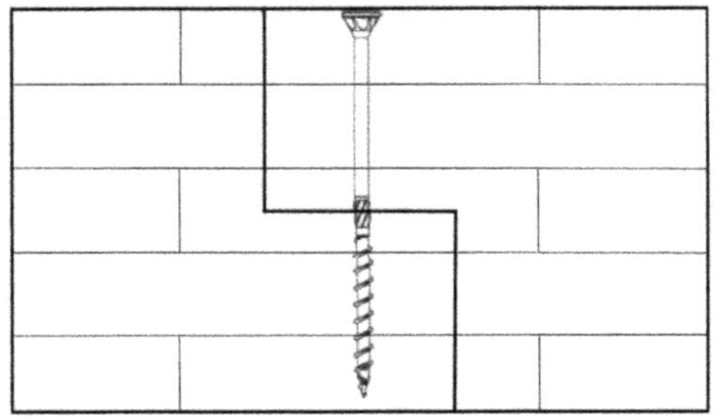

FIGURE 4.93 90° shear screws in five-ply CLT panel lap joint, front view.

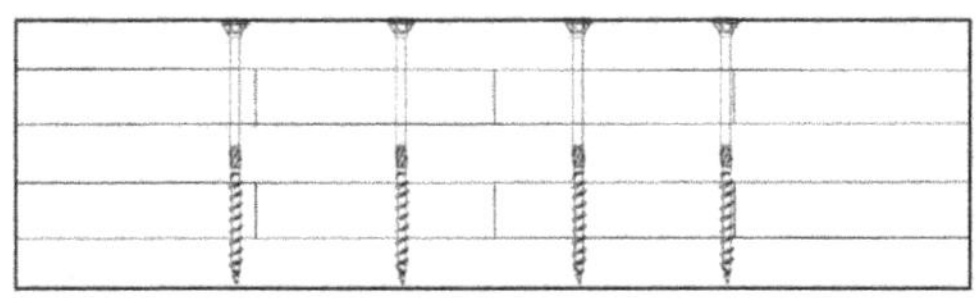

FIGURE 4.94 90° shear screws in five-ply CLT panel lap joint, front view.

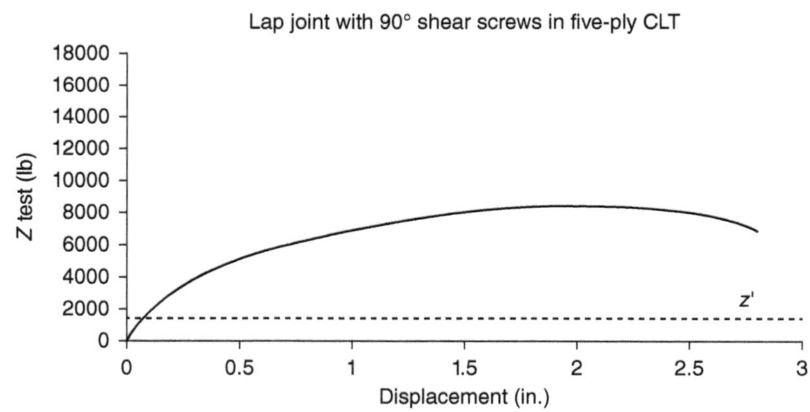

FIGURE 4.95 Monotonic loading curve for 90° shear screws in five-ply CLT panel lap joint, with adjusted lateral design value, Z', indicated.

4.6.3 45° + 90° Screws (see Figs. 4.96 to 4.99)

Without stiffness compatibility, connection design strength is determined from the stiffer screw crosses, with the 90° shear screws providing ductility. Looking back at Table 4.6, the minimum thread penetration length for the screw crosses is 4.6 in. < 6.5 in., which means that the inclined screws are not likely to fail through tensile fracture, and instead will continue to contribute to the load carrying capacity of the joint by acting as inclined dowels after withdrawal failure. (In this case, the test was stopped before this behavior was recorded for the screw cross joint, although a typical shape of this load curve is shown in Fig. 4.38.)

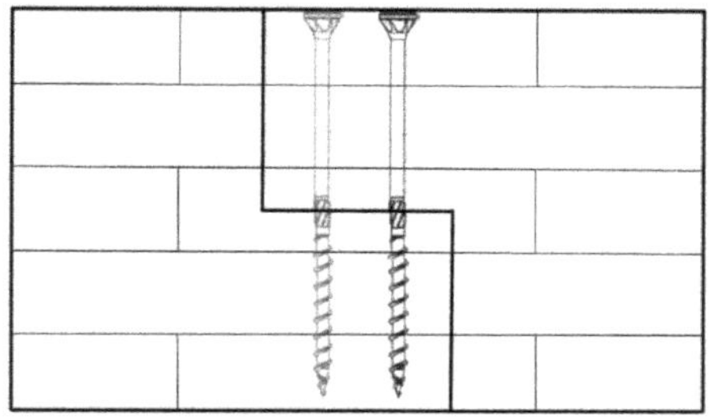

FIGURE 4.96 45° and 90° screw combination in five-ply CLT lap joint, front view.

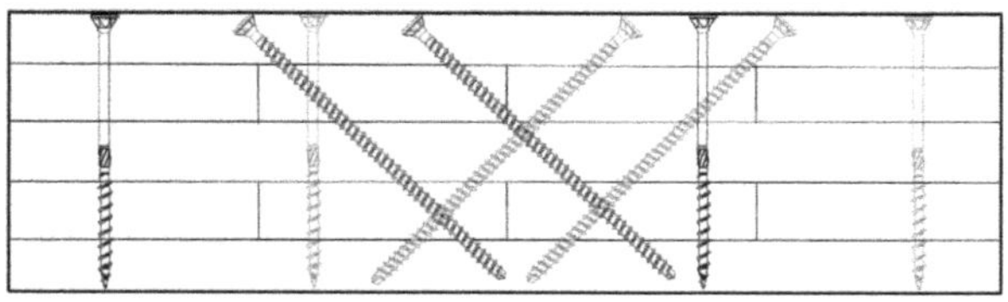

FIGURE 4.97 45° and 90° screw combination in five-ply CLT lap joint, side view.

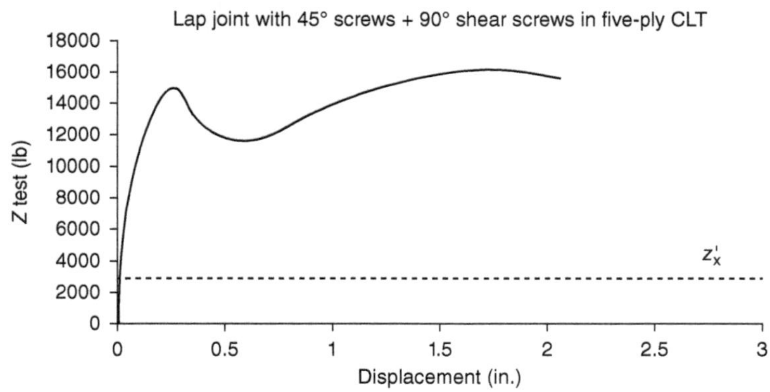

FIGURE 4.98 Monotonic loading curve for 90° shear screws in five-ply CLT panel lap joint, with adjusted lateral design value, Z_x', indicated.

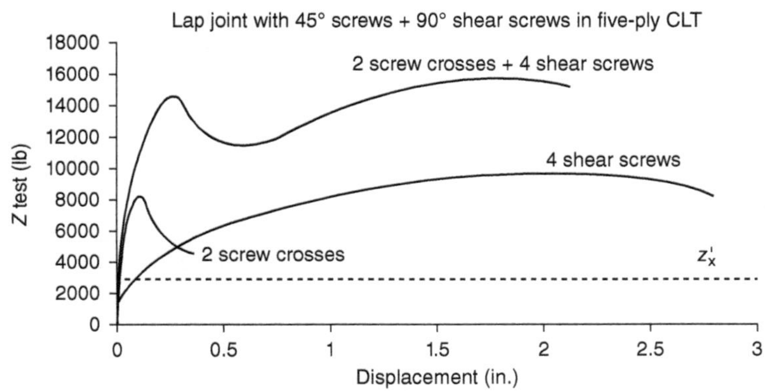

FIGURE 4.99 90° shear screws, 45° screw crosses, and both combined in one joint maximize stiffness and ductility.

4.7 Compression Reinforcement Example (Figs. 4.100 and 4.101 and Table 4.9)

$$C'_r = \min: \begin{bmatrix} n \cdot R'_c + (F_{c,90} \cdot A_{c,1} \cdot C_b)* \\ A_{c,2} \cdot F_{c90} \end{bmatrix}$$

where C'_r = adjusted compressive resistance of panel under steel plate
R'_c = adjusted axial resistance of one compression reinforcement screw

$$R'_c = \min: \begin{bmatrix} W^{\circ}_{90} \cdot l_{ef} \cdot C \\ B_a \end{bmatrix}$$

B_a = allowable buckling strength of one reinforcement screw
$F_{c,90}$ = allowable compression design value ⊥ to the grain (from CLT manufacturer)
$F_{c,90} \approx 425$ psi
$A_{c,1}$ = bearing area of steel plate on panel surface
$A_{c,2}$ = effective bearing area at depth of screw tips
C_b = bearing area factor (NDS 2015 Table 3.10.4)

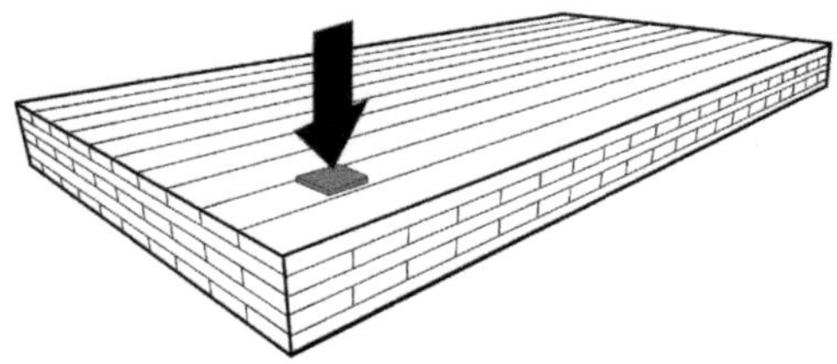

FIGURE 4.100 Localized compression administered through steel plate.

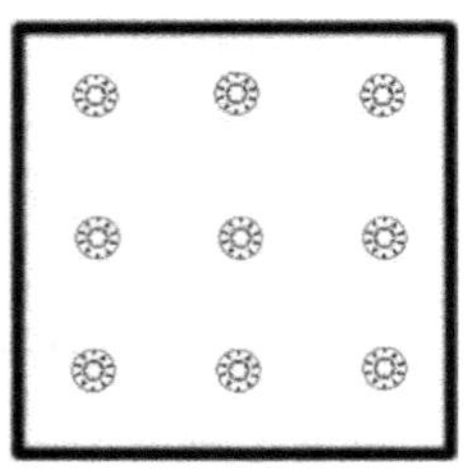

FIGURE 4.101 Layout of compression reinforcement screws under steel plate.

D	*G* = 0.42, *G* = 0.49
5/16"	850 [3750]
3/8"	1200 [5300]

Estimated factor of safety = 3.0+

TABLE 4.9 Estimated Allowable Buckling Resistance Design Values, Ba, for ASSY® Fully Threaded Screws in CLT, lb [N]

Screws:

⅜ in. × 6¼ in. (10 mm × 160 mm) ASSY® VG CSK
$D = 5/16$ in.
$D_r = 0.244$ in. (6.2 mm)
$l = 6¼$ in. (160 mm)
$l_{head} = 0.255$ in. (6.5 mm)
$n = 9$
spaced 1.75 in. (44.5 mm) apart o.c.

CLT:

$G = 0.42$
$t = 6⅞$ in. (175 mm)

$W_{90} = 188$ lbs/in. (33 N/mm) [ICC ESR-3178]
$l_{ef} = 6.25$ in. – 0.255 in. – ⅜ in. = 5.6 in. (142 mm)
$C_D = 1.0$

$R'c = 188$ lb/in. · 5.6 in. · 1.0 = 1053 lbs (4684 N) ($<B_a$)
$A_{c,1} = 25$ in.^{2} (16,130 mm^2)
$A_{c,2} \approx 56$ in.^{2} (36,130 mm^2)

$C_b = 1.1$

$$C'_r = \min: \begin{bmatrix} n \cdot R'_c + F_{c,90} \cdot A_{c,1} \cdot C_b \\ A_{c,2} \cdot F_{c90} \end{bmatrix}$$

$n \cdot R'_c + F_{c,90} \cdot A_{c,1} \cdot C_b$
$= 9 \cdot 1053$ lb + 425 psi · 25 in.^{2} · 1.1
= 21,165 lb (94,147 N)

$A_{c,2} \cdot F_{c90}$
= 425 psi · 56 in^2
= 23,800 lbs (105,868 N)

$$C'_r = \min: \begin{bmatrix} \textbf{21,165 lbs (94,147 N)}^* \\ \text{23,800 lbs (105,868 N)} \end{bmatrix}$$

*Compare:

Unreinforced 25 in.^{2} bearing area:
$C'_r = 425$ psi · 25 in^2 = 10,625 lbs (47,262 N)

References

APA (2017). *PRG 320-2017: Standard for Performance-Rated Cross-Laminated Timber.* Tacoma, WA: APA—The Engineered Wood Association.

American Wood Council (2012). *Technical Report 12.* Leesburg, VA: American Wood Council.

American Wood Council (2014). *National Design Specification for Wood Construction with Commentary, 2015 Edition.* Leesburg, VA: American Wood Council.

American Wood Council (2016). *Manual for Engineered Wood Construction 2015 Edition.* Leesburg, VA: American Wood Council.

Below, K., Mohammad, M., Ni, C., Malczyk, R., & Popovski, M. (2014). Design considerations and input parameters for connections and assemblies. In E. Karacebeyli & C. Lum (Eds.), *Technical Guide for the Design and Construction of Tall Wood Buildings.* Pointe-Claire, QC: FPInnovations.

Bejtka, I., & Blass, H. J. (2002). *Joints with Inclined Screws.* Proceedings of CIB-W18, Meeting 35, Paper 35-7-5.

Blass, H. (2003). Joints with dowel-type fasteners. In S. Thelandersson & H. J. Larsen (Eds.), *Timber Engineering.* Chichester: Wiley.

Blass, H., & Bejtka, I. (2003). Verbindungen mit geneigt angeordneten Schrauben. *Baunen mit Holz,* 105 (10).

Blass, H. J., & Schädle, P. (2011). Ductility aspects of reinforced and non-reinforced timber joints. *Engineering Structures,* 33 (11), 3018–3026.

Borg, M. (2000). *Behaviour of Timber Connections.* North Vancouver, BC: Timber Engineering Ltd.

Bradner, R. (2013). Production and technology of cross laminated timber (CLT): A state-of-the-art report. In R. Harris, A. Ringhofer, & G. Schickhofer (Eds.), *COST Action FP 1004.* Graz, Austria, pp. 2–36.

Brandner, R., & Schickhofer, G. (2014). *Properties of Cross-Laminated Timber (CLT) in Compression Perpendicular to the Grain.* Proceedings: International Network on Timber Engineering Research, Meeting 47.

Bradner, R., Flatscher, G., Ringhofer, A., Schickhofer, G., & Thiel, A. (2016). Cross laminated timber (CLT): Overview and development. *European Journal of Wood and Wood Products,* 74 (3).

Brühl, F., Schänzlin, J. & Kuhlmann, U. (2014). Ductility in timber structures: Investigations on over-strength factors. *Materials and Joints in Timber Structures: Recent Developments in Technology. RILEM.*

Closen, M. (2012). *Self-Tapping Screw Assemblies under Monotonic and Reverse Cyclic Load* (T). Vancouver: University of British Columbia.

Dietsch, P. & Brandner, R. (2015). Self-tapping screws and threaded rods as reinforcement for structural timber elements—A state-of-the-art report. *Construction and Building Materials,* 97, 78–89.

Ehlbseck, J., & Ehrnhardt, W. (1995). *Joints with Self-Tapping Screws (original Article C8).* Step. Centrum Hout, Almere.

Ehlbseck, J., & Werner, H. (1995). *Joints with Bolts and Dowels (original Article C6, C7).* Step. Centrum Hout, Almere.

European Committee for Standardization (2004). *Eurocode 5: Design of Timber Structures.* Brussels: BSI.

Flatscher, G., Bratulic, K., & Schickhofer, G. (2014). Screwed joints in cross laminated timber structures. *WCTE 2014.*

Fragiacomo, M. (2014). Seismic behaviour of cross-laminated timber buildings: Numerical modeling and design provisions. *European Conference on Cross-Laminated Timber.*

Gerber, A. (2016). *Timber-Concrete Composite Connectors in Flat-Plate Engineered Wood Products* (T). Vancouver: University of British Columbia.

Hossain, A. (2016a). *Withdrawal Resistance of ASSY® Self-Tapping Screws in SPF Cross-Laminated-Timber* (private report). Vancouver: University of British Columbia.

Hossain, A. (2016b). *Full Scale and Mid-Scale Tests on Shear Connections with Self-Tapping Screws for Cross-Laminated Timber Panels. Vancouver:* University of British Columbia.

Hossain, A., Danzig, L., & Tannert, T. (2016). Cross-laminated timber shear connections with double-angled self-tapping screw assemblies. *Journal of Structural Engineering,*142 (11).

Hossain, A., Popovski, M., & Tannert, T. (2016). *Cyclic Performance of Shear Connections with Self-Tapping Screws for Cross-Laminated Timber.* 16th World Conference on Earthquake, paper No 2913.

International Organization for Standardization. (1983). *Timber Structures: Joints Made with Mechanical Fasteners: General Principles for the Determination of Strength and Deformation Characteristics.* Geneva: ISO.

Jockwer, R. (2014). *Structural Behaviour of Glued Laminated Timber Beams with Unreinforced and Reinforced Notches* (Doctoral Thesis). ETH Zurich.

Jockwer, R., Stieger, R., & Frangi, A. (2014). *Design Model for Inclined Screws under Varying Load to Grain Angles.* Proceedings: International Network on Timber Engineering Research, Meeting 47.

Jorissen, A., & Fragiacomo, M. (2011). General notes on ductility in timber structures. *Engineering Structures,* 33 (11), 2987–2997.

Karacebeyli, E., & Popovski, M. (2003). *Design for earthquake resistance.* In S. Thelandersson & H. J. Larsen (Eds.), *Timber Engineering.* Chichester: Wiley.

Kevarinmäki, A. (2002). *Joints with Inclined Screws.* Proceedings of CIB-W18, Meeting 35, Paper 35-7-4.

Kevarinmäki, A. (2005). Puurakenteiden uusia liitostekniikoita. *Rakenteiden Mekaniikka,* 38 (2).

Krenn, H., & Schickhofer, G. (2009). *Joints with Inclined Screws and Steel Plates as Outer Members.* Proceedings CIB-W18 Paper 42-7-2.

Lam, F. C., Zhang, C., & Lee, G. (2013). *SWG/Wurth Self-Tapping Wood Screws-Data Report.* No. 2011-002 (private report). Vancouver, BC: UBC TEAM.

Mayo, J. (2015). *Solid Wood.* New York: Routledge.

Mestek, P., & Dietsch, P. (2011). *Design Concept for CLT-Reinforced with Self-Tapping Screws.* Proceedings of CIB-W18, Meeting 44, Paper 44-7-2.

Mohammad, M., Quenneville, P., Salenikovich, A., Zarnani, P., & Munoz, W. (2014). Failure modes in CLT connections. *WCTE 2014.*

Montgomery, W. G. (2014). *Hollow Massive Timber Panels: A High-Performance, Long-Span Alternative to Cross Laminated Timber* (Thesis). Clemson University.

Munoz, Mohammad, Salenikovich, A., & Quenneville, P. (2008). *Need for a Harmonized Approach for Calculations of Ductility of Timber Assemblies.* Proceedings of CIB-W18, Meeting 41, Paper 41-15-1.

Nie, X. (2015). *Failure Mechanism of Rolling Shear Failure in Cross-Laminated Timber.* Vancouver: University of British Columbia.

Perez, F., & Wallwork, T. (2016). Connections between CLT elements and future challenges for CLT in practice. In A. Falk, P. Dietsch, & J. Schmid (Eds.), *Proceedings of the Joint Conference of COST Actions FP1402 & FP1404 "Cross Laminated Timber—A Competitive Wood Product for Visionary and Fire Safe Buildings.* Stockholm: KTH Royal Institute of Technology, Division of Building Materials.

Pirinen, M. (2014). *Ductility of Wood and Wood Members Connected with Metal Fasteners* (Master's Thesis). Aalto University.

Popovski, M. (2000). *Seismic Performance of Braced Timber Frames* (Thesis). University of British Columbia.

Prat-Vincent, F., (2011). *Evaluation of the Performance of Joist-to-Header Self Tapping Screw Connections* (Master's Thesis). McGill University.

Ringhofer, A., Brandner, R., & Schickhofer, G. (2013). Withdrawal resistance of self-tapping screws in unidirectional and orthogonal layered timber products. *Materials and Structures,* 48 (5), 1435–1447.

Sadeghi, M. (2015). *Continuity Connection for Cross Laminated Timber (CLT) Floor Diaphragms.* Fredericton, NS: University of New Brunswick.

Sadeghi, M., & Smith, I. (2014). Edge connections for CLT plates: In-plane shear tests on half-lapped and single spline joints. *WCTE 2014.*

Silva, C., Branco, J. M., Ringhofer, A., Lourenço, P. B., & Schickhofer, G. (2016). The influences of moisture content variation, number and width of gaps on the withdrawal resistance of self-tapping screws inserted into cross laminated timber. *Construction and Building Materials,*125, 1205–1215.

Smith, I., & Frangi, A. (2014) *Use of Timber in Tall Multi-Storey Buildings.* Zürich: IABSE.

Smith, I., Asiz, A., Snow, M., & Chui, Y. (2006). *Possible Canadian/ISO Approach to Deriving Design Values from Test Data.* Proceedings of CIB-W18 Meeting 39, Paper 39-17-1.

Smith, I., Gong, M., & Landis, E. (2003). *Fracture and Fatigue in Wood.* Chichester: Wiley.

Sullivan, K. (2017). *Behaviour of Cross-Laminated Timber Diaphragm Panel-to-Panel Connections with Self-Tapping Screws* (Master's Thesis). Oregon State University.

SWG ASSY 3.0 WOOD SCREWS (ESR-3179). (2016). ICC Evaluation Service.

SWG ASSY VG PLUS WOOD-DRILLING SCREWS (ESR-3178). (2016). ICC Evaluation Service.

Tomasi, R., Crosatti, A., & Piazza, M. (2010). Theoretical and experimental analysis of timber-to-timber joints connected with inclined screws. *Construction and Building Materials*, 24(9), 1560–1571

Tomasi, R., Piazza, M., Angeli, A., & Mores, M. (2006). A new ductile approach design of joints assembled with screw connectors. *Unpublished Technical Report*.

Uibel, T., & Blass, H. J. (2006). *Load Carrying Capacity of Joints with Dowel Type Fasteners in Solid Wood Panels*. Proceedings of CIB-W18 Meeting 39, Paper 39-7-5.

Uibel, T., & Blass, H. J. (2007). *Tragfähigkeit von stiftförmigen Verbindungsmitteln in Brettsperrholz*. Karlsruhe: Univ.-Verl. Karlsruhe.

Uibel, T., & Blass, H. J. (2007b). *Edge Joints with Dowel Type Fasteners in Cross Laminated Timber*. Proceedings of CIB-W18 Meeting 40, Paper 40-7-2.

Wrzesniak, D., Fragiacomo, M., & Jorissen, A. (2014). Alternative approach to avoid brittle failure in dowelled connections. *Materials and Joints in Timber Structures: Recent Developments in Technology*. RILEM.

Zarnani, P., & Quenneville, P. (2014). *Resistance of Connection in Cross-Laminated Timber (CLT) under Block Tear-Out Failure Mode*. Proceedings of International Network on Timber Engineering Research, Paper 47-7-3.

CHAPTER 5

Hygrothermal Performance of CLT Assemblies: Recommendations for Design, Construction, and Maintenance

Mariapaola Riggio, Lech Muszyński

Oregon State University, Wood Science and Engineering Department

Introduction

The hygrothermal performance of building assemblies is a result of their response to heat, air, and moisture (HAM) transfer phenomena. Controlling the hygrothermal behavior of each assembly is crucial for overall building service life, energy efficiency, and the comfort and health of occupants. In this chapter, the hygrothermal behavior of cross-laminated timber (CLT) assemblies is discussed with the aim of defining general design principles for durability. Practical recommendations for the design, construction, and maintenance of CLT structures are presented with the specific objective of maximizing service life and durability as related to hygrothermal performance.

Durability considerations regarding insect attacks are not covered in this chapter. Design of energy-efficient CLT buildings is only partially discussed, and the reader is invited to refer to specialized publications for further information on this topic. It is also beyond the scope of this chapter to provide guidelines or values for hygrothermal modeling and simulation. It should be noted, however, that dynamic hygrothermal modeling should support design of efficient CLT building envelopes. These models must consider the multiple, local factors that affect hygrothermal performance of CLT assemblies in a building.

HAM phenomena and relevant variables can be considered at different scales: at the material level, such as the hygrothermal parameters determined by the natural characteristics and composition of wood; at the product level, such as differences in the manufacturing process and product-specific characteristics of CLT panels; at the component/assembly level, such as the specific location of the assembly within the building and its performance requirements; and at the scale of the building and site, such as type of occupancy, climate zone, and construction type. In this chapter, these considerations are addressed in order of increasing scale, from the material and product level (Sec. 5.1) to the assembly level (Sec. 5.2) and to the scale of the whole building (Secs. 5.3 and 5.4).

It is worth mentioning that the term "cross-laminated timber" encompasses a range of products with very different features. Species, number and thickness of layups, adhesives used, edge-glued (or not) lumber, and type of finishes or treatments result in variations in performance. Section 5.1 aims to provide the reader with a general understanding of the factors influencing hygrothermal behavior of raw wood materials as well as CLT products. More specific data, including specific values for design and calculation, can be retrieved from manufacturers through product datasheets and specialized literature.

The utilization of CLT in mass timber buildings can be broken into two major systems: (1) fully panelized (cellular) systems, in which CLT is used for all structural walls, and floors, and (2) frame/post-and-beam superstructures, coupled with CLT floors and roofs. In the second case, vertical elements of the lateral load-resisting system typically include one or more of the following solutions: braced framing, shear walls, and shear cores.

In the second system, the presence of linear structural components allows for flexibility with façade infill. However, this system also necessitates attentive design detailing to control hygrothermal performance. The possibility of differential movement, thermal bridging, and other detriments to the performance of the façade increases due to the respective behaviors of the different materials used for the structure and enclosure.

Building enclosures play a crucial role in the hygrothermal performance of a building and represent a buffer from the external environment. Principles and recommendations for the design of CLT enclosure assemblies regarding weather protection, moisture control, airtightness, and thermal control are presented in Sec. 5.2. Additional functions of CLT in a building enclosure, such as support, fire resistance, and acoustic performance, are covered in other chapters of this book.

Schematic examples of building assembly details for multilayered enclosure systems are included in Sec. 5.2. These are intended to highlight some key design elements and criteria. Of course, these details do not cover all possible alternatives. Project- and site-specific considerations may precipitate very different design resolutions, dependent on a wide range of factors such as climate, local regulations, the characteristics of the CLT product used, and more.

Multiple factors influence envelope design, such as the multifunctional requirements of building enclosures, the variety of external and internal environmental loading conditions (e.g., site conditions, site climate, building occupancy), and the interrelation of different building components and materials. An integrated design approach to timber envelope design is highly recommended to avoid conflicting solutions regarding different performance requirements. The involvement of experts from various disciplines, including structural engineers, building physics experts, fire engineers, and wood science experts, should take place early in the design phase.

In Sec. 5.3, design principles are presented that account for differential movement between parts of a mass timber building. These details aim to prevent potential negative effects on structural integrity, enclosure performance, and serviceability. A holistic approach to durability, encompassing design, construction, and maintenance, is described in Sec. 5.4. Finally, Sec. 5.5 illustrates techniques for the inspection and monitoring of CLT components over time.

5.1 Hygrothermal Behavior of Wood and CLT: Fundamentals

When comparing the hygrothermal performance of CLT with other structural materials, we must focus on the physical properties of wood, such as its hygroscopicity, permeability, moisture and heat conductivity, and specific heat. We also must consider that these properties may vary depending on individual features of CLT panels, such as grade of lumber used for lamellas, the number and thickness of the lamellas, and wood species.

Since CLT panels are overwhelmingly composed of dimension lumber with limited disturbance to the natural structure of the wood tissue, the properties of the solid wood determine most of the properties of the panels. Therefore, it is important to discuss these properties first.

Wood is often described as a natural composite that is porous, permeable, and hygroscopic. It is important to keep in mind that like most natural materials, wood is highly inhomogeneous and that most of the properties discussed below would vary greatly between wood of the same species grown in different regions or in different stands within the same region and between trees on the same site. Even within a single tree, many properties would vary depending on the position along the log the wood sample was taken from and on its position between the pith and the bark. In addition, knots and other natural features generated by living trees disturb the regular cellular structure of wood and affect its bulk properties. Since lamellas of an individual CLT panel are likely to include lumber from a number of trees and cuts from various positions within the log, we cannot assume narrow values of the properties of interest as determined on small clear specimens. A level of statistical distribution of the properties discussed below must be expected and accounted for in modeling and assigning characteristic or design values.

The main components of wood tissue are three types of organic polymers: cellulose (about 50% by weight), hemicelluloses (20% to 25%), and lignin (23% to 33% in softwoods). All three polymers have large numbers of hydroxyl (–OH) groups, which are potent sorption sites for water molecules and are the source of wood's natural affinity with water, or hygroscopicity. The hydroxyl groups can readily attract (adsorb) and release (desorb) water molecules in response to the changes of concentration of these molecules in the ambient environment. At a constant ambient temperature, the concentration of adsorbed water molecules tends to reach an equilibrium with the concentration of water molecules in the surrounding air. The moisture content of wood at this equilibrium state is commonly called *equilibrium moisture content* (EMC). In practice, rather than concentration of moisture in wood, an "oven-dry base" (or OD) moisture content of wood, expressed as a ratio of mass of water present in a given piece of wood to the oven-dry mass of that same piece, is used. Similarly, instead of concentration of water molecules in the gas, a measurement of relative humidity (RH) is commonly used. These EMC values are determined for different RH and temperature conditions and readily available in tabular or graphical format (sorption isotherms) (FPL, 2010).

Wood tissue may also contain a level of organic extractives (5% to 30%), which is highly dependent on the species, and a small amount of inorganic materials (within 1%) sometimes referred to as "ash." Many extractives are hydrophobic and their presence in larger quantities may somewhat mitigate the hygroscopicity of bulk wood.

As important as the properties of the components are on a molecular level, the physical properties of wood at the scale of lamellas are largely determined by the internal structure of wood. Wood is a complex natural composite with multilevel hierarchical structure, with smallest components less than a micron across. For the current discussion, it is sufficient to focus on the cellular structure of softwoods, which are the species mostly used for CLT in North America (ANSI/APA PRG 320-2018). Most of the material is organized in concentric rings of highly elongated cells, with an aspect ratio of about 100:1, and roughly square to hexagonal cross sections, oriented along the axis of the stem. This cellular structure is the main reason of the strong anisotropy of many wood properties.

The cell walls are organized in rigid microporous layers and contain a void in the center, customarily referred to as a lumen. The density of the cell wall material is surprisingly uniform across many species (approximately 1.52 g/cm^3), so the density of solid wood is dependent on the cell wall thickness or the proportion of the cell wall material volume to the volume of lumens. The density of commercially significant softwoods in North America ranges from 0.31 to 0.62 g/cm^3, with a mean value of 0.44 g/cm^3 (±0.08 g/cm^3). A related property is porosity, expressed as a unitless ratio of volume of all lumens to a reference unit volume of bulk wood, ranging from 0.59 (for the densest species) to 0.80, with a mean value of 0.71 (±0.05). These values are not given directly but can be easily derived from data published in the FPL *Woof Handbook* (2010).

Thermal properties of wood, like thermal conductivity and specific heat, are largely determined by its porosity: the less airy pores, the greater the density of wood and the greater its thermal conductivity. Another major factor is the moisture content. Generally, thermal properties of moist wood can be approximated from rules of mixtures, where wood is treated as three-component mixture of wood-cell material, air, and water, with the thermal properties of each component generally available. The effect of other components is negligible. In such calculations, it is important to remember that at moisture contents above a value referred to as the fiber saturation point (FSP), liquid water replaces air in the cell lumen. At the maximum possible moisture content (value dependent on species density), when water saturates all cell walls and fills all cell lumens, the mixture has only two components (mostly water) and is highly conductive. Since an overwhelming proportion of cells is oriented along the axis of the log, thermal conductivity of dry wood, relying on the continuity of the cell wall as thermal bridges, is somewhat an anisotropic property and can be greater along the grain by a factor of 1.5 to 2.8 than conductivity across the grain (FPL, 2010). However, at the stage of full saturation of lumens discussed above, the water provides a much more potent bridge, conducting heat in all directions without distinction.

Moisture conductivity of wood is a much more complex phenomenon. It is dependent not only on the porosity of wood but also on the size of the pores and on whether individual pores are interconnected into continuous systems of passages of various scales. Moisture conductivity is vitally important for the living tree's hydrology, and the cell lumens of neighboring cells are typically connected by a system of pits. This allows for movement of liquids mostly along and, to some degree, across the trunk. While sapwood, the external part of the log cross section, is generally highly permeable, in many species the physiological changes in heartwood lead to partial or near complete blockage of cell lumens, substantially reducing its natural moisture conductivity. In addition, in some softwood species, during the first drying, the capillary action of retreating liquid water found in lumens at the "green state" leads to permanent blockage of most pits (this phenomenon is called pit aspiration).

At a constant temperature, moisture conductivity can be described in terms of parallel diffusion of water molecules through the microporous cell wall material and the diffusion of water molecules through the air, filling cell lumens and the system of connecting pores from areas of higher moisture concentration to areas of lower concentration. Though both processes are fairly slow, the diffusion of water vapor through the gas in the system of pores is faster by a couple orders of magnitude. Due to the longitudinal orientation of cells, both modes of transport are faster in the longitudinal direction than in the transverse direction of the wood tissue.

Practical differentiation between these two parallel processes is notoriously difficult. Therefore, in practical measurements, wood is typically treated as a homogeneous medium for which total diffusivity is determined without distinguishing of the phases. However, in such an approach, the diffusivity strongly depends on the porosity of the wood since in denser species a smaller volume of voids (lumens) is available for the faster diffusion through gas filling the cell lumens.

This phenomenon is further complicated when moisture transport occurs in changing temperature conditions. Temperature changes may generate local pressure gradients due to the restricted expansion of liquids and gases trapped in the tight system of pores. Pressure gradients trigger bulk flow of gases and liquids from regions of higher pressure toward areas of lower pressure. This is yet another mass transport mechanism generally described by Darcy's law of bulk flow of fluids.

It is important to keep in mind that despite great progress in modeling the fundamental aspects of moisture transport phenomena in wood, most functional models used in practice utilize substantially simplified approaches. These simplified approaches may not provide equally accurate results for all practical problems. While this is a valid approach and calibrations are proposed for many typical situations, one should treat model simulations with due caution.

Because of the slow rate of moisture transport through wood tissue, the process of adsorption and desorption from larger volumes of wood in normal room conditions may take a substantial amount of time. The deciding parameter is the shortest path of a moisture molecule from the core to the surface of the element or from the surface to the core. Therefore, a thin sheet of veneer will dry or gain moisture from the air much faster than a piece of two-by-four, even if their total volumes may be equal. In a variable indoor environment, massive timber elements may serve as a temporary moisture buffer, slowly releasing moisture to a warm, dry room and slowly absorbing moisture from a humid room. This property, sometimes referred to as moisture storage capacity, might be quantified in ways analogous to the thermal capacity of a material.

Moisture content also affects the dimensional stability of wood. The presence of water molecules bound to the sorption sites within the cell wall material affects the spacing between the microstructural components of the cell wall. On the macroscale, this may be observed as the shrinking of wood tissue when the moisture content is reduced (water molecules removed) or swelling when the moisture content is increased (water molecules added). We generally assume all these sorption sites are saturated at the FSP (about 22% to 30%, depending on the wood species). Consequently, we assume that moisture content of the cell wall material will not increase if more moisture is added to a wood sample, even as the moisture content of the entire sample increases due to an increased amount of water present in the cell lumens. It follows that no dimensional changes are expected if moisture content in wood exceeds the FSP. Experimental data for small clear specimens of many wood species show that the change in volume due to moisture-related shrinkage or swelling is roughly equal to the volume of water removed or added to the specimen (FPL, 2010). In softwood species commercially available in North America, the volumetric shrinkage from FSP to an oven-dry state may range from 7% to 16% of the moist volume, with the mean value at 12% ± 2%. However, due to the cellular structure of wood, different amounts of linear shrinkage/swelling are expected in wood along the grain and in the radial and tangential directions. The amount of shrinkage and swelling may differ substantially between various species of wood and even between the heartwood and sapwood of wood of the same species. However, for most species, the dimensional shrinkage recorded in tangential direction (mean value about 7% ± 1%) is about twice as large as in radial direction (mean value about 4% ± 1%). For most practical considerations, the shrinkage and swelling in the direction along the grain may be safely neglected. Shrinkage values from FSP to oven-dry for many wood species, shrinkage coefficients for small moisture changes in wood in service, as well as useful formulas for practical calculation of any moisture changes below FSP are published in the FPL *Wood Handbook* (FPL, 2010).

5.1.1 Characteristics of CLT Panels Affecting the Hygrothermal Behavior

CLT panels composed of cross-laminated solid wood lamellas would inherit most of the hygrothermal properties characterized above for solid wood. However, some of them will be modified to a certain degree due to the arrangement of the lamellas within the volume of the panel and of the increased complexity of the internal structure compared to the wood itself.

Generally speaking, the CLT manufacturing process does not affect the chemical composition, hygroscopicity, cellular structure, density, or porosity of wood. However, some manufacturers may use thermally or chemically modified lamellas. These treatments do affect both the chemical composition and the hygroscopicity of lamellas to some extent. These modifications are aimed at improving the durability of the composite against biodegradation and may affect the entire volume of the panel. Much more often, just a thin buffer zone close to the surface of individual lamellas is impacted. While it is important to be aware of such options, these modifications comprise a relatively small volume of global CLT production.

In CLT wall and floor assemblies, the most interesting aspect of heat and moisture transfer is that in the direction perpendicular to the wide surfaces, that is, in the direction perpendicular to the grain of the lamellas in all layers, regardless of their in-plane orientation.

In this direction, the presence of adhesive layers may constitute a series of additional barriers to the transfer of moisture or gases through the layup, although we are not aware of any currently published data that account for this effect.

Five cross-laminated layers, commercially fabricated CLT leaving the production line is considered practically airtight. However, the presence of loose knots, checks, wane, and small gaps between the lamellas that are not edge-bonded constitutes a system of systematic breaches in the tight structure of solid wood. In CLT panels produced without edge-bonding of lamellas, the width of the gaps between lamellas in the surface layers will increase in dry conditions and decrease in humid conditions. This is due to the unconstrained surface of the face lamellas shrinking and swelling in width. Similar behavior may be observed in surface checks in edge-bonded panels.

Even if the total volume of these macro-voids is small compared to the overall porosity of wood in the wall or floor volume, they are thousands of times larger than the regular cell lumen structure. Due to this fact, they constitute local "bridges" for heat, air, and moisture transfer within a layer of the wall or floor layups. The effect of these local "bridges" diminishes in five-, seven-, and nine-layer CLT layups. The impact of these local bridges on the overall hygrothermal performance of CLT assembly is negligible compared to the edge joints between panels, which, if not properly sealed, offer much easier paths for convective heat and moisture transfer.

Despite the slow pace of moisture transfer through the substantial thickness of a CLT panel, checks, gaps between boards, and general discontinuities can affect wettability of CLT panels exposed to bulk water (as, for instance, unprotected CLT floors during construction). It has been observed that macroscopic discontinuities in the laminas, such as gaps between non–edge-glued boards and wane wells, may facilitate accumulation of liquid water at the interface of the upper plies of CLT floor panels (Schmidt et al., 2019). Additionally, CLT panels are more prone to wicking at their edges and wherever end-grain faces are exposed (e.g., cuttings for windows and door openings).

If wetting of CLT occurs via exposure to bulk liquid, the water uptake is much faster than the drying process, which relies on evaporation from the surface of the panels. If this evaporation process is hindered, for instance, by the presence impermeable coating or membranes, undesirable high moisture concentrations can linger in a panel's internal laminas for a long period of time after direct exposure (Wang, 2016a; Lepage et al., 2017).

The cross lamination of layers makes CLT panels remarkably dimensionally stable in in-plane directions, allowing the design and finishing of perimeters, openings, and fitting nests with submillimeter precision. In fact, cross lamination of layers was introduced as a major improvement from unidirectionally bonded massive glulam panels, which had been previously used in construction. These unidirectionally bonded panels were notorious for large cumulative shrinkage or swelling effects in the direction perpendicular to the grains in all layers. However, the cross lamination does not prevent the CLT panels from swelling or shrinking in thickness, which should be taken into account in design and construction (see Sec. 5.3).

It has been observed that in CLT panels exposed to extreme moisture fluctuations, the constraint of transverse swelling and shrinkage of layers due to cross lamination can cause stresses, leading to gradual physical deterioration such as checking and interfacial shearing. This type of damage is referred to as the "zipper effect" (Gamper et al., 2013).

Warping of CLT panels may occur if the wide surfaces are exposed to different physical conditions (e.g., top surface wet, bottom surface dry). Warping of massive CLT panels has been a concern because a warped wall or floor would not fit properly in a construction system designed with tight tolerances, and they are virtually impossible to straighten on-site. However, recent research show that the actual scale of warping in CLT layups is minuscule and it takes a substantial time of exposure to reach critical conditions (Gereke et al., 2011). In typical CLT construction projects, the time between when panels leave conditioned factory storage and their installation in the structure should be short enough to avoid measurable warping (see Sec. 5.4).

5.2 Control of Moisture, Air, and Heat Flows in CLT Assemblies: Design Principles

CLT can be used in wall and floor systems and, if part of a building enclosure component, in roof systems and above-grade wall assemblies. CLT is not appropriate for below-grade walls and base floors.

Typically, a building enclosure is made up of multilayered systems, which varyingly control the flow of heat, air, and moisture from the exterior environment to the interior of the building and vice versa (Fig. 5.1). Each of these control functions in the enclosure can be assigned to a specific layer, but some layers can be used to fulfill several functions at the same time (e.g., vapor control and airtight layer). This results in fewer work steps. In contrast, multiple products used in combination can serve one single function, for instance, to create a continuous airtight layer. Both cases require special attention, through correct detailing and installation, to ensure continuity.

To extend service life of CLT assemblies, the following principles should guide the design of building enclosures: (1) prevent rain penetration in the enclosure assembly, (2) minimize air pressure differentials across the assembly, (3) minimize temperature drop and thermal bridges within the enclosure assembly, and (4) minimize penetration

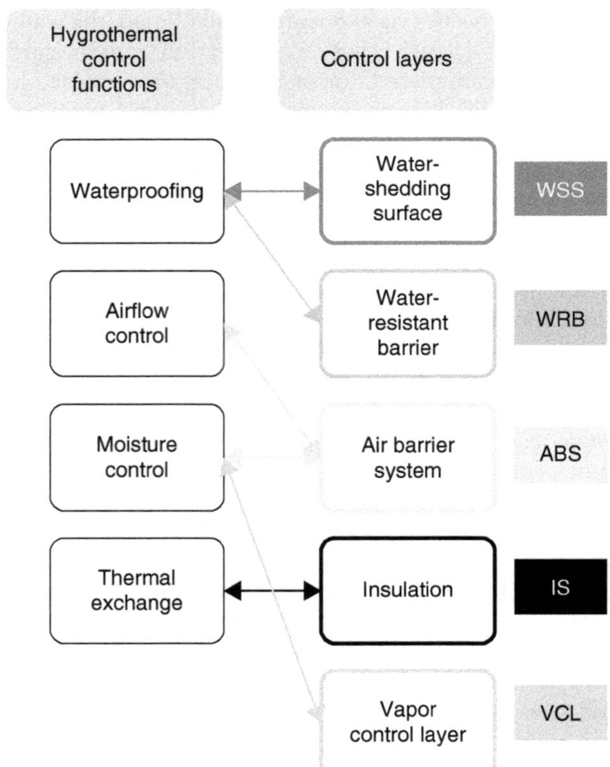

FIGURE 5.1 HAM control functions and related control layers. (*Source: Adapted from Gasparri & Aitschison 2019.*)

of indoor humidity into the enclosure assembly through diffusion, air infiltration, and exfiltration (Straube, 2012). These principles are further explained in the following sections.

5.2.1 Water Control

This subsection illustrates water control design principles for external moisture sources. Typical external moisture sources in a building are precipitation, ice dams, and ground water from the soil. Wind affects the deposition of precipitation on a building. Other sources can be irrigation systems and vegetation, such as in a roof garden or a green façade. In CLT buildings, the built-in moisture accumulated due to exposure of the panels during construction and the consequent water absorption and entrapment in a component or assembly should be carefully considered and avoided. Section 5.4 provides guidelines and relevant information about on-site moisture management practices recommended for CLT buildings.

Building height and shape affect the pattern of precipitation-related wetting and redistribution. Enclosures for mid- and high-rise CLT buildings are subject to higher environmental loads, such as wind-driven rain on the upper stories and increased rundown water at the lower floors (FPInnovations, 2014; Ott et al., 2015).

Depending on the source, different moisture transfer mechanisms and driving forces can occur (Glass et al., 2013). Liquid water flow includes gravity-driven water, such as intrusion through a leak; capillary action, which occurs when porous materials are in contact with wet interfaces; surface tension, causing water to cling to the underside of horizontal or nearly horizontal surfaces; and kinetic energy, propelling water drops into unprotected holes. Diffusion and convection of water vapor resulting from vapor pressure difference between indoor and outdoor environments and air pressure difference due to air leakage are treated respectively in separate subsections of this chapter. If CLT gets wet, drying should occur in one of the following ways: drainage of free liquid water, desorption of bound vapor, and evaporation of bound liquid through venting or ventilation in particular (Glass et al., 2013).

CLT panels can act as moisture buffer, as wood materials have an inherent capacity to store moisture, as long as safe moisture levels are not exceeded. This material characteristic can have positive effects both on the indoor comfort (Hameury & Lundström, 2004; Nore et al., 2019) and on the durability of a building component, allowing for short-term storage of moisture. On the other hand, long-term moisture accumulation may result in deterioration.

Design strategies to control liquid water flow are guided by defensive constructive solutions minimizing exposure to environmental loads. These strategies work through three mechanisms: controlling moisture entry, limiting moisture accumulation, and removing accumulated moisture. It is worth mentioning that these strategies may conflict with each other (Glass et al., 2013). For instance, a solution that is effective at preventing moisture from entering an assembly also may prevent moisture from leaving an assembly. Conversely, a solution effective for removing moisture is likely not effective enough in preventing moisture from entering. For this reason, it is important to predict and control moisture sources in a building during construction and in service.

Defensive constructive solutions are often referenced as distance, deflection, drainage, and drying (Beebe & Kam-Biron, 2016). Distance from ground should be provided by raising the timber structure above grade as well as by adequately detailing CLT wall to slab interfaces to prevent moisture absorption through capillarity (Fig. 5.2).

FIGURE 5.2 Waterproofing of a CLT base wall: (top) reinforce concrete sole plate with damp-proof membrane (left) and liquid-applied waterproofing (right); (bottom) damp proofing of the base wall with bitumen sheeting. (*Courtesy of Xlam Dolomiti.*)

Deflection reduces moisture loads by directing precipitation water away (outward and downward) from a building enclosure. This is the most effective defense against precipitation water intrusion and must be thoughtfully incorporated into enclosure systems. The effectiveness of deflection solutions is influenced by the presence and depth of projections, building height, and local driving rain wind pressure. Details such as flashing and drip edges are key to deflecting water in some exposed locations.

Precipitation water may penetrate the enclosure through imperfections or damaged areas. Drainage is the only transport mechanism with sufficient capacity to dissipate this moisture. A well-designed drainage path (or multiple paths to ensure some redundancy) is fundamental in CLT assemblies to direct water downward through the assembly and outward from the building. Flashing and waterproofing should be used at vulnerable positions in connections and drainage planes to completely seal openings and discontinuities. Differential movement of building materials should be considered in all drainage details. Any deformation of flat roof assemblies has a detrimental effect on the drainage capacity of the roof. The slope to the drains of CLT decks should be a minimum of 2% in a loaded configuration (Kaufmann et al., 2018): this can be achieved by either a tapered insulation package, built-up tapered wood framing, or a sloped concrete screed (Riggio et al., 2020). Additionally, precambered roof timber beams can prevent sagging of large-span roofs and structurally create a positive slope for drainage. The overall geometry of a roof should be designed to avoid local water stagnation (e.g., at horizontal valleys).

Water control design solutions for CLT buildings must include opportunities for drying in both assembly directions, both to the exterior and the interior. Therefore, attention should be paid to the sequence and characteristics of control layers in an assembly. In the following paragraphs, exterior wall assembly systems are classified on the basis of different levels of implementation of drainage and drying strategies (Straube & Burnett, 1999) (Fig. 5.3).

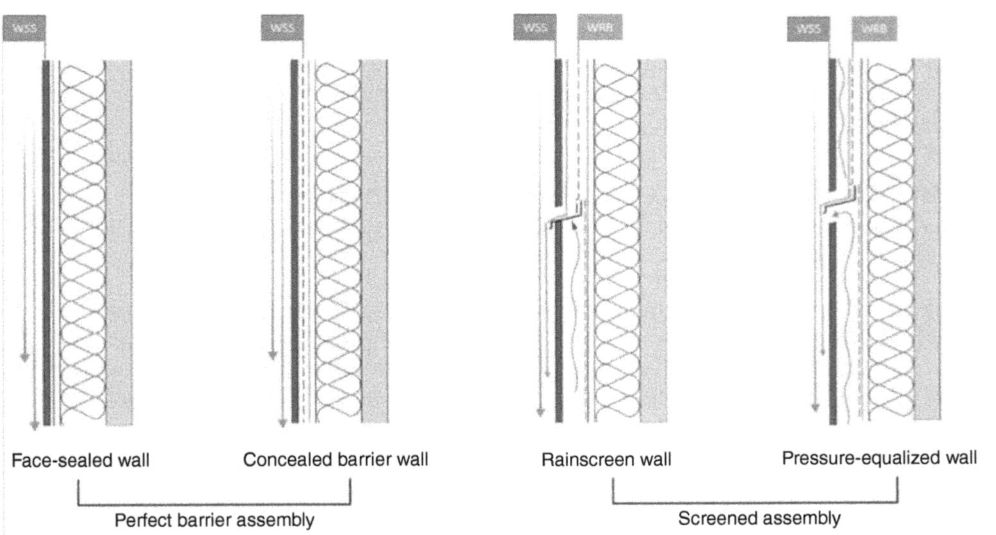

FIGURE 5.3 Types of multilayered, exterior wall assembly systems.

A perfect barrier assembly is a rain control approach that employs a single water control layer and a water-shedding surface (WSS) that prevents liquid water from traveling further into the assembly. This WSS is either the cladding (face-sealed assembly) or a waterproof barrier concealed behind the cladding (concealed barrier wall). This type of assembly is generally not recommended for CLT construction in wet climates since it provides limited moisture control.

Drained wall systems (or rainscreen assemblies) incorporate a second path of drainage with a water-resistive barrier (WRB) within a cavity behind the cladding. Water intercepted by the WRB must be dissipated to the exterior by means of drainage, evaporation, or a combination of both. Drainage cavities may have different depths, depending also on the material used for the WSS (i.e., deeper for porous cladding materials). In smaller cavities, water will be retained by surface tension. In this case, membranes must have increased water resistance.

Drying mechanism can be enhanced with pressure-equalized rainscreens, which include ventilation openings at both the top and the bottom of the internal cavity. This reduces the air pressure differential between the exterior environment and interior of the wall system, mitigating the risk of moisture and water vapor from the exterior air being driven into the system. This is particularly recommended for CLT construction in climates with high moisture loads.

All CLT roof systems require the same durability considerations as façade assemblies: effectively deflecting and draining water and providing opportunities for drying. Flat roofs for timber structures can be designed using conventional assemblies; however, "protected membrane" assemblies (inverted roof assemblies) are recommended for roof decks or roofs, which expect high foot traffic and other surface loads. Every roof system must carefully interlace with the exterior wall system to maintain continuity of drainage paths and critical control layers. Dryability of roof framing can be enhanced by interior ventilation cavities that allow air to move over the roof diaphragm. Flat roofs, however, create a challenge in this regard; ventilation is generally not effective due to the lack of differential pressure (Kaufmann et al., 2018; Piva, 2019). In this case, integration of mechanical ventilation systems or use of other design solutions to enhance dryability should be taken into consideration. While most of the above-mentioned principles are applied to other building systems as well, some are specific to CLT assemblies.

Water Control Layers in CLT Assemblies: Materials and Systems

In this section, solutions for WSS systems and WRB systems used to control liquid water entry in a CLT building enclosure are discussed.

WSS WSSs are the first plane of protection against precipitation water entry. These include cladding, roofing, and details in the wall and roof assemblies, such as flashing.

WSS functionality is based on the fact that the number and size of openings should be minimal and driving forces acting on the assembly controlled. General design approaches for managing forces that drive water through the cladding are provided by Brown et al. (1999). The choice of materials will affect the number and size of openings and the design of joints and interface details around penetrations. In any case, all these details must be carefully designed.

Drip edges should be incorporated in the underside of projecting horizontal surfaces, such as windowsills, balcony, and soffits, and on flashing. Openings and joints should be shielded with overlapping materials, sealant, preformed gaskets, or deflectors. Properly flashed movement joints, such as cross-cavity flashings at each floor, are required to accommodate initial and seasonal wood and building movement. They will also provide drainage and ventilation behind the cladding.

Different cladding materials can be used for rainscreen walls. Porous claddings will allow water to accumulate in the cladding itself. This moisture will be driven inward by solar heating and eventually past the WSS. In this case, it is necessary to provide a safe outlet for moisture. Stored water may evaporate into capillary breaks beyond the cladding or dissipate to the exterior by venting. With the use of porous cladding, a more robust WRB is also recommended. Similarly, porous roofing materials may require a vented space between the roofing and the membrane.

Examples of absorptive WSS are so-called reservoir cladding materials, such as stone, brick veneer, and stucco. Many authors advise against the use of these materials in hot, humid climates, such as the eastern United States, to avoid moisture accumulation due to inward vapor pressure (Derome, 2010; Glass et al., 2013). In tall CLT buildings and in exposure conditions including wind-driven rain, it is recommended to carefully consider the use of absorptive WSS materials and open-joint rainscreen assemblies (Finch, 2016). While wood is a hygroscopic material, some bulk treatments, such as acetylation (Mantanis, 2017) and thermal treatments (Esteves & Pereira, 2009), make it more hygrophobic and less susceptible to biotic decay. This can qualify wood as a durable cladding solution if fire requirements are met.

Since the exterior surface of the building enclosure is also exposed to solar radiation, weathering resistance of the material used for cladding and roofing should be specified for the designed service life as well as any planned, periodic maintenance activities.

Furring strips should be preferably arranged vertically and, in any case, not block the drainage paths. They should be of a durable material. If pressure-treated wood is used as furring strips for cladding attachments or to create vented spaces within roof assemblies, fasteners must be selected with appropriate corrosion resistance compatible with the preservative used (Zelinka & Rammer, 2009).

WRB WRBs are the second plane of protection in a building enclosure assembly. Objectives of the WRB are (1) to intercept both free and bound water passing the WSS and (2) to dissipate this water effectively to the exterior.

A waterproof membrane is typically used to intercept free water passing the WSS. Different factors are considered in the selection of a WRB membrane: resistance to penetration of water vapor, maximum pressure caused by standing water that the product can tolerate (i.e., in decks and roofs), tensile strength for forces exerted both longitudinally and transversally, maximum elongation percentage the product can tolerate before failure, nail tear strength, UV stability, and flexibility at low temperatures. WRB systems with different properties may be used for different elements in the enclosure system.

Vapor-permeable water-resistant membranes also allow the wood to dry out toward the exterior. Some of these types of membranes are not resistant to standing water; in this case, they cannot be used on roofs. WRBs must be resistant to standing water to waterproof horizontal or subhorizontal assemblies, such as roofs, decks, and windowsills.

In many cases, the same material can function as the WRB and the air barrier. These materials are typically applied on the external face of CLT wall panels. A WRB applied in contact with the CLT wall has the advantage of (1) being more easily installed on a continuous surface, and (2) potentially providing better protection of the CLT wall from rain during construction. Disadvantages of this installation are that (1) the WRB can be damaged by insulation fixings, and that (2) it leaves the insulation unprotected from rain and exposed to wind actions. In the case that a moisture-sensitive insulation material is used, and especially for open-joint cladding in a rainscreen system, a breathable WRB should be placed outboard the insulation (see assembly details at the end of this section). It can be noted that the later installation type is the most common one in most countries in Europe.

Self-adhered or mechanically attached WRB membranes can be used (Finch, 2016), but the former may raise fire performance concerns, particularly when exposed in ventilation cavities behind the cladding. The currently available liquid-applied membranes are not suitable; their durability can be impaired by dimensional changes and crack openings. Continuity is an essential feature of the WRB. Design measures to manage moisture-driving forces should be incorporated in the WRB. This includes flashing and sealants at penetrations (Fig. 5.4) and extending the WRB across the rough opening to the window and door frames.

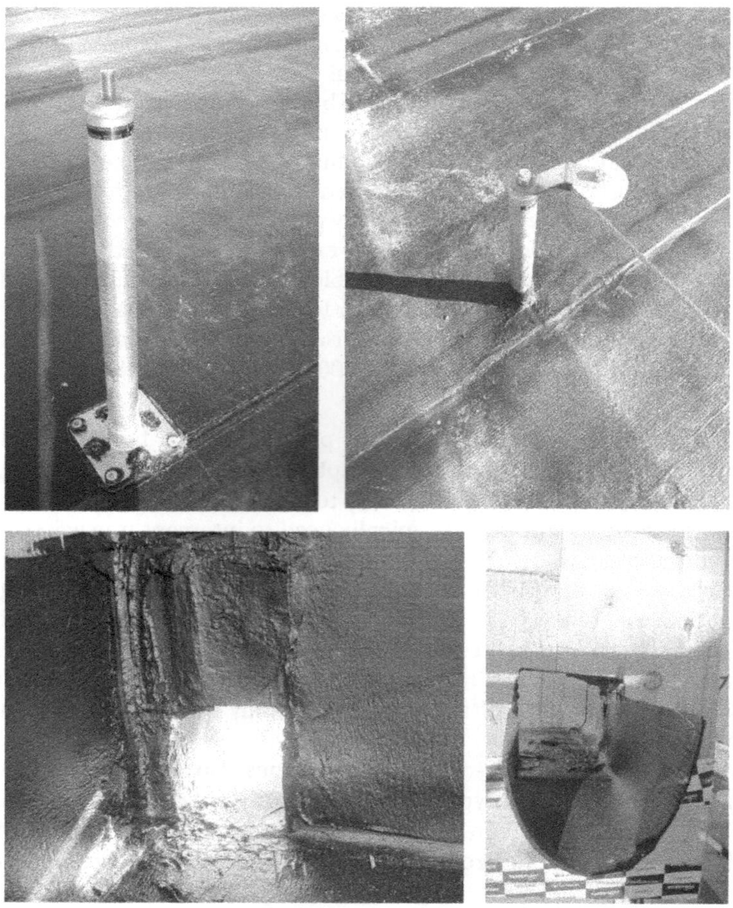

FIGURE 5.4 Continuity of the water-resistive barrier: sealing at roof penetrations and at drains. (*Courtesy of Xlam Dolomiti.*)

Where the WRB is also part of the air barrier system (ABS), it can be made airtight by using sealants, tapes, gaskets, or through other methods.

One of the most common low-slope roof types is the exposed membrane roof, also known as a conventional roof. This is a type of face-sealed assembly and, as such, is vulnerable to incidental moisture trapping. To achieve the best performance for these types of roofs, a fully adhered or ballasted WRB, with high thermal and UV stability and mechanical resistance, should be applied over a cover board, protecting the external insulation (Straube, 2012) (Fig. 5.5).

Finch (2016) recommends a peel-and-stick membrane with torch-welded laps versus pressure-adhered laps whose tolerance for defects in application is very low.

Inverted or protected membrane roof (PMR) systems locate the WRB below the insulation. This protects the membrane from UV radiation, temperature swings, and mechanical damage. Straube (2012) recommends including a sloped drainage gap over the membrane and under the insulation of the PMR.

FIGURE 5.5 Waterproofing of a CLT flat roof and parapet: conventional roof assembly with ballasted water-resistive barrier. (*Courtesy of Xlam Dolomiti.*)

5.2.2 Airflow Control

There are three main areas concerning a building's hygrothermal performance related to the control of inward/outward air movement within and through assemblies: (1) moisture control, (2) comfort and health, and (3) energy efficiency (Straube, 2012).

Ventilation or external airflow behind the cladding and/or roofing is useful because it accelerates drying. On the other hand, airflow within the building enclosure can cause increased moisture due to infiltration during hot, humid weather, causing interstitial vapor condensation in air-conditioned assemblies, and water vapor from exfiltrating air in winter due to heated interior and cold exterior ambient temperatures. Condensation supports also mold growth, affecting indoor air quality. Increased energy efficiency demands require perfectly airtight enclosures to avoid heat loss and gain. However, increased airtightness can cause moisture and air quality issues if not matched by a ventilation system that can supply fresh air, dilute pollutants, and control indoor humidity.

Airflow occurs when there is a pressure difference between two points and a continuous flow path connecting them. Factors that generate pressure differences are wind, stack effects due to differences in air density with temperature, convective loops, wind washing and pumping, and mechanical systems (Straube, 2012). Wind forces typically create a positive pressure on the windward face of a building and negative pressure on the leeward walls. Low-slope roofs tend to have mostly negative (uplift) pressure, while roofs with slopes above about 25 degrees experience negative pressure on the leeward face and positive pressure on the windward (Straube, 2008). Wind and stack effects are higher for high-rise buildings due to the higher differential air pressure.

To reduce stack effects, enclosure assemblies and buildings should be horizontally and vertically compartmentalized (Straube, 2008). For instance, vertical compartmentalization of the cavity in rainscreen wall assemblies is recommended in tall buildings at the corners and at some intermediate locations.

Convective loops can potentially form within a highly air-permeable insulation or around insulation through small gaps. Mechanical equipment such as exhaust fans and blowers can generate negative pressure if they exhaust more air than what is supplied (Straube, 2012). Airtightness can be tested with a blower door test (ASTM E1827-11). This should be carried out before inner planking is installed.

Airflow Control Layers in CLT Assemblies: Materials and Systems

An ABS is required to control airflow through the building enclosure components that separate two different environments.

One of the basic requirements of an ABS is, of course, air impermeability. ABSs comprise different assemblies using different materials (e.g., for the opaque and transparent enclosure system), which combined provide an uninterrupted plane of primary airflow control. Therefore, air impermeability performance of both materials and assemblies has to be specified. Performance specifications for materials and assemblies used in the air control layer are included in Section C402.5.1.2 of the 2018 IECC. Accordingly, air permeance of materials comprising part of the ABS shall not exceed 0.004 cfm/ft^2 under pressure differential of 0.3 in. w.g. tested in accordance with ASTM E2178. The average air leakage of opaque assemblies that comprise the ABS shall not exceed 0.04 cfm/ft^2 under pressure differential of 0.3 in. w.g. tested in accordance with

ASTM E2357, 1677, or 283. The whole building shall not have an air leakage rate of more than 0.25 cfm/ft^2 at a pressure differential of 0.3 in. w.g. (75 Pa), with calculated surface area the sum of above- and below-grade building envelope, as verified by whole-building pressurization test per ASTM E779 or ASTM E1827 or by independent third party.

The airtightness of CLT panels leaving the production line depends marginally on the type of layup (i.e., edge gluing, number of layers) and the moisture content, as described in Sec. 5.1. However, additional gaps can open within CLT panels caused by movement in the elements and cracks due to shrinking and swelling, following excessive exposure to liquid water during and after construction. In time, these cracks may create numerous air leakage pathways around and through the panels themselves. Routing of pipes and cables in the panels further reduces airtightness of CLT. For these reasons, airtightness of building enclosures relying exclusively on CLT panels locally sealed may vary, depending on the product and the installation, and can change with time.

For CLT panels with certified airtightness, continuity of the ABS can be achieved by locally sealing discontinuities, such as all the joints between panels, seam, connections, and corners, using sealing strips or surface-applied tapes (TRADA, 2017) (Fig. 5.6). However, in many cases, the use of an additional airtight vapor-permeable layer is recommended to ensure redundancy of the ABS and for the reasons mentioned above. In all cases, ABS continuity must be ensured through penetrations, windows, doors, and floor interfaces; around corners; and at the connection between walls and roof, parapets, and so on. It must link across dissimilar materials and joints. In case the ABS is locally punctured by connectors and fasteners, its continuity must be restored by local taping. Foils used in an airtight layer must be glued to joints since overlapping is not sufficient.

The ABS should perform for its designed service life, considering all the imposed stresses and material resistance to fatigue, temperature cycles, movements, and so on. (Straube, 2012). It must be structurally strong to resist air pressure forces under full design wind loads, effects of mechanical ventilation systems, or stack effects. The ABS, including the fastening methods, must be stiff enough to avoid deflection due to pumping or deformations. They must also be able to withstand damage due to excessive stretching in the case of hygrothermal movement of the CLT structure.

The ABS vapor permeability requirements depend on the specific use, such as location of the ABS within the building enclosure. It will also rely on the type of assembly and the environmental loads, including seasonal changes. In some assemblies, the products or elements used for water penetration control, like a sheet membrane, can also be used for airflow control.

Combining external insulation with an ABS, placed inboard the insulation layer and outboard the CLT panel, minimizes chances of moisture accumulation in the CLT panels due to either infiltration during hot, humid weather or exfiltration in cold climates. CLT structural connections can interfere with and puncture air barrier membranes installed on the interior face of the panels. Therefore, placing an air barrier on the external side of the CLT panel also has the advantage of making fewer penetrations. In this way, continuity is more easily achieved. While this location makes inspection and repair more difficult, it also protects the air barrier layer from wide temperature ranges and exposure to solar radiation, thus ensuring its service life. External air barriers formed with rigid

FIGURE 5.6 Tapes and gaskets applied to joints, penetrations and connections to provide continuity of the ABS—note CLT is used here as airtight layer. (*Courtesy of Xlam Dolomiti.*)

insulation boards, such as foam plastic, foil-faced boards, or spray foams, have several limitations in CLT enclosures. The main limitation is their impermeability to vapor. Self-adhered vapor-permeable air membranes applied on the external faces of CLT panels have multiple advantages, including fast and easy installation. A separate, vapor-permeable air barrier on the exterior of a fibrous insulation in the façade acts as a wind barrier, reducing the risk of wind washing. This barrier should be water-resistant if moisture-sensitive insulation is used and UV-resistant if the cladding system allows for local UV irradiation. In exterior insulation and finishing systems (EIFSs), plaster functions as a wind-driven barrier.

Tapes, sealants, and gaskets provide continuity of the ABS. The choice may depend on the type of connection, type of surface of application, ambient conditions, and stresses they are subject to. Use of flexible transition strips, such as silicone or EPDM rubber, can accommodate dimensional movements in the wood construction. Since increased airflow control will limit drying in the enclosure, it is fundamental that any source of wetting in a CLT assembly, including built-in moisture, is eliminated once the ABS is installed.

5.2.3 Heat Flow Control

Control of inward/outward heat flow through a building enclosure is important for a variety of reasons: reducing energy consumed, enhancing indoor comfort, and increasing durability of the enclosure by avoiding cold weather condensation and facilitating drying within the enclosure assembly.

In the heating period, the transient heat flow developed as a result of the difference between indoor and outdoor temperatures must be minimized by a building enclosure with the highest possible thermal resistance [or low heat transmittance coefficient U value (Btu/h·ft^2·°F) (W/m^{2}K)]. Therefore, the thermal performance of a building in cold climates depends on the resistive qualities of the enclosure. The contribution of CLT to the control of heat flow is considered in the total calculation of the required insulation thanks to the good thermal resistance of these massive wood panels (see Sec. 5.1).

On the other hand, in hot climates/seasons, thermal performance is achieved by rejecting/extracting the heat or storing it before it can enter in the internal space. In this case, the thermal capacity of the enclosure plays a major role in reducing and delaying by several hours the internal effect of temperature fluctuations outside. Opportunities offered by the inherent thermal mass of CLT panels are discussed later in this section.

Thermal performance is measured differently for each heat flow mechanism: conduction, convection, and radiation. ISO 6946 provides the method of calculation for the thermal resistance (R-value) and thermal transmittance (U-value) of building components and building elements, with the assumption of steady-state conditions. This calculation requires an individual evaluation of the thermal resistance of each layer, obtained as the ratio between the thickness of the layer and its thermal conductivity. The thermal conductivity of CLT depends on the wood used and its moisture content. Twelve percent moisture content (MC) is commonly assumed, as described in Sec. 5.1. CLT panels can significantly contribute to the thermal resistance of the building enclosure assembly depending on the thickness, that is, the number of lamellas. To increase the thermal resistance of an assembly, insulation materials are added, as discussed in the following section. While thermal insulation in timber frame structures is often installed between posts, therefore providing different thermal resistance in elements disposed in parallel, in CLT assemblies, the CLT panel and the insulating layer work "in series," thus minimizing thermal bridging through the enclosure.

A thermal bridge occurs when heat flows at a much higher rate through one part of an assembly than another. This is caused by discontinuities in the thermal barrier, such as when more conductive building materials or elements penetrate the enclosure thermal layer. Thermal bridges can raise durability and health concerns by triggering condensation, which causes mold growth. They can also reduce the thermal performance of the enclosure.

Geometrical thermal bridges are a result of the shape of the thermal envelope and are typical of architectural details such as corners, parapets, balconies, overhangs, intersections between walls and floors or walls and roofs, and window and door perimeters. Some of these thermal bridges can be eliminated with continuous thermal insulation installed outboard the CLT panels. CLT balconies can be built as cantilevered elements or as a continuation of the CLT floor or be separate elements, which are externally supported (e.g., bolt-on, prefinished balconies, which are thermally separated with a thermal break). One of the advantages of the latter balcony type is that air, water, and heat control layers can be more easily continuously laid. In any case, geometrical thermal bridges have a much lower impact on CLT assemblies than in other types of construction due to the high thermal resistance of wood.

Construction thermal bridges are usually highly localized and repetitive along the length of the structure. They are associated with the presence of heterogeneous structural components in the thermal envelope (e.g., framing) and connectors in the assembly. This type of thermal bridge can be minimized or eliminated through careful detailing (Brambilla et al., 2018). In CLT-based enclosures, continuity of the structural element, namely, the panel, reduces the presence of construction thermal bridges. In hybrid CLT buildings, however, the incidence of thermal bridges can be higher than in all-wood construction; concrete or steel structures penetrating the thermal envelope will need a structural thermal break component.

Potential risk of point thermal bridging is found, for instance, at the cladding connections. To avoid this, cladding can be attached to the furring with short screws, and then the strapping can be screwed through the insulation. Thermally improved clips, long screws, or strapping with thermal breaks can also be used for this purpose (Fig. 5.7).

Mass that functions as storage absorbs heat in higher ambient temperatures and releases it when temperatures are lower, limiting temperature fluctuations in interiors. The benefits of using CLT as a thermal mass to store and release heat have not yet been studied extensively. Findings reported by Glass et al. (2013) show that the effectiveness of thermal mass varies depending on climate, solar heat gains, and internal heat generation in the building. Thermal mass can be advantageous, especially in mixed climates and during seasons when outdoor temperature fluctuates above and below indoor temperature with cycles of 24 hours, by reducing heating and cooling peak loads and shifting the time of peak loads (Glass et al., 2013).

The effectiveness of CLT panels acting as thermal mass, both in the building enclosure and in interior floors/ceilings and walls, depends on the density, amount,

FIGURE 5.7 Examples of thermally improved strapping to avoid thermal bridging at the cladding connections (top, *Courtesy of Xlam Dolomiti*); (bottom) thermally improved strapping is used also to support the semirigid insulation. (*Courtesy of Andersen Construction.*)

location, and exposed surface area of the mass. It is also dependent on the rate of heat transfer into the mass. In general, benefits of thermal mass are limited for multistory, highly insulated buildings. Indeed, thermal mass has a disproportionately lower influence on overheating compared to the importance of shading and night flushing (Kaufmann et al., 2018).

Heat Flow Control Layers in CLT Assemblies: Materials and Systems

Building insulation materials can be classified into three main families: synthetic, mineral, and organic products. From the hygrothermal point of view, the choice of insulation(s) should be guided by the following criteria beyond thermal performance: vapor permeance, water tolerance, durability, and rigidity. These criteria should be considered in relation to vapor permeance and airtightness of the other layers in the assemblies to ensure durability.

The use of insulation products with low vapor permeance may slow down drying in case of incidental wetting during construction and in service. They can also interfere with vapor flow through the assembly, causing an accumulation of moisture and condensation. Low permeance and therefore risk of moisture trapping is one of the reasons why synthetic insulation materials are less suitable for most CLT assemblies. On the other hand, moisture-resistant insulation, such as extruded polystyrene, may be an appropriate solution in some building areas. For instance, it is used to protect a wall assembly against water spray 12 to 20 in. above ground level or in other areas vulnerable to water intrusion, such as terraces and inverted roofs. Some natural insulations can be highly compatible with CLT assemblies as a result of their breathability. Examples include wood fiber panels (varying densities and thermal properties and the need to be protected by a WRB) and cork panels (examples in Benedetti, 2010).

Semirigid fibrous insulation boards (e.g., Rockwoll) can be an appropriate solution for CLT assemblies, provided that the CLT panel provides rigid support. The relative flexibility of the insulation has the advantage of (1) accommodating differential movement in the assembly and (2) maintaining continuity at transitions in the enclosure. It is worth mentioning that, even for moisture-resistant materials, any water absorbed by the insulation will reduce its thermal resistance.

The CLT Handbook provides examples of insulation thicknesses for CLT building assemblies (Glass et al., 2013). It is worth mentioning that both insulation thickness and type that are optimal for cold climates are not always equally effective in protecting against heat in warmer climates (Benedetti, 2010).

The placement of insulation plays a key role in the hygrothermal performance and energy efficiency of a building enclosure assembly. CLT enclosure assemblies should be always insulated from the exterior. External insulation provides a high level of decrement delay that protects the CLT structure and the ABS from temperature extremes. This reduces hygroscopic movement of the panels as well as the risk associated with moisture accumulation. It also facilitates the construction of a continuous insulation layer, limiting thermal bridging at corners and at some types of intersections. An additional external layer, such as wood fiberboard panels with a water-resistant surface, is sometimes added to the exterior insulation, if allowed from a fire safety perspective. This is able to act as a windproofing/second water-bearing layer and protects the fibrous insulation.

In some cases, split insulation (external and internal) is used to fill in the gaps of furred CLT walls. The correct ratio and type of inboard and outboard insulation are critical to limit the risk of interstitial condensation (Straube, 2012). As a rule of thumb, the proportion of insulation on the interior side of a vapor control layer (e.g., a CLT panel) should not exceed an equivalent thickness of approximately 20% of the total thickness of the insulation, including the contribution of the CLT panel, to avoid the risk of condensation (Kaufmann et al., 2018).

Investigations have demonstrated the advantages of ventilated façade assemblies versus so-called EIFSs in preventing indoor summer overheating. The main benefit provided is the increase of the phase shift of the thermal wave thanks to both the thermal mass of the screen layer and the ventilated cavity effect (Salvalai, 2018). A similar benefit is provided by ventilated roofs (Ringhofer & Schickhofer, 2014).

5.2.4 Vapor Diffusion Control and Relationship of Air, Vapor, and Thermal Control

Vapor diffusion is the movement of water vapor molecules caused by the thermal self-motility (Brownian motion) of these particles. Vapor diffusion moves in porous building components containing air pockets from areas of more to less vapor concentration and, according to the drop in temperature, from warm to cold areas.

Vapor diffusion can occasionally cause damage due to excessive wetting. However, it is also an important drying mechanism. Therefore, the sequence and respective vapor diffusion properties of layers in a building enclosure should be carefully considered to maximize the drying capability of incidentally wet materials in the assembly. This should be done without making the assembly vulnerable to condensation.

Condensation control depends on a combination of air, heat, and vapor control. In fact, the variables that affect condensation include the temperature of the air, the temperature of the surfaces, and the amount of water vapor in the air. Condensation occurs because the capacity of air to store water vapor decreases rapidly as the temperature drops. The dew point temperature measures the temperature at which air cannot hold additional moisture. Moisture condenses on surfaces that are colder than the surrounding air dew point temperature.

Vapor Control Layers and Condensation Control in CLT Buildings

The vapor control layer consists of a material or materials that retard (vapor retarder) or stops (vapor barrier) the diffusion of water vapor in enclosure assemblies. ASTM E96 defines three classes of vapor permeance of vapor control layer: Class I—0.1 perm or less, Class II—0.1 < perm <1.0 perm, and Class III—1.0 < perm <10 perm. Continuity of the vapor retarder is not necessary in most cases, unlike the air barrier. Different components of the enclosure, such as walls and roofs, generally have different vapor control requirements. Components in timber building enclosures should be as vapor-permeable as possible (e.g., in walls) and as minimally vaporproof as necessary (e.g., in roofs).

The vapor diffusion-resistance factor (μ) describes the vapor resistance of a building material as compared to the resistance of air. A high μ-value indicates a greater vapor diffusion resistance. The μ-value varies with the moisture content of a material; for many materials, it drops as moisture content rises. The Sd-value is often used to express the vapor resistance of a material in terms of "equivalent air layer thickness." Membranes can be classified accordingly from highly breathable to vapor stop barriers.

Polymer-based membranes used as vapor control layers and barriers can be of two main types: monolithic and microporous products (see, for instance, Rothoblaas, 2017). The first category includes highly performing membranes whose vapor permeability depends on the chemical formulation of the functional layer, and it is activated through a chemical reaction between water vapor molecules and the used polymers. In microporous products, the physical structure of the membrane and, specifically, the presence of micropores in the vapor check film define the permeance of the product. In addition to conventional vapor barriers with a constant diffusion resistance, vapor barriers with a variable diffusion resistance have appeared on the market in recent years.

Determination of layer sequence in the enclosure assembly should be optimized for the planned occupancy and climatic conditions. As a general rule, vapor control layers are located inboard the insulation on the "warm side," or high pressure side, of the assembly. This helps protect from wetting due to condensation. However, the direction of vapor flow will often change between summer and winter. The use and location of any vapor control layer must carefully account for the risks associated with the variation of external and internal hygrothermal loads (Gasparri et al., 2018).

During the winter, hot air migrates to the outside where partial pressure is lower, and, if it finds points of discontinuity, higher vapor flow occurs through convection. This causes condensation at the cooler outer layers. In a cold climate/season, a well-constructed layer composition gradually decreases vapor resistance toward the outer layers, allowing excess water vapor to dissipate. This thereby creates an equilibrium with the outdoor conditions. Insulating on the exterior of moisture-sensitive materials, minimizing thermal bridges, and avoiding air leakage are fundamental strategies to avoid interstitial condensation.

The relatively low vapor permeability of solid wood and the massive nature of CLT panels control the flow of vapor through the assembly in most situations. In fact, the use of an additional vapor diffusion control layer in CLT assemblies is not always recommended and could increase the risk of condensation in some situations. Therefore, it should be evaluated accordingly during the design phase (Alsayegh et al., 2013). CLT walls are expected to perform well in cold climates without the addition of vapor retarders when insulation is to the exterior (Lepage, 2012; Glass et al., 2013; McClung et al., 2014). During the summer, hot air is pushed from the outside inward through inverse diffusion. To contrast high outdoor vapor drive in hot, humid climates, it is critical that indoor drying is not hindered; examples of layer sequences for this type of climates are reported, for instance, in Goto et al. [2011 (reported in Glass et al., 2013)]. In any case, to allow for a breathable assembly, vapor barriers/retarders should never be placed on both sides of the assemblies.

Vapor barriers with a variable diffusion resistance can be a solution for optimizing vapor and condensation control both in cold and in hot, humid climates. They promote drying since their resistance to vapor diffusion decreases with increasing humidity levels. They can be coupled with CLT panels, which themselves can be considered a natural vapor barrier with variable diffusion resistance (Glass et al., 2013; Kaufmann et al., 2018). This beneficial property should be taken into account in the design of the enclosure assembly.

5.2.5 Assembly Details

Figures 5.8 and 5.9 schematically illustrate the sequence of hygrothermal functional layers of some commonly used ventilated rainscreen wall assemblies and roof assemblies, respectively.

Figures 5.10 to 5.15 illustrate potential solutions for detail connections of CLT assemblies in the building envelope.

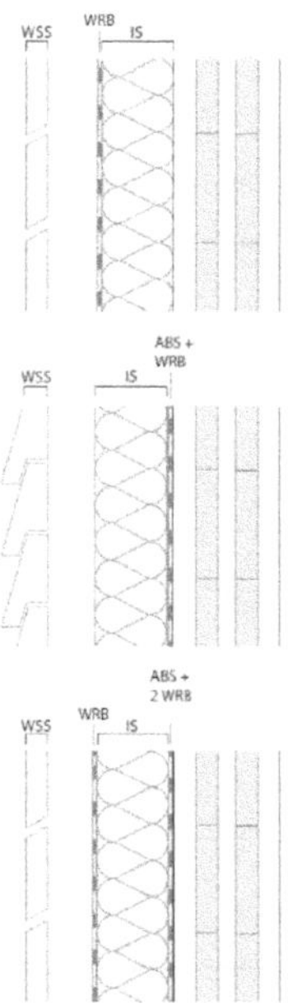

VENTILATED RAINSCREEN WALL ASSEMBLIES (sequence from the exterior layer)—vapor-permeable interior lining not illustrated

WSS. Open joint cladding;

Ventilation gap

WRB. Flame-retardant, permanent UV resistant, vapor-permeable, windproofing, water-resistive barrier

IS. thermal insulation;

CLT wall, here exposed as a vapor control layer and airtight layer (only in case of certified airtight CLT, with edge-glued lamellas—continuity of the ABS provided by sealants, tapes, and gaskets, as required)

WSS. Close joint cladding;

Ventilation gap;

IS. thermal insulation (Note: if possible, protect the insulation with a windproofing layer also in this case—no need of permanent UV resistance in case of close joint cladding)

ABS + WRB. vapor-permeable (Sd < 0.1 m), airtight, water-resistive barrier

CLT

WSS. Open joint cladding;

Ventilation gap

WRB. Flame-retardant, permanent UV resistant, vapor-permeable, windproofing, water-resistive barrier

IS. thermal insulation;

ABS + WRB. vapor-permeable (Sd < 0.1 m), airtight, water-resistive barrier

CLT

FIGURE 5.8 Examples of rainscreen wall assemblies. (*Adapted from Kaufmann et al., 2018.*)

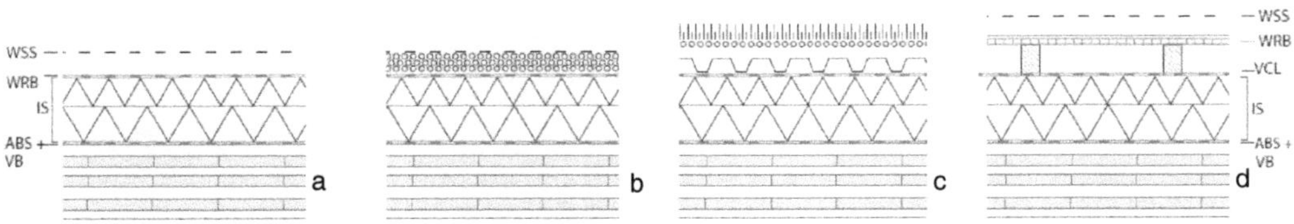

FLAT ROOF ASSEMBLIES (sequence from the top) – optional interior lining not illustrated

WSS. Protective layer (detail b- ballast on geotex mat), (detail c- green roof with root resistant barrier, drainage layer, filter sheet, protection mat and substrate);

WRB. Wind-tight, sealing layer –Note: UV and thermal resistance up to 250 degree F for use in layers subject to high temperatures or accumulation of heat

Rear ventilation layer with vapor-permeable underlay (only in detail d);

IS. Compression-resistant, slopped, thermal insulation;

ABS + VB. moisture-variable vapor barrier/airtight layer;

CLT slab

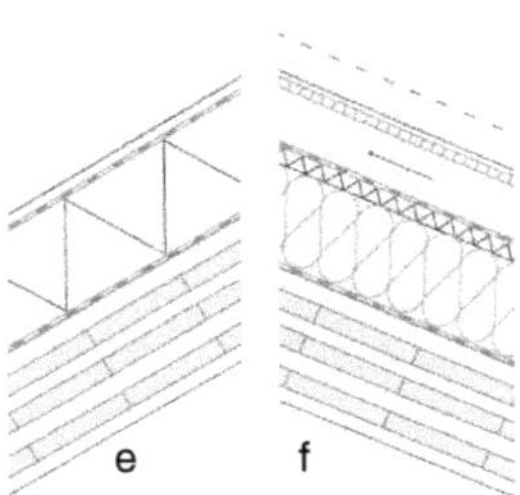

SLOPED ROOF ASSEMBLIES (sequence from the top)

WSS. Roof covering

WRB. Sealing layer – Note: UV and thermal resistance up to 250 degree F for use in layers subject to high temperatures or accumulation of heat

Rear ventilation layer with vapor-permeable underlay (only in detail f);

IS. Compression-resistant, thermal insulation;

ABS + VB. Vapor barrier/airtight layer;

CLT slab

FIGURE 5.9 Examples of roof assemblies. (Source: *Adapted from Kaufmann et al., 2018.*)

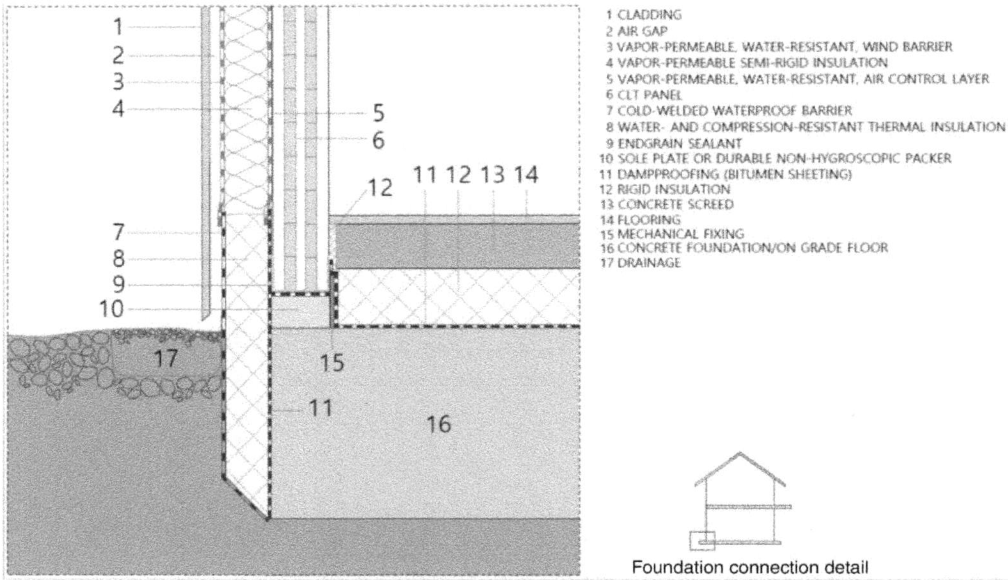

FIGURE 5.10 Example of a wall-to-base slab connection (case of not airtight CLT panel).

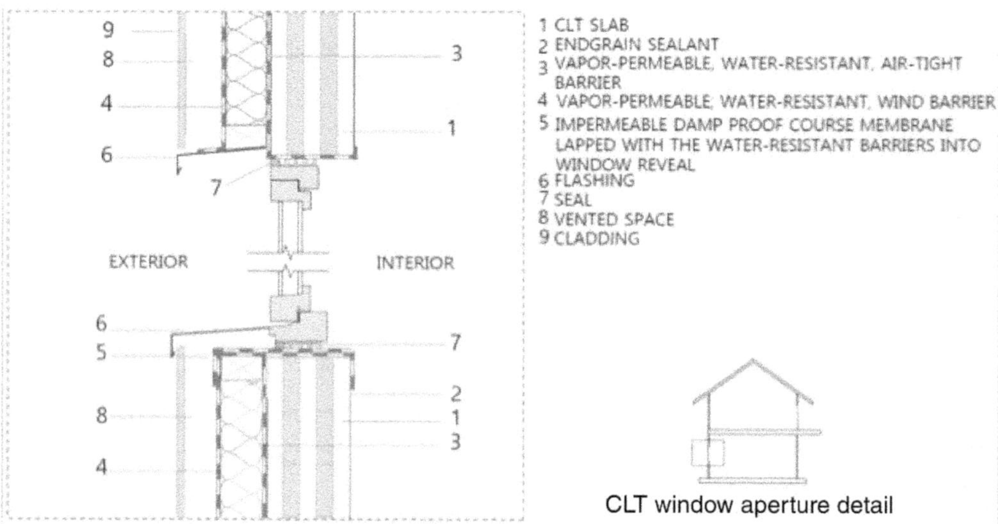

FIGURE 5.11 Example of a window aperture detail (case of not airtight CLT panel).

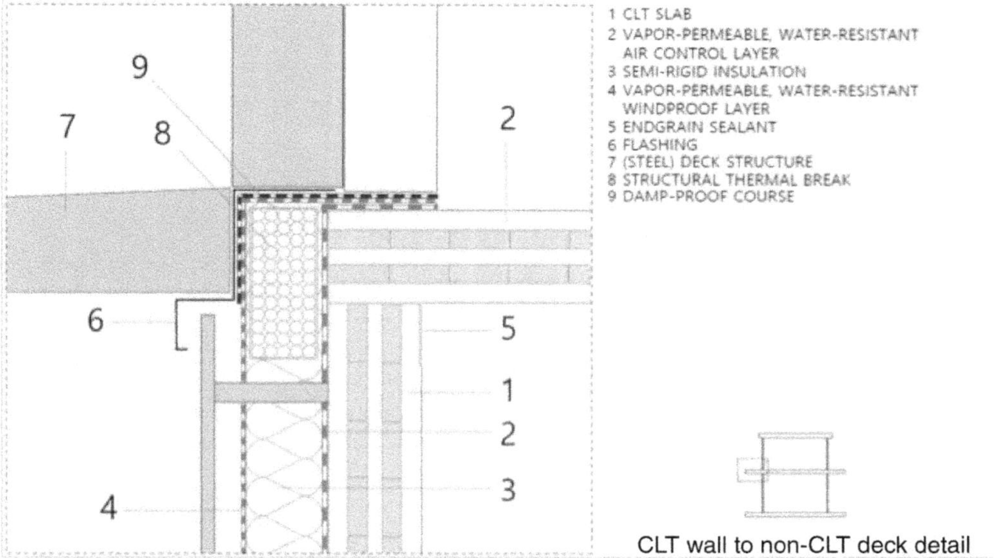

FIGURE 5.12 Example of a bolt-on deck detail (case of not airtight CLT panel).

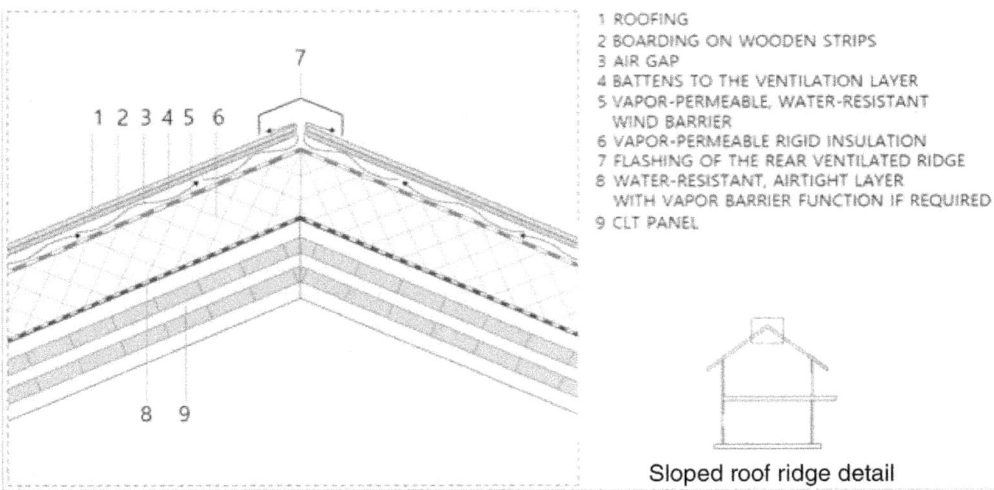

FIGURE 5.13 Example of a ventilated roof ridge detail.

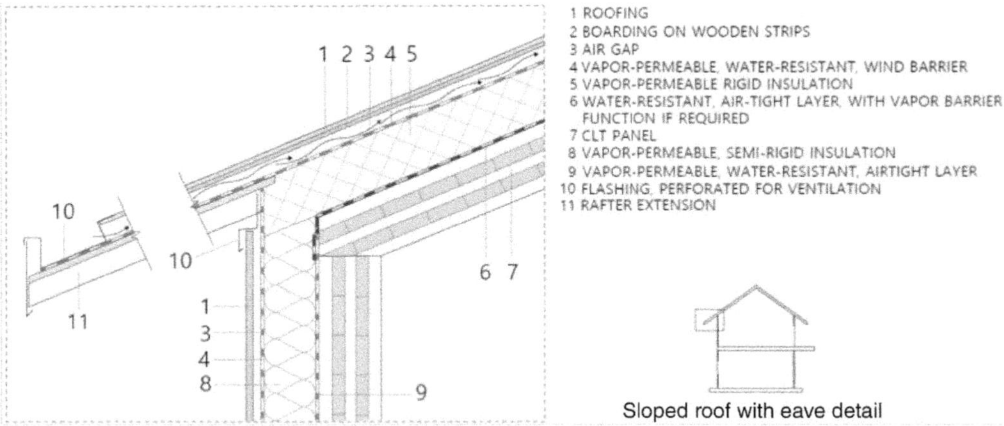

FIGURE 5.14 Example of a ventilated roof eave detail.

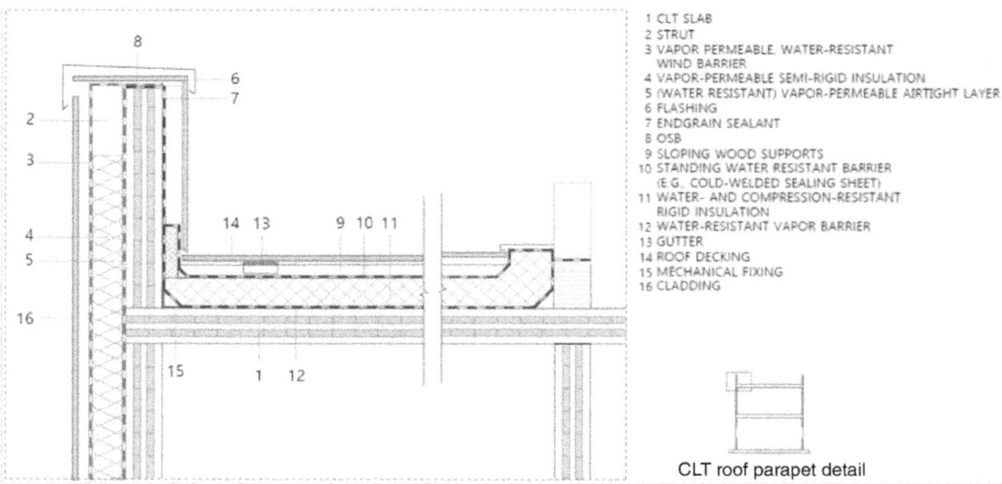

FIGURE 5.15 Example of a deck and parapet detail.

5.3 Accounting for Building Movement

Moisture-induced dimensional changes of wood and of CLT panels specifically have been discussed in Sec. 5.1. Designs should take into account differential movement between parts of a mass timber building to prevent potential negative effects on structural integrity, enclosure performance, and serviceability. As a general rule, connection design and detailing should accommodate instead of restrain differential movement. In fact, restrained movements may cause internal stresses and damage, such as checking and splitting in the wood, which could undermine a connection and member capacity.

Differential movement may occur between different wood components or between wood and other materials, such as concrete and steel. When CLT is used in combination with other wood products, it is important to consider the different dimensional stability of the specific products according to their grain/lamination directions. As an example, published values for within-plane dimensional movements of wood-based panel products are reported by Carll and Wiedenhoeft (2009). The seasonal MC changes in service are in general much smaller than from initial construction; however, in some climates, seasonal fluctuations of RH levels cause several percent MC changes.

Downward movement is the predominant type of cumulative dimensional change in tall wood buildings. This is a result of a combination of shrinkage due to in-service drying, gap closure, and both instantaneous and time-dependent (i.e., creep) shortening due to the applied loads (Gasparri et al., 2016). Depending on the construction type, downward movement and the combined effects of the above-mentioned factors can vary. For instance, it is more pronounced if CLT panels are used in a platform system. Contributing factors are the transverse shrinkage of the CLT floor panels, which is higher than the in-plane shrinkage (Sec. 5.1), and, if the floor panels are directly loaded from the vertical systems above, compression across the grain, which leads to higher deformation than along the grain (Brandner, 2018). Deformations caused by compression perpendicular to grain are more pronounced in circumstances of high MC or large MC fluctuations (Brandner, 2018). This is also the case for long-term deformations under sustained loads, whose rates increase with increasing or alternating moisture (Muszyński et al., 2005). This effect is well known as mechano-sorptive creep (Grossman, 1976).

Potential risk of high deflections, due to a combination of elastic, viscoelastic, and mechano-sorptive deformation, in large-span, low-slope roofs and green roofs should be considered in the design. To avoid sagging and consequent water ponding, the roof load-bearing system should be stiff enough to provide robustness against creep effects. Concurrently, RH and MC values should be kept within a safe range to avoid these effects.

Cumulative vertical movement should be accounted for in the design of the building enclosure, such as by including flexible airtight and watertight joints in connected elements of the façade assembly, in order to maintain the continuity of the control layers. Similarly, flexible connections for piping and provisions in the mechanical systems should be included in the detailing to allow for movements.

If posttensioned systems are used in combination with CLT panels, the effects of various factors influencing and/or influenced by dimensional movement, such as the potential tension loss in the tendons, should be carefully taken into account when considering long-term performance of the system (Nguyen et al., 2019).

5.4 Durability of CLT Buildings: A Holistic Approach

Though all materials can be degraded by mechanisms triggered by environmental loads, there are unique considerations for wooden structures. These are related to the specific chemical/physical properties of the wood material (discussed in Sec. 5.1), coupled with the fact that persistent high moisture content in the material is an essential element for all biological agents of wood (Nore et al., 2014; Wang et al., 2018). CLT presents some peculiar characteristics, making its hygrothermal behavior different from that of traditional light-frame systems, as discussed in Sec. 5.1. In this section, considerations related to durability of CLT systems are therefore presented, specifically considering hazards associated with moisture loads.

In order to design and assess the durability of a building, concepts of limit states can be applied, similarly to structural design. In structural engineering, loads and load resistance are considered, and limiting conditions, such as deflections, are specified. Similarly, environmental loads and systematic resistance to these loads should be defined to ensure that designed loads do not exceed the system resistance as related to specific limiting conditions, such as rot and mold (Morris & Winandy, 2002; Norén et al., 2017). To achieve the expected performance, two complementary approaches can be adopted: (1) defensive strategies to reduce the loads by applying the 4-Ds rule described in Sec. 5.2 to control moisture entry (distance and deflection), limit moisture accumulation (drainage), and remove accumulated moisture (drying) and (2) to increase the resistance, specifically the durability of the material.

The environmental loads acting on a building are coming both from the outdoor environment (e.g., wind, external temperature, solar irradiation, precipitation, groundwater, and outside humidity) and from indoor sources (e.g., heat, water vapor, air pressure difference due to internal moisture loads, building use, and occupant behavior). Environmental loads are influenced by climate modifiers, such as trees and overhangs. Consequently, different parts of the enclosure should be detailed to take into consideration the different exterior microclimates, that is, the portion of the exterior environment that is close to and affected by the building itself. Factors such as the location or the climate zone in which the building is constructed, as well as the nature of its occupancy, will dictate the type, magnitude, duration, and frequency of the environmental loads.

Longer construction times can expose wood on-site to the elements for extended periods of time, potentially resulting in excess water absorption. Once hazards due to environmental loads are recognized, defensive design strategies can be adopted to minimize the associated risks.

In order to increase a system's resistance to an action, that is, increase the resistance of a CLT building against given hygrothermal loads, some general design principles should be taken into consideration: (1) provide redundancy in the system, such as a second water control layer in the building enclosure, to avoid wetting of a CLT assembly in case of leakages in the primary control layer; (2) ensure robustness, namely, the ability of a system to recover from unexpected conditions, such as ensuring that a CLT assembly can dry out if accidentally wetted; and (3) design tolerance, for instance, allowing for movements in a building to accommodate for dimensional changes in the timber.

Effective design for durability holistically considers upstream and downstream factors affecting the final product, namely, manufacturing, construction, and maintenance. Manufacturing variables can influence the durability of CLT panels in various ways.

Factors such as wood species, lumber quality, layup, and bonding method determine the type and presence of discontinuities within and between lamellas. Checks, wane, and inter-board gaps can facilitate moisture penetration and entrapment on construction sites. Also, checking could be reduced by using boards with the same moisture content. This helps to avoid internal stresses due to differential, constrained dimensional changes in the alternating, orthogonal directions. Moisture resistance also varies among species, thus influencing the way and rate CLT panels take upliquid water and dry out (Van Acker et al., 2014).

North American species typically used in CLT laminations have low to moderate natural resistance to fungal and termite attack (Wang et al., 2018). Additional protective measures may be needed for some applications, in wet/humid climates, or in markets with severe termite risk. Chemical treatments are available to improve the durability of wood, but their use is still not widespread in CLT systems. Conventional pressure treatment is not feasible for CLT due to the size of the panels. While oil-borne products used for exterior and industrial applications are not suitable for CLT panels, some water-borne preservatives are used for pressure treatment of the lamina prior to manufacture of the panel in some regions with high risk of insect attack (AS/NZ 1604.5). Borate-treated lamina has shown poorer bonding performance than untreated wood, and more research is needed to formulate optimized species, treatment, adhesive, and processing combinations (Stirling & Morris, 2017). Boron or boron/copper rods can be used in locations where there is a high and predictable risk of moisture ingress. These locations should be accessible for easy inspection and maintenance. Some authors recommend treating wood surfaces in contact with the rods with borate or a film-forming coating to prevent leaching (Glass et al., 2013). Film-forming coating should not be applied in the area close to the rod since that will stop diffusion.

The use of surface-applied penetrating treatments has been deemed ineffective against decay due to the superficial penetration (Stirling & Morris, 2017). However, many of them can be considered as a measure against mold. In this regard, some boron-based fire retardants can provide also some resistance against biotic attack decay and insect attack. New processes, such as dip diffusion, glue line additives, and low-uptake sprays, have been recently proposed as viable treatment strategies for CLT panels (Wang et al., 2018).

Physical barriers designed to protect CLT from moisture intrusion and prevent fungal attack are receiving considerable interest (Wang et al., 2018). These barriers can include paint films; nonporous, water repellent, vapor-permeable coatings (Lepage et al., 2017); and membranes. Proper application, careful handling of panels during installation, and maintenance are key for the effectiveness of these approaches.

When using preservative-treated wood, corrosion-resistant metal fasteners, including nuts and washers, should be specified. The International Building Code (IBC) gives provisions for stainless steel, hot-dipped zinc-coated galvanized steel, silicon, bronze, or copper (IBC, 2018). Fastener coating types and weights should be chosen in accordance with the treated wood and connector manufacturer's recommendations. Compatibility of wood treatments with building enclosure membranes, adhesives, and sealants should also be verified.

The durability of materials in a system should be planned in accordance with the level of deterioration risk. This includes considerations of the type of hazard, level, and time of exposure to a certain environmental load both during construction and in service and of the performance required for a specific component.

Different levels of durability and service life can be attributed to different elements in a system. As a result, certain components, namely, those with lower performance requirements and a higher risk of deterioration, are designed to be easily replaced after reaching their planned service life without affecting the durability of the whole system. This role of "sacrificial" elements was traditionally attributed to elements such as corbels and wall plates (distance elements) and fins (deflection). In CLT structures, this "selective durability" attribution could be applied, for instance, in the design of balconies, canopies, and other more exposed enclosure elements. Design for durability should in any case facilitate and plan for maintenance (see also Sec. 5.5).

Poor installation alone is one of the major causes of leakage in building enclosures for any type of construction (Norén et al., 2017). Shortcomings during some critical installation phases can be avoided by proper detailing and by specifying installation sequence. Installation/construction sequence should be planned, whenever possible, during the design phase to avoid moisture traps in areas temporarily exposed during construction. Moisture trapping connections should be avoided both in service and during construction.

Prefabrication can be an effective way of limiting the exposure of CLT panels and other elements during construction (Fig. 5.16). The installation of wall and roof panels, which have already been sealed to be watertight following the erection of the wood

FIGURE 5.16 Prefabrication of CLT systems. (Top) Preassembled windows and sealant tapes allow installing parts of the WRB and ABS in a controlled environment. (*Top and middle photos, Author Giulio Boem, Courtesy of Essepi.*) (Bottom) Preassembled self-adhered membrane to protect end grain of base wall panels. (*Courtesy of Brad Nile, Andersen Construction.*)

structure, can be an excellent strategy for reducing the exposure of the internal assemblies to external environmental loads. This can be particularly relevant for large or tall buildings (Gasparri et al., 2015, 2019).

Prefabricated enclosure components need to be somewhat resistant to moisture, as they can be exposed to inclement weather during construction. The enclosure details should also account for potential vertical expansion/contraction and lateral drift of the wood structure (Finch, 2016). Key factors in reducing the risk of wetting and deterioration are material choice, thoughtful design of interfaces and joints for ease of sealing, and preinstallation of windows. Water-absorptive materials, such as insulation, should be adequately protected in the prefabricated assembly to avoid accidental wetting. The use of prefabricated wall systems does not necessary prevent other parts of the construction from wetting. For instance, it may be wetted from above since rainwater might still percolate through the seams of CLT floor panels.

Prefabricated CLT floor modules incorporating a precast concrete screed have proven effective in some projects, such as the LCA tower in Austria (Kaufmann, 2016). Low-permeance materials included in the prefabricated assembly can retard drying of CLT panels accidentally wetted during construction. Therefore, holistic on-site moisture management, considering all possible ways of water intrusion, also becomes important when using prefabricated assemblies. This is especially the case in wet climates.

It is worth mentioning that while guidelines exist for on-site moisture management of timber construction in general (e.g., Wang, 2016b) and CLT specifically (STA Advice Note 14, Part 4), the opposite problem, that is, protection of CLT during construction in very hot and dry climates, is still not adequately covered. High external temperatures and low air RH can lead to excessive drying of wood, causing unwarranted checking and, potentially, warping of the panel. Prefabrication of the enclosure can be a solution to this problem by thermally insulating the structure during construction.

Independent from the choice to use prefabrication processes, it is crucial to maintain the MC of CLT panels as closely as possible to the manufactured MC/in-service EMC during shipping, storage, and construction. CLT MC at the time of manufacture should be 12% ± 3% (APA PRG320). Higher humidity during construction can increase these values. While this is sometimes unavoidable, prolonged direct exposure to liquid water should be prevented or at least limited.

Factors to consider while planning appropriate on-site protection include the climatic conditions at the construction site, including local climate, microclimate, and possible variable exposure due to predominant solar and wind orientation; schedule constraints, such as seasonal weather in the different construction phases; wetting and drying ability of the materials and assemblies; risks associated with exposure to elements, such as discoloration, mold, and checking, and the impacts on specific building components (floors, walls, etc.); and costs of protection measures versus costs of potential remediation, like forced drying, resurfacing, repair, replacement, and so on.

Planning construction schedules so that the erection of CLT panels occurs during the dry season is the simplest preventive strategy. This strategy can be optimized by sequencing construction phases and installations to minimize exposure of structures to the elements.

Shop-applied vapor-permeable sealants on all the exposed end grains, like panel edges, connections, cutouts, and notches, as well as coating of the panel faces, may make the surfaces water-repellent and reduce water absorption for short exposure periods. However, they can also reduce dryability of CLT. The use of coatings on the upper face of CLT floor panels may be inadequate for two main reasons: it does not prevent water from entering through discontinuities in the exposed faces (e.g., wane, gaps between boards), and it may reduce bonding of the concrete topping (as reported in the case of Brock Commons—see Naturally:wood, 2018).

In the case of protection using physical barriers, it must also be considered that improper handling on-site or dimensional changes of the panels can damage the barriers, thus altering the wetting/drying capacity of the panel. Attention should also be paid to avoiding membrane damage by connectors.

Strategies for moisture protection should be implemented during transportation; each lift of CLT should be delivered in a waterproof, breathable package, and a tarp should be added to protect the load against precipitation and accumulation of debris during transportation (Wang, 2016b). Whenever possible, CLT panel delivery should be coordinated for just-in-time installation.

Distance, deflection, drainage, and drying solutions should also be implemented for water management during storage and construction. To avoid capillary absorption, CLT panels should be stored off the ground on-site by placing panels on stickers with adequate clearance to allow for air to flow beneath the packages. Any water accidentally trapped in the wraps during handling should be drained.

Spatial orientation of CLT members within a construction system can affect susceptibility to water ingress and distribution. The top and bottom edges of CLT walls are the most vulnerable areas. Gravity can transfer precipitation entering from the top edge further into the panel; bottom edges can wick moisture via capillary action if the end grain is not physically separated from other wet surfaces. Because of the presence of macroscopic gaps in the lamellas edge, sealing alone may be insufficient for protecting CLT walls from gravity-driven moisture.

Deflection is the first and most effective water exclusion strategy. Entire canopy structures were temporarily erected over the buildings during construction in projects like Limnologen in Sweden (Serrano, 2009) or the ETH House of Natural Resources in Switzerland (Leyder & Chatzi, 2015). This led to a drier and more efficient construction site. An example of deflection using a canopy structure is presented in Fig. 5.17 (top).

Partial deflection can be provided by temporary covers (Fig. 5.17—bottom) or preinstalling temporary or permanent protective membranes on an assembly. These latter materials tend to reduce dry ability of CLT, implying that (1) the panels must be factory dry at the time of application (2) any preinstalled membrane should be perfectly lapped on-site and sealed at penetrations to avoid accidental wetting. Tape on joints between panels and carefully sealing at penetrations and membrane laps create continuity of the WSS. These measures allow floors and walls to be built as self-sheltering elements. Openings in external walls should be temporary closed by plywood sheets or a robust plastic layer to protect the structure from wind-driven rain.

If, despite providing a constructive shelter, water leaks in, it should be quickly removed. Similarly, snow should be removed as soon as possible, ideally before it melts and becomes a moisture source. Temporary drainage should be arranged in any locations where water could accidentally accumulate, such as floors, around staircases, and larger holes. When perimeter drainage is not possible in flat slabs, drain holes should be provided in other locations. Water diversion should be provided at joints and other sensitive areas by gutters. With on-site installation of windows, openings cutouts should be protected. For instance, plywood boxes can be used around windows, and internal components can be protected from wind-driven rain by temporarily covering the window with a plywood panel or other barrier.

FIGURE 5.17 On-site water control: (top) canopy used to deflect precipitation from the whole structure (*Courtesy of L. Muszyński*); (bottom) temporary covers used for partial deflection (*Courtesy of Xlam Dolomiti*).

If despite these precautions CLT elements get wet, the source of excessive moisture influx has to be identified and removed before moving to the drying procedure. In order to prevent drying damage (e.g., excessive checks on visible surfaces or distortions of elements before the assembly), the drying procedure should be carefully planned, and, as much as possible, controlled drying conditions should be provided (controlled or monitored temperature, RH, and airflow rate in the dried sections of the building). This may be achieved by dividing large buildings into separate sections, where the environment may be easier to control. In addition, the moisture content of the drying elements should be measured regularly at different locations and depths of CLT panels using a resistance-type moisture meter (Sec. 5.5). The aim is to provide conditions for gradual removal of the excess moisture back to the standard level of 12% ± 3%). Rapid drying of CLT should be avoided.

The best practice is to design gradual or staged drying procedures providing mild drying conditions that keep up with the changing moisture content of the drying elements. It is generally considered safe to raise the temperature slowly from the ambient outside air temperature to the target interior temperature at a rate of roughly 3°F to 5°F per day during controlled drying. Heat can be added by an outdoor gas-fired furnace with a heat exchanger and exterior combustion exhaust. The interior RH is typically maintained by a heating, ventilating, and air-conditioning (HVAC) system and finished buildings is about 55% with relatively wide tolerances (about ±10%) in areas where only the temperature is actively controlled. The target RH ramp for the drying procedures should be established based on the target equilibrium wood moisture content using sorption isotherms, as described in Fig. 5.18 (see FPL, 2010 Chapter 4, for practical applications). Air humidity during the drying procedure may to some extent be controlled by using portable humidification units, when needed, to reach the target humidity gradually. It should also be kept in mind that the airflow may substantially affect the pace of drying. Faster airflows will accelerate the drying process for the same temperature and RH parameters of the air.

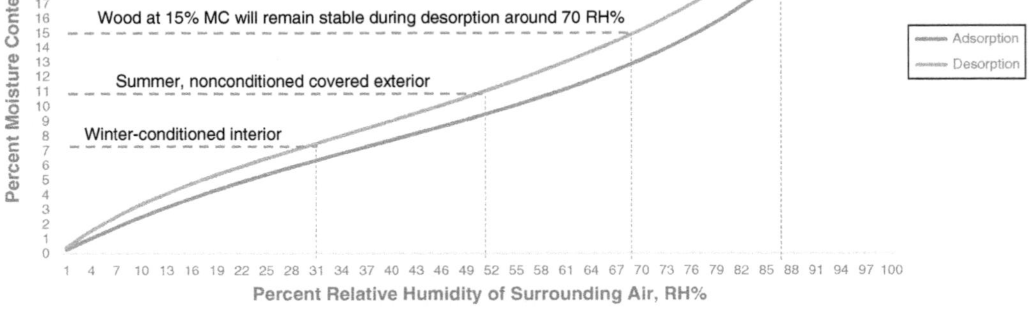

FIGURE 5.18 Sorption isotherm at a temperature of 20°C (68°F). (Based on Equation 4-5 in the Wood Handbook, FPL, 2010, Chapter 4, page 4-3. (*Courtesy of E. Schmidt.*)

It is worth mentioning that different materials/products and components in a timber system can have different dry rates. For example, interior plywood splines used to join CLT panels will take longer to dry due to limited airflow (Naturally:wood, 2018). Timber elements should only be enclosed once all timber elements reach 14% MC or below.

5.5 Hygrothermal Performance of CLT Buildings: Control and Maintenance

5.5.1 Maintenance and Inspection Plan

Maintenance needs to be considered at the design stage. Periodic inspection and maintenance intervals should be planned according to service life of each component. High-risk areas, such as openings and penetrations in the roof and decks, should be prioritized in the inspection plan. Water stagnation on the building enclosure can be prevented by simple measures such as regular (e.g., twice a year) inspection and cleaning of the roof drainage system. Large-span roofs should be checked after exceptional/long-duration snow loads or periodically, in case of high permanent loads, against possible deflection, which can cause water ponding.

Details of water-bearing pipes should be designed so that leaks can be quickly discovered. Regular maintenance of elastic joints and sealants is required in areas where small amounts of water can leak out over a long period. Examples of such areas are around apertures for taps, grouting of tiles, and joints around shower trays and bathtubs.

When fire engineering allows, exposed panel faces provide for easier inspection, especially in case of a leakage. In addition, easy accessibility is fundamental in facilitating reapplication of coatings and other surface treatments, when required. The integration of a monitoring system can facilitate inspection tasks, providing some continuous data on the conditions, such as wood moisture content and water leaks, at the most vulnerable locations.

Measurements, like wood moisture content, can be taken discontinuously (manually) or continuously (automatically, using permanently installed sensors). The most convenient and sustainable approach in the long term should be decided on during the design phase and inform the maintenance plan.

Methods to measure wood moisture content and systems for detecting moisture, heat, and air leakages are explored in the following sections.

5.5.2 Wood MC Measurement

Moisture content measurements with nondestructive methods, like resistance or capacitance methods, only depict local conditions. Moisture distribution can vary substantially, even in close locations. A large number of permanent sensors would be ideal for depicting this variability; however, this is not always possible for technical, aesthetic, and economic reasons. Some sampling criteria can therefore guide selection of MC measurement locations. Measurement points should be selected considering the anticipated vulnerability of a location to exposure and/or moisture trapping during both construction and in service.

Vulnerability should be considered at three scales: (1) building system (position, aspect, exposure, structural role, etc.); (2) assembly/detail, including the potential for moisture trapping, detail complexity, waterproofing method, and orientation in space; and (3) the material/subassembly scale, such as at the upper layers of floor/roof panels and end grain at the panel edges (Schmidt & Riggio, 2019; Schmidt et al., 2019).

The most common approaches to estimate MC are based on the relationships between moisture content and electric properties of wood (FPL, 2010). The resistance (or conductance) method and the capacitance method are the two possible approaches. They use the relationship between MC and the direct-current conductance or the dielectric properties of wood, respectively.

An alternative approach to the electric methods is to sense moisture content by applying the relationship between temperature RH and wood MC (hygrometric or sorption method). This method, which measures the microclimate close to the measuring point, is based on the fact that wood MC influences a small volume of air in its proximity in a predictable manner.

Resistance Method

Common commercially available resistance-based sensors use two-pin-type electrodes inserted into the wood. Electrodes can be ram-in, nail-type electrodes, or screw-type electrodes. Insulated electrodes measure MC of the wood only at the tips of the pins. Pins can be of different lengths, thus providing the possibility of measuring MC at different depths/lamellas of a CLT panel and depicting moisture gradients within a panel. Uninsulated electrodes, in contrast, will sense the highest moisture content along their entire length.

Resistance-based readings are calibrated for some reference species of wood and according to the wood temperature. This is done using correction tables or built-in reference functions provided by the sensor manufacturer. With a constant surrounding climate, acceptable accuracy can be obtained deriving the temperature in the CLT lamellas from the surface temperature. In CLT panels with strongly fluctuating surrounding climate, like those exposed to direct solar radiation though glazing, temperature correction using thermistors within the panel and in a location/depth close to the moisture content sensor leads to more accurate readings. The presence of some preservatives, flame-retardant treatments, or other chemical or surface treatments may affect the accuracy of the measurement. These circumstances may require special calibration of the instrument or an alternative measurement method, like the hygrometric method described below.

Resistance-based meters are available both for discontinuous, manual readings and for permanent monitoring. The former type is a handheld meter with ram-in-type interchangeable pins. The readings on the display must be manually recorded in an inspection report. The measurements have a limited visual impact on the inspected member, that is, small holes left by the pins.

For permanently installed, resistance-based sensors, the ohmmeter has typically multiple channels to determine moisture content at different locations and depths, and it is connected to a data logger for data storage and transmission. The measured data can be stored locally or sent wirelessly to a cloud database. The latter option allows for higher design flexibility, more continuous control, and the possibility of early warning options.

In each case of discontinuous or continuous measurements, it is important to consider some procedural recommendations. The meter manufacturer's instructions should specify the measurement direction as parallel or perpendicular to grain. Accordingly, pins are inserted in alternate directions on a CLT panel's face, depending on the grain direction of the measured lamina/depth. If site constraints do not allow strict adherence to this recommendation, some negligible errors can be expected at moisture levels above 20%. James (1988) reported that readings across the grain may occasionally be as much as 2% lower in moisture content than readings parallel to grain.

Any material discontinuity, such as gaps between boards, checks, splits, fissures, and knots, between the two pins must be avoided when inserting the electrodes. This is not easily controllable since discontinuities in the internal lamellas are not visible. Lack of signal is a possible indication that the electrodes are placed across a gap. Heavy environmental fluctuations can cause internal checking of the panel and loss of signal in a permanently installed sensor (Riggio et al., 2019). Recommendations for permanent installation of moisture sensors are provided in Dietsch et al. (2014) and Schmidt and Riggio (2019). Guidelines for use of handheld resistance-type meters for periodic inspection are given in Riggio et al. (2014).

Capacitance Method

Dielectric meters are sometimes referred to as "capacity-type" or "radiofrequency-type" meters. These types of meters can accurately measure MCs up to 30%, within 1% accuracy. They are equipped with surface contact electrodes. Field penetration of this type of electrode is typically 10 to 20 in., depending on the electrode used. With any type of electrode, however, the meter registers the highest values near or in contact with the electrode.

Measurement of dielectric properties is affected by parameters such as wood density, temperature, grain orientation, and AC frequency. Density correction may be carried out by using special tables or charts supplied by the meter manufacturers or, if possible, appropriately setting the meter for the specific wood species. Wood preservatives, flame retardants, and chemical and/or surface treatment may affect the accuracy of the measurement and require specific calibration of the instrument. The effect of temperature on this type of meter is negligible.

The following procedural recommendations are given (Riggio et al., 2014): (1) the measurement area should be free of any inhomogeneity or discontinuity, such as knots, bark, and checks, and of surface moisture; (2) measurements should be taken far from metallic elements like connectors; and (3) the electrode must be pressed firmly against the specimen in order to avoid air gaps or poor contact. Since readings with the capacitance meters are limited to the first lamellas, this method is not recommended to evaluate thorough equilibration of CLT panels before completing the enclosure.

Hygrometric Method This method is based on the assumption that air encapsulated in small cavities inside solid wood specimens remains in equilibrium with moisture content of the cavity walls. If the temperature and RH of the encapsulated air is measured, the moisture content of the surrounding wood can be calculated based on the sorption isotherm (Fig. 5.18). In practice, it requires boring a small hole into the CLT panel to insert a thermistor and a hygrometer. The hole is then hermetically closed with rubber or silica sealants to ensure that the ambient air does not leak into the cavity and does not disturb the readings. Readings are taken once the hygroscopic equilibrium between the wood and the air is reached. Despite being somewhat more laborious than the methods discussed above, the distinct advantage of this method is more stable and more reliable readings in long-term measurements, the fact that it does not depend on tight contact between the sensor and the material, and that it is less sensitive to interference from external conditions.

5.5.3 Water Leak and Condensation Detection

Adhesive tapes with flat copper or stainless-steel conductors are types of linear detection sensors that can be placed in areas prone to water intrusion. They can detect moisture and leaks over a contiguous region, and they can also be used in conjunction with thermistors to predict condensation buildup. As they are typically self-adhesive, they permit easy installation and wide coverage. They can be placed on pipes and other plumbing fixtures, in wall cavities, and under roof WRB membranes. These linear sensors are applied in a grid pattern on a roof surface; grid density defines resolution of the leak detection. Precautions should be taken to avoid damaging the tape. These sensors should not be grounded or connected to a conductive surface. Maintenance requires keeping them free from dirt and debris that can cause false alarms

5.5.4 IR Thermal Imaging

Thermography is a technique used to visualize temperature distributions on the surface of an object based on its thermal properties. Because of the low thermal conductivity of wood, thermography allows for superficial inspection (approx. 1/16 in.). The thermal gradient calculated from the temperature distribution makes it possible to map some types of inhomogeneity on the inspected surface.

There are two main methods to acquire thermal images: active and passive heating procedures. During active heating, the inspected element is exposed to a source of thermal radiation (e.g., an infrared lamp), and images are taken a few seconds afterward. The process of heat conduction after excitation makes this approach more effective in case of slight variations of thermal properties on the surface of the inspected member, like localized moist areas. The passive method uses environmental heating sources, such as the sun or the building system. It is advisable to map areas with very high moisture content and heat leaks.

Thermography can be used as a preliminary noncontact screening procedure to select areas for more detailed analysis (Riggio et al., 2015). For instance, it allows for indicative localization of (1) thermal bridges, (2) air leaks, and (3) moist areas, typically identified as cold spots. All thermographic measurements are affected by surface roughness/porosity and environmental parameters, such as dust deposition, air temperature, wind speed, and unconsidered heating sources. This can distort results.

Acknowledgments

We thank the reviewers of this chapter: Eugenia Gasparri, University of Sidney; Paolo Lavisci, WoodSolutions; Jeff Morrell, National Centre for Timber Durability and Design Life—USC; and Franco Piva, Ergodomus Timber Engineering. We thank Paola Brugnara, Rothoblaas, for her thoughtful suggestions. We thank Massimo Paoli and Albino Angeli, Xlam Dolomiti; Giulio Boem and Silvio and Jlenia Pedrotti, Essepi; and Brad Nile, Juliana Ruble, and Ryan Wagar, Andersen Construction, for the photographs illustrating this chapter. Noura Alhariri is coauthor of the illustrations in this chapter.

References

Alsayegh, G., Mukhopadhyaya, P., Wang, J., Zalok, E., van Reenen, D. (2013). Preliminary characterization of physical properties of cross-laminated-timber (CLT) panels for hygrothermal modelling. *Advances in Civil Engineering Materials*, 2(1):472–484.

American National Standard Institute ANSI/APA PRG 320-2018. Standard for Performance-Rated Cross-Laminated Timber, APA, 2018.

AS/NZ 1604.5:2012. Specification for Preservative Treatment. Part 5: Glued Laminated Timber Product. Australian/New Zealand Standard, 2012.

ASTM E1677-11, Standard Specification for Air Barrier (AB) Material or System for Low-Rise Framed Building Walls, ASTM International, West Conshohocken, PA, 2011.

ASTM E283-04, Standard Test Method for Determining Rate of Air Leakage through Exterior Windows, Curtain Walls, and Doors under Specified Pressure Differences across the Specimen, ASTM International, West Conshohocken, PA, 2012.

ASTM E2178-13, Standard Test Method for Air Permeance of Building Materials, ASTM International, West Conshohocken, PA, 2013.

ASTM E1827-11, Standard Test Methods for Determining Airtightness of Buildings Using an Orifice Blower Door, ASTM International, West Conshohocken, PA, 2017.

ASTM E96/E96M-16, Standard Test Methods for Water Vapor Transmission of Materials, ASTM International, West Conshohocken, PA, 2016.

ASTM E2357-18, Standard Test Method for Determining Air Leakage Rate of Air Barrier Assemblies, ASTM International, West Conshohocken, PA, 2018.

ASTM E779-19, Standard Test Method for Determining Air Leakage Rate by Fan Pressurization, ASTM International, West Conshohocken, PA, 2019.

Beebe, K., Kam-Biron, M. (2016). Five Ds of Moisture Management: Deflection, Drainage, Drying, Distance, and Durable Materials. SEAC 2016 Convention Proceedings, Seattle, WA.

Brambilla, A., Gasparri, E., Aitchison, M. (2018). Building with timber across Australian climatic contexts: A hygrothermal analysis. ASA 2018: Engaging Architectural Science: Meeting the Challenges of Higher Density (ANZAScA), RMIT University, Melbourne.

Brandner, R. (2018). Cross laminated timber (CLT) in compression perpendicular to plane: Testing, properties, design and recommendations for harmonizing design provisions for structural timber products. *Engineering Structures*, 171:944–960.

Brown, W.C., Chown, G.A., Poirier, G.F., Rousseau, M.Z. (1999). Designing exterior walls according to the rainscreen principle, Construction Technology Update No. 34, Institute for Research in Construction, National Research Council of Canada.

Carll, C., WIedenhoeft, A.C. (2009). Moisture-related properties of wood and the effects of moisture on wood and wood products. In *Moisture Control in Buildings: The Key Factor in Mold Preventions*, Heinz R. Trechsel and Mark T. Bomberg (eds.), 2nd ed. West Conshohocken, PA: ASTM International, Chapter 4:54–79.

Derome, D. (2010). The nature, significance and control of solar-driven diffusion in wall systems. ASHRAW Research Project 1235. Atlanta, GA: American Society of Heating, Refrigerating and Air-Conditioning Engineers.

Dietsch, P., Franke, S., Franke, B., Gamper, A., Winter, S. (2014). Methods to determine wood moisture content and their applicability in monitoring concepts. *Journal of Civil Structural Health Monitoring*, 5(2):115–127.

Esteves, B., Pereira, H. (2009). Wood modification by heat treatment: A review. *Bioresources* 4(1):370–404.

Finch, G. (2016). High-Rise Wood Building Enclosures. Thermal Performance of the Exterior Envelopes of Whole Buildings. XIII International Conference, ASHRAE.

FPInnovations (2014). Special Publication SP-55E. Chapter 6—Building enclosure design. In: *Technical Guide for the Design and Construction of Tall Wood Buildings in Canada*, E. Karacabeyli and C., Lum (eds.), 1st ed.

FPL (2010). *Wood Handbook: Wood as an Engineering Material*, Madison, WI: U.S. Department of Agriculture, Forest Service, Forest Product Laboratory

Gamper, A., Dietsch, P., Merk, M., Winter, S. (2013). Building climate—Long-term measurements to determine the effect on the moisture gradient in large-span timber structures. Gebudeklima - Langzeitmessung zur Bestimmung der Auswirkungen auf Feuchtegradienten in Holzbauteilen, *Bautechnik*, 90(8).

Gasparri E., Aitschison, M. (2019). Unitised timber envelopes. A novel approach to the design of prefabricatedmass timber envelopes for multi-storey buildings. *Journal of Building Engineering*, 26:100898.

Gasparri, E., Brambilla, A., Aitchison, M. (2018). Hygorthermal analysis of timber-based external walls across different Australian climate zones. World Conference on Timber Engineering, August 20 -23, 2018, Seoul, Republic of Korea.

Gasparri, E., Lam, F., Liu, Y. (2016). Compression perpendicular to grain behavior for the design of a prefabricated CLT facade horizontal joint. WCTE 2016—World Conference on Timber Engineering, Vienna: Vienna University of Technology.

Gasparri, E., Lucchini, A., Mantegazza, G., Mazzucchelli, E.S. (2015). Construction management for tall CLT buildings: From partial to total prefabrication of façade elements. *Wood Material Science & Engineering*, 10 (3):256–275.

Gereke, T., Gustafsson, P.J., Persson, K., Niemz, P. (2011). The hygroscopic warping of cross-laminated timber. In *Delamination in Wood, Wood Products and Wood-Based Composites*, V. Bucur (ed.), Springer, DOI: 10.1007/978-90-481-9550-3_14.

Glass, S.V., Wang, J., Easley, S., Finch, G., Cammalleri, V., Ross, L.A., Morris, P.I. (2013). Enclosure: Building enclosure design for cross-laminated timber construction. *CLT Handbook*, U.S. Edition. Chapter 10.

Goto, Y., Ghazi Wakili, K., Ostermeyer, Y. Frank, Th., Ando, N., Wallbaum, H. (2011). Preliminary investigation of a vapor-open envelope tailored for subtropical climate. *Building and Environment*, 46: 719–828.

Grossman, P.U.A. 1976. Requirements for a model that exhibits mechano-sorptive behavior. *Wood Science and Technology*. 10: 163–168.

Hameury, S., Lundström, T. (2004). Contribution of indoor exposed massive wood to a good indoor climate: In-situ measurement campaign. *Energy and Building*, 36(3): 281–292.

International Code Council—ICC. 2018. *International Energy Conservation Code*—IECC.

International Code Council—ICC. 2018. *International Building Code—IBC, 2018.*

ISO 6946:2017. *Building Components and Building Elements—Thermal Resistance and Thermal Transmittance—Calculation Methods*. International Organization for Standardization.

James, W.L. (1988). *Electric Moisture Meters for Wood*. USDA-FS FPL-GTR-6.

Kaufmann, F. (2016). Building systems—Constraint or changes for architects. World Conference on Timber Engineering, August 22–26, 2016, Vienna, Austria.

Kaufmann, H., Krötsch, S., Winter, S. (2018). *Manual of Multi-Storey Timber Construction*. Edition Detail. Chapter C: Construction—Part 3: The layer structure of building envelopes.

Lepage, R. (2012). *Moisture Response of Wall Assemblies of Cross-Laminated Timber Construction in Cold Canadian Climates* (Thesis), Civ. Eng. University of Waterloo, Ontario, Canada, p. 139.

Lepage, R., Higgings, J., Finch, G. (2017). Moisture uptake testing for CLT floor panels in a tall wood building in Vancouver, in *15th Canadian Conference on Building Science and Technology*, pp. 1–17.

Leyder, C., Chatzi, E. (2015). Structural health monitoring of an innovative timber structure, in *2nd International Conference on Performance-Based and Life-Cycle Structural Engineering* (PLSE 2015), Brisbane, Australia, pp. 1383–1392.

Mantanis, G.I. (2017). Chemical modification of wood by acetylation or furfurylation: A review of the present scaled-up technologies. *BioResources*, 12(2):4478–4489.

McClung, R., Ge, H., Straube, J., Wang, J. (2014). Hygrothermal performance of cross-laminated timber wall assemblies with built-in moisture: Field measurements and simulations. *Building and Environment*, 71:95–110.

Morris, P.I., Winandy J.E. (2002). Limiting conditions for decay in wood systems. 33rd IRG Annual Meeting, Cardiff, South Wales, UK, May 12–17, 2002.

Muszyński, L., Lagaňa, R., Davids, W., Shaler, S.M. (2005). Comments on the experimental methodology for quantitative determination of the hygro-mechanical properties of wood. *Holzforschung*, 59(2): 232–239.

Naturally:wood (2018). Brock Commons Tallwood House. Performance overview. Retrieved from https://www.naturallywood.com/sites/default/files/documents/resources/brock_commons_cs_-_performance_web.pdf

Nguyen, T.T., Dao, T.N., Aaleti, S., Hossain, K., Fridley, K.J. (2019). Numerical model for creep behavior of axially loaded CLT panels. *Journal of Structural Engineering*, 145(1):04018224.

Nore, K., Kraniotis D., Sortland, M.L. (2019). Wood as an exposed building material for indoor climate adaptation. In *Cold Climate HVAC 2018*, D. Johansson, H. Bagge, and A. Wahlström (eds.), CCC 2018, Springer Proceedings in Energy, Springer.

Nore, K., Mattsson, J., et al. (2014). Cross laminated timber vs. timber frame walls in water damage: Comparing drying and mould growth. In NSB 2014 (Full paper no. 157). June 15–19, Lund, Sweden.

Norén, J., Pousette, A., Sandberg, K. (2017). Moisture safety of tall timber facades. SP Rapport 2017:38. RISE Research Institutes of Sweden, Skellefteå.

Ott, S., Tietze, A., Winter, S. (2015). Wind driven rain and moisture safety of tall timber houses—Evaluation of simulation methods. *Wood Material Science & Engineering*, 10(3): 300–311.

Piva, F. (2019). Manuale delle costruzioni di legno (*Manual of timber construction*). Legislazione tecnica. (in Italian)

Riggio, M., Alhariri, N., Hansen, E. (2020). Paths of innovation and knowledge management in timber construction in North America: A focus on water control design strategies in CLT building enclosures. *Architectural Engineering and Design Management*, 16(1): 58–83.

Riggio, M., Anthony, R.W., Augelli, F., Kasal, B., Lechner, T., Muller, W., Tannert, T. (2014). In situ assessment of structural timber using non-destructive techniques. *Materials and Structures/Materiaux et Constructions*, 47(5):749–766.

Riggio, M., Sandak, J., Franke, S. (2015). Application of imaging techniques for detection of defects, damage and decay in timber structures on-site. *Construction and Building Materials*, 101, 1241–1252.

Riggio, M., Schmidt, E., Mustapha, G. (2019). Moisture monitoring data of mass timber elements during prolonged construction exposure: The case of the Forest Science Complex (Peavy Hall) at Oregon State University. *Frontiers in Built Environment.* doi: 10.3389/fbuil.2019.00098.

Ringhofer, A., Schickhofer, G. (2014). Multi-storey residential buildings in CLT—Interdisciplinary Principles of Design and Construction. In *WCTE 2014 Conference Proceedings*, pp. 1–7.

Rothoblaas (2017). Membranes and tapes for wooden buildings. Catalog. Rothoblaas ©.

Salvalai, G. (2018). Experimental analysis of different multilayer insulated façade. *TEMA Technologies Engineering Materials Architecture*, 4(2).

Schmidt, E., Riggio, M. (2019). Monitoring moisture performance of cross-laminated timber building elements during construction. *Buildings*, 9: 144.

Schmidt, E.L., Riggio, M., Barbosa, A.R., Mugabo, I. (2019). Environmental response of a CLT floor panel: Lessons for moisture management and monitoring of mass timber buildings. *Building and Environment*, 148: 609–622.

Serrano, E. (2009). Limnologen—Experiences from an 8-storey timber building. Limnologen potpourri—Praktische Erfahrungen von immeubles en bois de huit étages, 15 Internationales Holzbau Forum, 2009.

Stirling, R., Morris, P. (2017). Development of protective treatments for cross-laminated timber (CLT). Proceedings of the American Wood Protection Association, Las Vegas, NV, April 2017.

Straube, J. (2008). *Air Flow Control in Buildings.* Building Science Digest 014. Somerville, MA: Building Science Press.

Straube, J.F. (2012). *High Performance Enclosures: Design Guide for Institutional Commercial and Industrial Buildings in Cold Climates.* Somerville, MA: Building Science Press.

Straube, J., Burnett, E.F.P. (1999). Rain control and design strategies. *Journal of Thermal Envelope and Building Science*, (23):41–56.

TRADA (Timber Research and Development Association) (2017). Cross-laminated timber. Design and performance. Exova BM Trada.

Van Acker, J., De Windt, I., Li, W., Van den Bulcke, J. (2014). Critical parameters on moisture dynamics in relation to time of wetness as factor in service life prediction. Document No. 14-20555. International Research Group on Wood Protection, Stockholm, Sweden.

Wang, J. (2016a). Wetting and drying performance of wood-based assemblies related to on-site moisture management. Proceeding of WCTE 2016 World Conference on Timber Engineering, Vienna/Austria, August 22–25, 2016, no. 2, pp. 1–22.

Wang, J. (2016b). *Guide for On-Site Moisture Management of Wood Construction.* Natural Resources Canada, Canadian Forest Service; Homeowner Protection Office, Branch of B.C. Housing.

Wang, J.W., Stirling, R., Morris, P.I., Taylor, A., Lloyd J., Kirker, G., Lebow, S., Mankowski, M.E., Barnes, H.M., Morrell, J.J. (2018). Durability of mass timber structures: A review of the biological risks. *Wood and Fiber Science*, 50 (Special Issue):110–127.

Zelinka, S.L., Rammer, D.R. (2009). Corrosion rates of fasteners in treated wood exposed to 100% relative humidity. *Journal of Materials in Civil Engineering*, 21 (12):758–763.

Recommended Readings and Additional Resources

- Benedetti C. (ed.). (2010). *Timber Buildings. Low Energy Constructions*. Bu, Press.
  (It provides a rich catalog of assembly examples and details, with descriptions of hygrothermal performance and calculation examples.)
- Glass, S.V., Wang, J., Easley, S., Finch, G., Cammalleri, V., Ross, L.A., Morris, P.I. (2013). *Enclosure: Building Enclosure Design for Cross-Laminated Timber Construction. CLT Handbook.* U.S. Edition, Chapter 10.
  (The first comprehensive publication in the United States on the hygrothermal behavior of CLT. It includes also examples of assembly details.)
- Structural Timber Association (2017). *STA Advice Note 14. Robustness of CLT Structures.*
  Part 1—Key principles for moisture durability
  Part 2—Key principles for CLT wall to foundation interfaces
  Part 3—Key principles for good practice detailing for the external envelope of CLT
  Part 4—Construction process best practice
- Teibinger, M., Matzinger, I. (2013). *Construction with Cross-Laminated Timber in Multi-Storey Buildings. Focus on Building Physics. Guidelines.* Published by Holzforshung Austria.
- Kaufmann, H., Krötsch, S., Winter, S. (2018). *Manual of Multi-Storey Timber Construction.* Edition Detail.
  (A comprehensive publication on timber construction, with a chapter dedicated to the design of the building enclosures. Rich in illustrations.)
- https://www.dataholz.eu/en.htm
  (Online catalog of assembly details)

CHAPTER 6
Acoustics

Matt Mahon
LSTN Consultants

Denis Blount
Arup

This chapter is intended to provide an overview of the unique acoustic considerations to be made for the design of projects with cross-laminated timber (CLT) mass timber structure.

The chapter will begin with a broad overview of acoustics in buildings to frame the discussion of acoustic concepts. Following this, we outline the limited acoustic code requirements in the United States for relevant project types. The sections covering the specific acoustic applications of CLT will be of particular interest.

6.1 Acoustics in Buildings

In the design and construction of buildings, we must consider the impact that acoustics may have on occupants of the building and on people in the environment surrounding the building. The study and practice of architectural acoustics is broadly intended to address these impacts. For the characterization of the various acoustical impacts, architectural acoustics can be subdivided into the following basic elements:

- *Room Acoustics:* The way that an enclosed space's volume, shape, and finishes affect the reverberance, noise buildup, and reflection anomalies within that space.
- *Sound Isolation:* The way dividing constructions reduce noise transmission from adjacent spaces.
- *Footfall/Impact Noise Isolation:* The way noise from occupant footfall is generated and transmits through a floor/ceiling construction into spaces below.
- *Mechanical Noise and Vibration Control:* The way noise and vibration from building services are generated and propagate into occupied areas of a building.
- *Environmental Noise Control:* The way noise emitted externally from a building propagates into the surrounding environment

There are many references, not to mention design consulting firms, that can provide detailed explanation of acoustic fundamentals and guidance on architectural acoustics and noise control. That is beyond the scope of this book. Here, we will provide a brief overview of acoustic subjects to provide appropriate context for the discussion of the acoustics of CLT constructions.

6.1.1 Basic Acoustics

What we refer to as "sound" or "noise" is the human perception of pressure waves in air. These waves involve the temporary displacement of air molecules from their quiescent position (see Fig. 6.1). These perceptible pressure waves are fundamentally characterized by fluctuation in the pressure in air from the atmospheric pressure. Sound pressure waves can be described by their pressure *amplitude*, which correlates roughly with the human perception of *loudness*.

Sound pressure amplitude is measured in pascals (Pa) or newtons per meter squared (N/m^2). Audible sound occurs over an extreme range of pressure amplitudes, and humans' perception is not linear with amplitude. Due to this, sound pressure is typically expressed in sound pressure level (dB ref 2×10^{-5} Pa), which define sound pressure as a logarithmic ratio of a reference level (which is set at the approximate threshold of hearing such that 0 dB is essentially inaudible).

Audible sound consists of a series of fluctuations in pressure (see Fig. 6.2). In addition to the amplitude of sound, we can characterize sound using *frequency*, which describes the number of times an air molecule completes a cycle of displacement from its neutral position over a period of time. The unit used to describe frequency is hertz (Hz) defined as the number of cycles per second in a pressure wave.

Typical human hearing is taken to be from 20 Hz to 20,000 Hz, though high-frequency hearing typically diminishes with age.

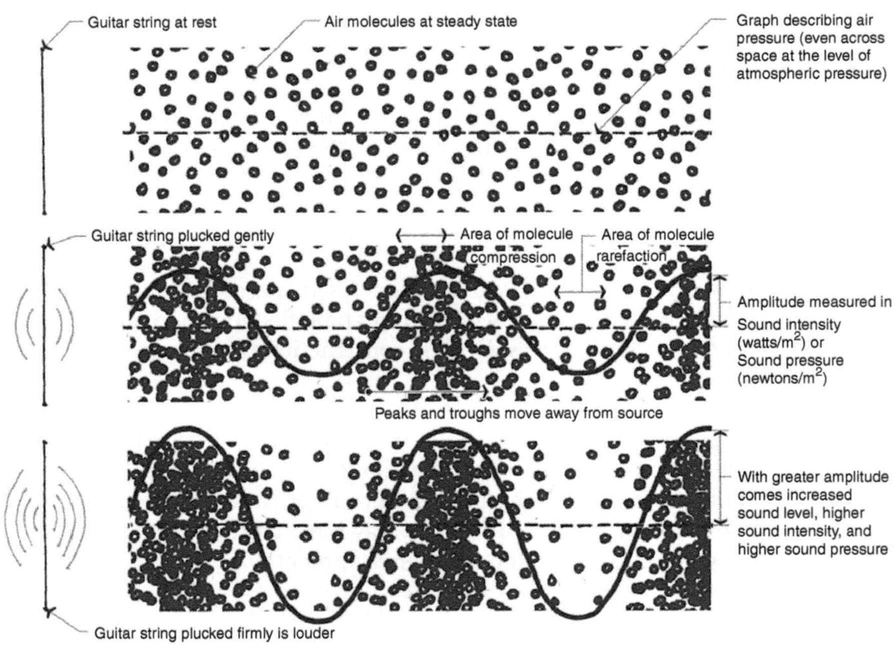

FIGURE 6.1 Sound wave exhibiting compression and rarefaction. (*Reprinted from Architectural Acoustics Illustrated by Michael Ermann 2015.*)

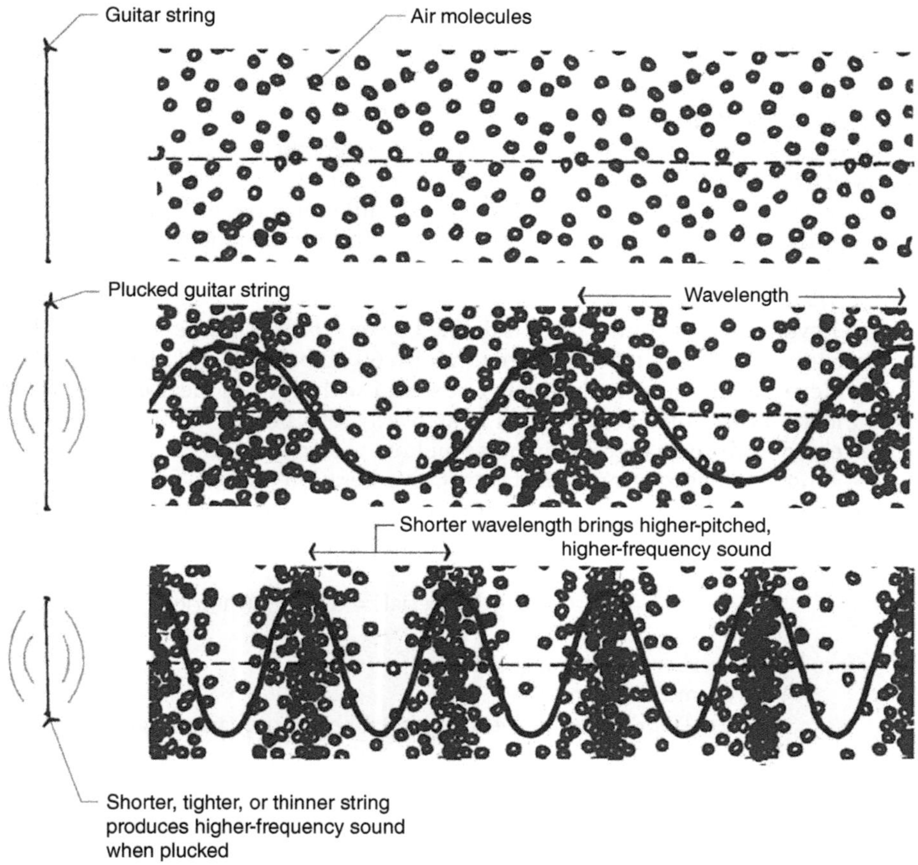

FIGURE 6.2 Sound pressure amplitude, frequency, wavelength, and period. (*Reprinted from Architectural Acoustics Illustrated by Michael Ermann 2015.*)

The *wavelength* and frequency of sound are related to the speed of sound in a medium, typically air,

$$c = \lambda * f$$

where c = speed of sound = 1100 ft/sec (340 m/sec)
λ = wavelength in ft (m)
f = frequency in hertz or cycles/sec

The frequency and wavelength of sound are important factors in acoustic analysis, as acoustic performance and responses vary based on these characteristics.

When a sound is emitted from a source, such as a human speaking, the sound wave spreads away from the source spherically. The reduction sound pressure experienced at a distance from the source is proportional to the square of the distance,

$$L_{p2} = L_{p1} - 20 \log \frac{r_2}{r_1}$$

where L_p = sound pressure level at distance from source in dB, ref 2×10^{-5} Pa
r = distance from source in any unit

6.1.2 Room Acoustics

When a sound is emitted within an enclosed room, the sound wave spreads spherically away from the source. For a room with sound-reflective surfaces, when the sound wave encounters a solid surface, it is reflected back off that surface.

As a listener within the room, you will hear the direct sound from the source as well as various reflections of the room's solid surfaces. We refer to the diffuse reflections as *reverberance* (see Fig. 6.3).

Reverberance within a space is defined by *reverberation time,* which is defined as the time it takes for a sound emitted to diminish 60 dB below its original level (see Fig. 6.4).

Reverberance affects the quality of sound within the space; for example, it may be difficult to carry on a conversation in an overly reverberant room. Increasing reverberance also leads to an increase in noise buildup within a space.

Empirical observation indicates that reverberation time is a factor of the volume of a space and the total sound absorption within the space. The following is the formula developed by Wallace Clement Sabine, the father of modern acoustics, in 1895:

$$T_{60} = 0.049\frac{V}{A}\ \text{(imperial)} = 0.161\frac{V}{A}\ \text{(metric)}$$

where T_{60} = reverberation time in sec
V = volume in ft³ (m³)
$A = S_1\alpha_1 + S_1\alpha_1 + S_1\alpha_1 + \cdots + S_1\alpha_1$ = total absorption area in sabins or ft² (m²)
S = surface area in ft² (m²)
α = absorption coefficient (unitless)

As this formula indicates, sound-absorbing materials or finishes may be added to surfaces to reduce the reverberance within a space. *Note that sound-absorbing*

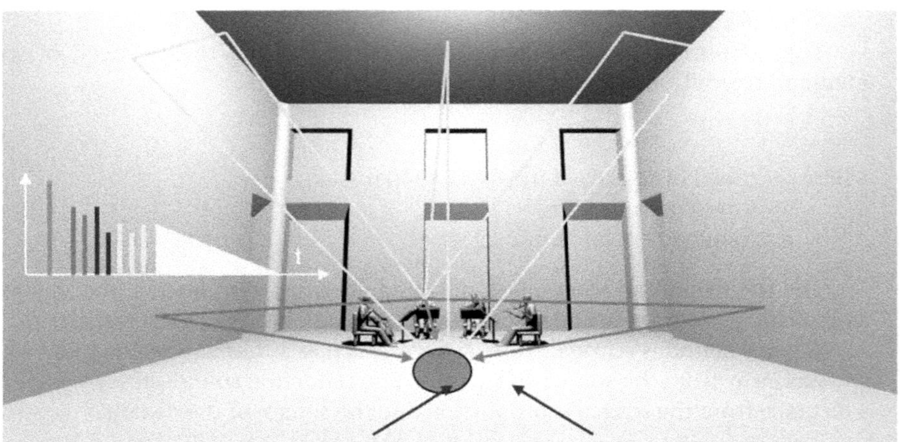

FIGURE 6.3 Reflections and reverberance. (*Image courtesy of Arup.*)

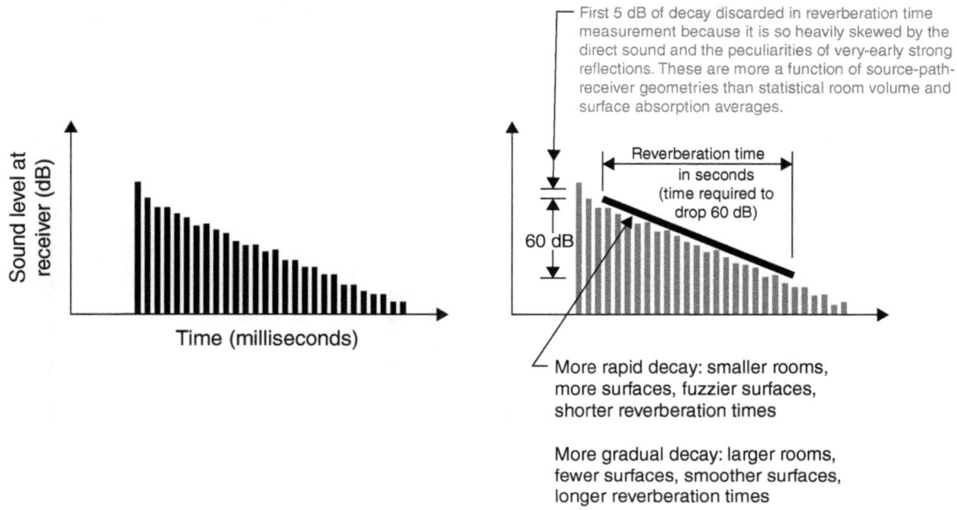

FIGURE 6.4 Calculation of reverberation time. (*Reprinted from Architectural Acoustics Illustrated by Michael Ermann 2015.*)

materials are not typically effective at reducing the noise transfer between spaces, a common confusion.

Sound absorption is defined by *absorption coefficients* over a surface area. These coefficients are typically defined in octave frequency bands. The coefficients are measured from a material sample mounted in an acoustic test lab, using the change in reverberation time to back-calculate the values that solve the reverberation time equation. *noise reduction coefficient (NRC)* is a single-number metric to describe the average mid-frequency (i.e., speech frequency) sound absorption.

In addition to reducing reverberance, sound-absorbing materials, changes in room volume, and shaping, can be used to avoid unwanted acoustic anomalies within rooms, such as discrete echoes, flutter echo, focusing, and the whisper gallery effect.

6.1.3 Sound Isolation

In the section above, we have described how sound may be reflected or absorbed by a room surface. A portion of the sound energy, however, may be transmitted through the boundary constructions of a room. Architectural acoustics is concerned that appropriate reduction of sound is provided between spaces due to the constructions that divide them.

Sound isolation performance is characterized by *sound transmission loss (TL)*, a measure of the reduction in energy experienced by sound as it passes through a barrier or partition. Higher TL results in greater sound isolation performance:

$$\tau = \text{Sound Transmission Coefficient} = \frac{\text{Transmitted Energy}}{\text{Incident Energy}}$$

$$\text{TL} = \text{Transmission Loss} = 10\log_{10}\frac{1}{\tau} = -10\log_{10}\tau$$

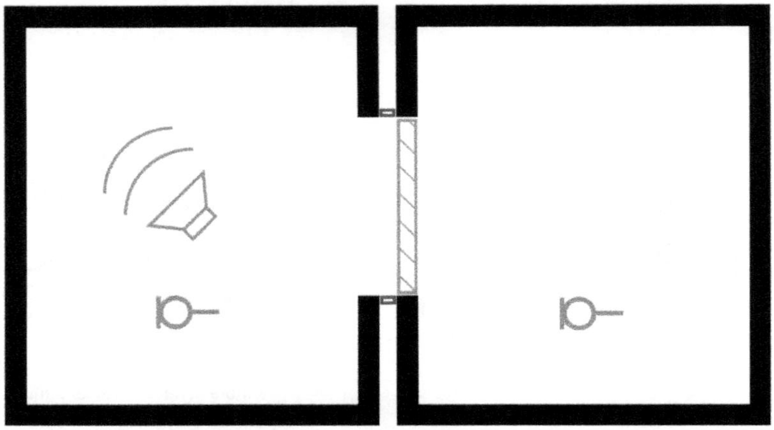

FIGURE 6.5 Lab test setup diagram. (*Image courtesy of Arup.*)

TL for a construction can be directly measured in a lab environment consisting of two adjacent rooms joined by an opening in which the test construction has been added (see Fig. 6.5).

The formula to determine TL is reproduced as

$$\Delta L_{\text{TL}} = \overline{L}_{p,s} - \overline{L}_{p,r} + 10 \log_{10} S_W - 10 \log \frac{.049V}{T_{60}} \text{(Imperial)}$$

where $\overline{L}_{p,r}$ = spatial average sound pressure level in the receiver room (dB)
$\overline{L}_{p,s}$ = spatial average sound pressure level in the source room (dB)
ΔL_{TL} = reverberant field TL (dB)
S_W = area of the transmitting surface (metric or imperial)
V = volume in ft^3
T_{60} = reverberation time in sec

Refer to ASTM E90 for more information on the laboratory measurement of airborne sound TL of building partitions. Refer to ASTM E492 or more information on the laboratory measurement of airborne sound TL of floor/ceiling constructions.

TL is typically measured in third-octave bands, which help sound transmission at different frequencies. It is typical to expect, as an example, that a given partition will transmit more low-frequency sound than high-frequency sound.

To simplify discussion of sound transmission performance, the industry uses a single number metric, *sound transmission class (STC)*. This metric is calculated per ASTM E413.

STC uses a third-octave band reference curve base that prioritizes reduction of speech frequencies. This means that STC is most applicable to characterize the sound transmission at speech frequencies. STC alone should not be used to describe transmission of sounds with significant low-frequency content, such as music or mechanical equipment (see Fig. 6.6).

If lab tests are not practical or relevant, field measurements can be made to characterize sound isolation in situ. Note that field measurements often present limitations

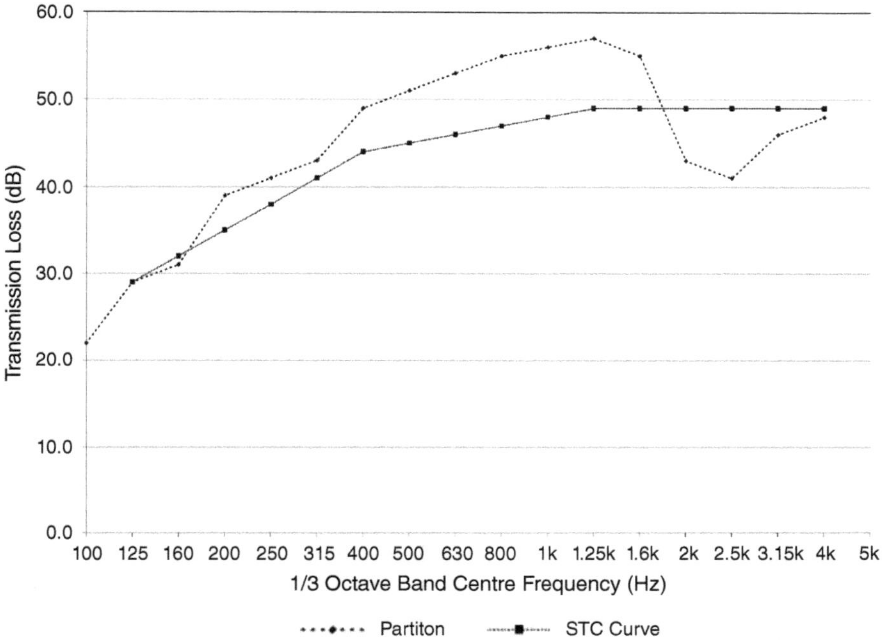

FIGURE 6.6 Graph with STC reference contour and deficiencies. (*Image courtesy of Arup.*)

due to the uncontrolled environment. Refer to ASTM E336 for more information on the field test.

TL performance of a partition or construction is typically a factor of a number of characteristics of the sound and the materiality of the construction, as summarized below:

- Frequency of sound: TL typically increases as frequency increases.
- Mass of construction: TL typically increases as the mass of a construction increases.
- Stiffness of construction: TL typically decreases as the stiffness of a construction increases.
- Damping in construction: TL typically increases as the damping in a construction increases.
- Multiple leaves in a construction: TL typically increases by adding multiple layers (leaves) separated by an air gap.

6.1.4 Footfall/Impact Noise Isolation

The preceding sections have focused on airborne sounds, that is, sounds generated and propagated through air. In addition to this, we must consider structure-borne sounds. Of particular note, footfall and other impulsive forces can impart vibration and noise in a building. Note that human perception and response to structural vibration due to footfall is principally a structural concern, which is not addressed in this section. Our focus here deals with the perception of noise generated from such impulsive forces.

When an impulsive force is transmitted into a floor, noise may be generated and radiate off of the ceiling below (and other surfaces). The energy generated by footfall varies based on a wide range of factors:

- Walker's weight
- Walker's speed
- Walker's footwear
- Floor coverings and construction
- Building structural system

The energy generated by footfall varies significantly in practice. In order to characterize the noise generated by footfall impact, a tapping machine is used as a standardized source of impact. A single number metric for footfall impact noise, *impact isolation class* (IIC), can be calculated by measuring the sound pressure levels in a range from 100 to 3150 Hz generated in a space below the tapping machine in various configurations and comparing those average levels against a reference frequency contour. ASTM E492 provides the methodology for IIC lab tests, and ASTM E1007 provides the methodology for IIC field tests.

A higher IIC rating indicates a floor/ceiling construction that generates and transmits less footfall impact sound into the space below. It should be noted that a tapping machine, while standardized, does not accurately reproduce the sound of people walking. This leads to some deficiencies in the metric that will be discussed further in this chapter.

Impact isolation performance of a floor/ceiling construction and the quality of sound generated/transmitted is typically a factor of a number of characteristics, as summarized below:

- Floor coverings
  - Hard floor finishes such as concrete, hardwood tongue and groove, ceramic tile, and stone transmit more high-frequency noise that is often described as "tapping" or "clicking."
  - Soft floor finishes, notably carpet on pad but also pad-backed vinyl tiles, can reduce that high-frequency noise.
  - Inserting a resilient underlayment between the floor finish and the structure can also acceptably reduce that high-frequency noise. Increasing the mass of the floor finish above the resilient underlayment can further improve the performance.
- Building structural system
  - Lightweight floor constructions such as timber joists generate significant low-frequency "thud" and "thump."
  - Massive floor constructions such as concrete slabs can reduce that low-frequency sound.
- Ceiling in space below
  - Including a dropped ceiling can reduce impact noise levels radiated from the structure above.
  - Suspending that dropped ceiling on resilient hangers can reduce the structure-borne noise transferred from the structure above in the ceiling below.

6.1.5 Mechanical Noise and Vibration Control (HVAC)

Modern building services design typically includes mechanical systems (alternately referred to as HVAC systems) that heat and cool spaces as well circulating outdoor fresh air to the interior of buildings. Various methods exist to accomplish these needs, but they often involve fans that force air over heat-exchanging elements and ductwork that distributes that tempered air.

Noise in mechanical systems is generated by turbulence in the air (notably at the fans but also as the air passes through ductwork) and directly by the rotating and reciprocating equipment. The noise can be transmitted via an airborne path or through structure off of which it can radiate (structure-borne).

Mechanical noise can be both beneficial and a nuisance. Within a building, mechanical noise can—to a point—provide noise masking that improves the perception of privacy. When the noise is excessive, though, it can be disturbing and uncomfortable. Outside of a building, noise emitted from mechanical system can become an environmental problem, with attendant comfort and code compliance implications.

The control of mechanical noise typically involves the following sources of noise and analysis/mitigation:

- Radiated noise from mechanical equipment (cooling towers, chillers, boilers, pumps, and fans)
  - Analysis of published radiated noise levels for selected equipment
  - Design of sound-isolating constructions and enclosures to prevent sound transmissions into noise-sensitive spaces
  - Reselection of equipment that produces lower noise levels
- Duct-borne noise from fans
  - Analysis of published ducted noise levels for selected fans
  - Analysis of the insertion loss along the duct network designed to supply (or return, exhaust, etc.) air from the fans to the spaces served and the room acoustics in the spaces served
  - Design of duct elements (sound attenuators, lined ductwork, plenum, etc.) to increase the insertion loss of the duct network
  - Reselection of fans that operate more efficiently and produce less noise
- Regenerated noise from the turbulent flow of air through
  - Review of duct layouts to minimize high-static pressure loss configurations which may also be a source of regenerated noise
- Structure-borne noise from mechanical equipment (cooling towers, chillers, boilers, pumps, and fans)
  - Review of equipment types and mounting locations
  - Recommendations of vibration isolation requirements, including structural mounting conditions, vibration isolators (isolator bases, types, and static deflections), and connection details

Refer to Chapter 48 of the ASHRAE Handbook for more information about noise and vibration control in mechanical systems.

6.2 Acoustic Code Requirements

There are few code requirements for acoustic design in the United States. This is not to say that acoustic design is not an important consideration for most building types but simply that the various building code authorities leave acoustic design to building owners and design teams to determine what is appropriate for each individual project.

Residential multifamily housing and hotels are building types where the code authorities have provided acoustic design requirements. The two most commonly used guidelines for multifamily dwelling acoustic design requirements are the International Building Code (IBC) and the U.S. Department of Housing and Urban Development (HUD) guide. The 2015 IBC requires, per Sections 1207.2 and 1207.3, the following minimum separation between dwelling and sleeping units:

- STC 50 airborne sound separation (45 in the field, minimum)
- IIC 50 (45 in the field, minimum)

HUD's *Guide to Airborne, Impact, and Structure-Borne Noise Control in Multifamily Dwellings* recommends various levels of acoustic isolation dependent on a "Luxury," "Average," or "Minimum" rating. The IBC requirements meet or exceed the "Minimum" HUD guidelines (Table 6.1).

For commercial construction and most other building types in the United States, there are no national/U.S. code requirements for acoustic design. Best practice suggests that STC 50–55 partitions be provided between tenant spaces (horizontal and vertical separation). Given that it is common for multilevel commercial construction in the United States to utilize concrete floor slabs, STC 50–55 is often achieved without any acoustical consideration. Given the proliferation of carpet flooring and hung ceilings, impact isolation between office units is often achieved without design consideration.

6.3 Acoustics for CLT Projects

CLT structures are a novel application of newly available engineered wood technologies. CLT structure differs substantially from more traditional structural systems and also other timber systems. The following sections broadly compare the acoustics of CLT with other structural systems and other mass timber systems and subsequently provide more detailed consideration of the acoustics for CLT projects.

6.3.1 Comparing CLT with Other Structural Systems

Concrete and Steel Composite

The most common structural systems for high-rise commercial and residential buildings are cast-in-place concrete and steel composite. Concrete and steel composite structures exhibit some unique acoustic properties:

- Floor structures are typically 4- to 12-in.-thick concrete or concrete on composite metal deck slabs. These high-mass floor slabs provide substantial airborne sound isolation between vertically adjacent spaces. The floor slabs do not, however, provide significant footfall/impact noise isolation. In residential constructions, it is often sufficient to leave the underside of structure exposed.
  - CLT floor structures are typically 6- to 12-in.-thick CLT panels. Despite comparable thickness, CLT panels are significantly less dense than concrete slabs, which results in decreased airborne sound isolation.
  - Bare concrete floors and CLT slabs with no ceilings below both provide poor impact noise isolation.
- To provide lateral support, core walls are commonly cast-in-place concrete sheer walls. These high-mass core walls can provide high levels of sound isolation from

Partition function between dwellings			Luxury Grade I		Average Grade II		Minimum Grade III	
APT. A	I	APT. B	STC	IIC	STC	IIC	STC	IIC
Bedroom	Above	Bedroom	55	55	52	52	48	48
Living room	Above	Bedroom[1,2]	57	60	54	57	50	53
Kitchen[3]	Above	Bedroom[1,2]	58	65	55	62	52	58
Family room	Above	Bedroom[1,2,4]	60	65	56	62	52	58
Corridor	Above	Bedroom[1,2]	55	65	52	62	48	48
Bedroom	Above	Living room[5]	57	55	54	52	50	48
Living room	Above	Living room	55	55	52	52	48	48
Kitchen	Above	Living room[1,2]	55	60	52	57	48	53
Family room	Above	Living room[1,2,4]	58	62	54	60	52	56
Corridor	Above	Living room[1,2]	55	60	52	57	48	53
Bedroom	Above	Kitchen[1,5]	58	52	55	50	52	46
Living room	Above	Kitchen[1,5]	55	55	52	52	48	48
Kitchen	Above	Kitchen	52	55	50	52	46	48
Bathroom	Above	Kitchen[1,2]	55	55	52	52	48	48
Family room	Above	Kitchen[1,2,4]	55	60	52	58	48	54
Corridor	Above	Kitchen[1,2]	50	55	48	52	46	48
Bedroom	Above	Family room[1,5]	60	50	56	48	52	46
Living room	Above	Family room[1,5]	58	52	54	50	52	48
Kitchen	Above	Family room[1,5]	55	55	52	52	48	50
Bathroom	Above	Bathroom	52	52	50	50	48	48
Corridor	Above	Corridor	50	50	48	48	46	46

NOTES:

[1]The most desirable plan would have the floor-ceiling assembly separating spaces with equivalent functions, e.g., living room above living room, etc.; however, when this arrangement is not feasible, the assembly must have greater acoustical insulating properties.

[2]This arrangement requires greater impact sound insulation than the converse, where a sensitive area is above a less sensitive area.

[3]Or dining, family, or recreation room.

[4]The airborne STC criteria in this table apply as well to vertical partitions between these two spaces.

[5]This arrangement requires equivalent airborne sound insulation and perhaps less impact sound insulation than the converse.

TABLE 6.1 HUD Guidelines

core areas into program areas, notably reducing elevator noise transmission into program areas.

- CLT hybrid structures may make use of concrete sheer walls, which provide similar benefits. CLT structures may, however, use 6- to 12-in.-thick CLT panels as sheer walls for lateral support. Again, CLT panels are significantly less dense, resulting in decreased airborne sound isolation as well as increased transmission of elevator vibration energy, which can reradiate as noise into program spaces.

Stick Frame Timber

The most common structural system for low-rise and mid-rise multifamily residential buildings remains the "stick frame" timber in many parts of the United States. Stick frame timber is somewhat infamous for its acoustic properties:

- Floor structures are typically plywood subfloor supported by joists constructed from dimensioned lumber, wood I-joists (TJI), or open-web timber or hybrid steel trusses. Resiliently hung drop gypsum ceilings are typically provided in residential applications to conceal services and building structure.
  - CLT floor slabs typically provide higher sound isolation than the plywood subfloor alone, but if a drop ceiling is omitted from the CLT assembly, the benefits of the additional mass provided by CLT may be lost due to the lack of the airspace (provided by the wood joist structure in stick-framed construction).
  - The increased mass of CLT floor slabs improves low-frequency impact noise isolation, which is particularly difficult to mitigate in stick-framed construction (given that stick-framed structures are lighter weight and less stiff).
- Stick frame timber walls are typically wood framing with gypsum board attached to each side.
  - CLT walls may be monolithic CLT panels, which, due to limited mass and single layering, can underperform double-leaf gypsum board walls.

6.3.2 Cold-Formed Steel

Increasingly common for mid-rise commercial and residential projects, cold-formed steel structural systems share many traits with stick frame timber:

- Floor structures are lightweight plywood subfloors supported by steel C-shaped framing or trusses. Drop ceilings are typically provided.
  - CLT floor slabs typically provide higher sound isolation than plywood subfloor alone, but if a drop ceiling is omitted, the benefits may be lost.
  - The increased mass of CLT floor slabs improves low-frequency impact noise. Steel framing is typically less stiff than timber, so the benefits are less dramatic than with stick frame timber.

6.3.3 CLT Compared with Other Mass Timber Systems

CLT is just one of a number of new engineered mass timber systems being developed and deployed today. Other mass timber systems notably include nail-laminated timber (NLT) and dowel-laminated timber (DLT). Acoustically, further research is needed to fully understand and quantify the differences between these systems.

6.3.4 CLT Floor/Ceiling Constructions

Mass timber structures have inherent acoustic challenges given that mass timber floor assemblies have significantly less mass than concrete structures, resulting in significantly less airborne sound isolation between floors. Lightweight stick-framed wood structures (e.g., wood joist, Trus Joist, I-Joist, etc.) used in buildings throughout the United States take advantage of the following:

- The airspace between joists to provide airborne sound isolation performance
- Additional mass toppings to improve airborne sound isolation
- Resilient connections and/or resilient underlayments to further improve airborne and impact sound isolation

While the three sound isolation strategies used for lightweight timber structures noted above are just as effective for mass timber buildings, they are also less desirable given that exposing the wood structure (either as a floor and/or ceiling finish) is an attractive opportunity for mass timber structures. The ability to expose the wood structure as a finish can result in a more sustainable and more cost-effective solution (i.e., less material), not to mention that people generally like wood finishes given wood's "warm," "tactile," and "natural" connotations (see Fig. 6.7).

Unfortunately, employing the acoustic design strategies used for lightweight wood structures results in "covering up" the mass timber. The practical compromise is to expose one side of the mass timber and have an "acoustic assembly" either above or below the mass timber structure.

One approach is to use a concrete topping slab with a resilient underlayment to add mass and resilience on top of the mass timber structure. This approach must add adequate mass to meet the airborne sound isolation targets and adequate resilience to meet the impact isolation targets. While practical, this approach adds a significant amount of weight to the building that is not necessarily required for structure. Additionally, the concrete topping slab involves a wet trade on the construction site that may not otherwise be necessary.

FIGURE 6.7 Bellevue Church (Bellevue, Washington). Designed by atelierjones, acoustics by Arup. *(Image courtesy of Arup.)*

There is ongoing research investigating the use of sheet goods, such as gypsum cement board, plywood, homasote, and so on, to incorporate additional mass without the use of a wet trade.

Resilient hung ceilings are sometimes less attractive than the massive topping slab approach since a significant airspace is required to achieve adequate airborne sound isolation, thereby reducing floor-to-floor clear heights, and resilient flooring is often still required to provide adequate impact sound isolation performance, thereby covering up the CLT on both sides of the assembly.

In addition to CLT, there are a variety of other mass timber products, each with their own acoustic considerations (see Fig. 6.8):

- NLT—unlike CLT, all of the dimensioned lumber elements run in the same direction, which can often be more structurally efficient. Without crossing the timber members, NLT has "seams" between members that may deteriorate the airborne sound isolation performance when compared to CLT. The acoustic test data for NLT are even more limited than CLT, so much research is needed to fully understand and quantify the pros and cons of NLT versus CLT.

FIGURE 6.8 Images via rethinkwood.com (*CLT, NLT, DLT, glulam*).

- DLT—similar to NLT, except instead of using nails to fasten the dimensioned lumber, inset dowels are used (explained more here). DLT has similar acoustic design considerations as NLT and is the only mass timber product that does not use glue or nails.
- Glulam—typically used for beams and columns, so this is typically not an acoustic design consideration.

6.3.5 CLT Flanking Noise

Flanking sound is transferred via an indirect path between spaces (e.g., over/under a wall partition instead of through). Flanking sound control is an important acoustic design consideration wherever sound isolation is needed. Flanking control is especially critical for the detailing of mass timber constructions given that mass timber relies on additional "acoustic elements" to improve sound isolation through the structure and therefore has a greater risk that flanking paths are introduced (see Figs. 6.9 and 6.10).

Several example details are provided below to illustrate the acoustic design concepts and possible solutions.

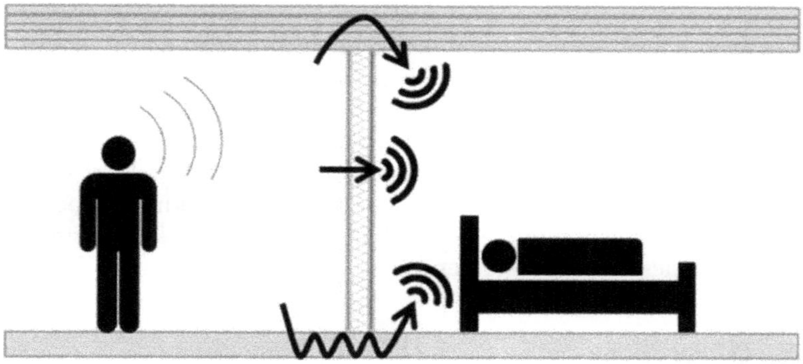

FIGURE 6.9 Airborne sound transfer and flanking illustration. (*Image courtesy Arup.*)

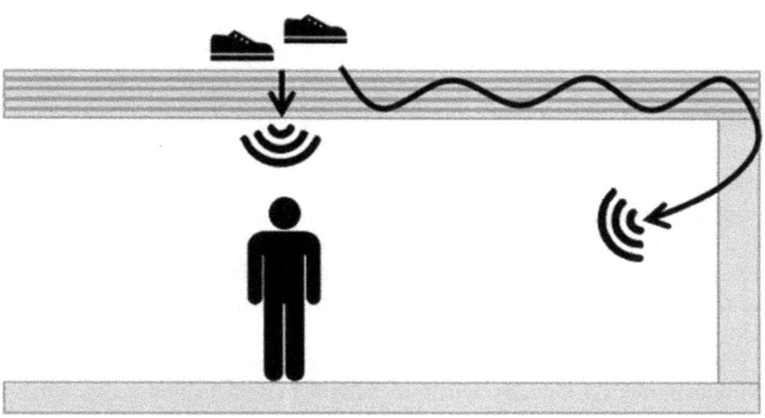

FIGURE 6.10 Impact sound transfer and flanking illustration. (*Image courtesy Arup.*)

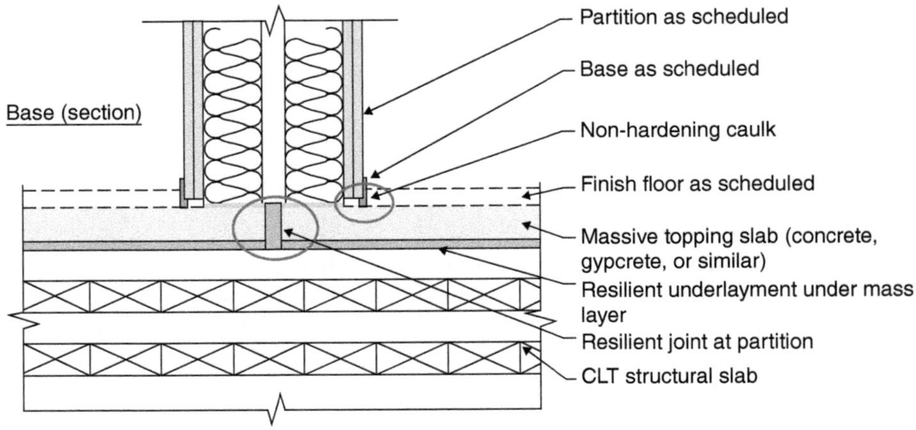

FIGURE 6.11 Example demising partition section detail. (*Image courtesy Arup.*)

Flanking noise control design can be intricate work and focuses on how the sound isolation strategies for the walls, floor, and ceiling come together to maintain the target separation at all intersections and interfaces (see Fig. 6.11).

Unfortunately, there is no "one size fits all" approach to flanking noise control design since the design considerations depend on (1) the target level of sound separation (e.g., the STC or IIC rating), (2) the particular acoustic design strategies used (e.g., massive topping slab vs. resilient hung ceiling vs. fully acoustically isolated systems etc.), and (3) the space adjacencies under consideration.

6.3.6 CLT Partitions

CLT structures typically minimally make use of CLT panels in floor slabs. CLT panels may also be used as load-bearing shear walls. The common alternative to this is post-and-beam framing to support CLT floor slabs or hybrid steel/concrete systems. The following sections concern acoustic considerations of CLT panels used in partition constructions.

Bare CLT Partitions

The sound isolation performance of bare CLT partitions is largely a factor of the thickness (and density) of the CLT panels used. Increasing surface generally mass results in improved sound isolation; thus, increasing the thickness or density of the CLT panel is likely to increase its TL.

Due to the relatively low density of wood and thickness of panels, single CLT panels suffer from decreased performance due to shear impedance.

Compared with higher-density single-panel constructions like concrete panels or CMU masonry, CLT underperforms at all frequencies due to decreased surface mass.

Single-panel CLT walls without added materials cannot practically achieve the required sound isolation requirements for demising partitions between residential units in a multifamily residential property and should not be used as such.

CLT Partitions with Laminated GWB

A common method to increase the effective surface mass of the CLT panel is to laminate gypsum wall board onto one or both sides of the CLT panel. This is also done in some instances to encapsulate the flammable wood material with the flame-retardant gypsum board.

It may be possible to achieve the required sound isolation requirements for demising partitions between residential units in a multifamily residential property by laminating. It is not recommended, as the multiple layers of gypsum board required would be inefficient compared to the techniques described in the following section.

CLT Partitions with Furred GWB

The sound isolation performance of CLT panel partitions can be greatly increased by adding layers of gypsum board to each or either side of the CLT panel on framing that provides an air gap between the gypsum board and the CLT panel. Further benefit can be derived in a cost-effective manner by including fiberglass or mineral wool batt insulation in the airspace and decoupling the framing from the CLT panel.

For demising partitions between residential units, anticipate either of the following minimum constructions:

- One layer of ⅝-in. gypsum board mounted on a 1-in. hat channel attached to both sides of a five-ply CLT panel
- Two layers of ⅝-in. gypsum board mounted on 2-in. Z-channels with batt insulation in the cavity attached to one side of a five-ply CLT panel
- One layer of ⅝-in. gypsum board mounted on a 3⅝-in. wood or metal stud with batt insulation in the cavity framed independently (with a 1-in. air gap) on one side of five-ply CLT panel

Where demising partitions align with glulam post-and-beam construction, note that the furred gypsum construction should continue around to box out the posts and beams, as indicated in Fig. 6.12.

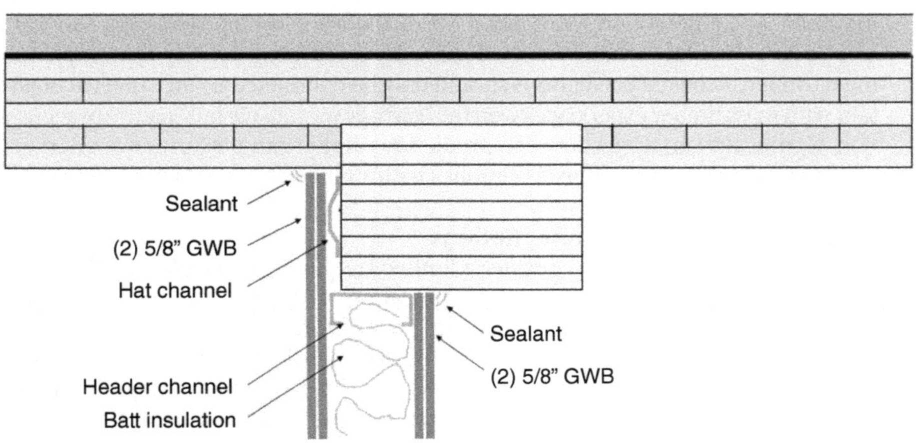

FIGURE 6.12 Furred gypsum construction. (*Image courtesy Arup.*)

6.3.7 Lab Tests and Other Analysis Methods

ISO/ASTM Tests for Sound Transmission and Footfall/Impact

Acoustic testing of mass timber assemblies in accredited laboratories is a critical step to the widespread adoption of mass timber. Lab tests provide a standardized and repeatable method to objectively characterize the acoustic performance of various assemblies, which is needed for owners and project teams to confidently design solutions for their buildings.

Several ASTM and ISO standards exist to measure and quantify the acoustic performance building partitions. The airborne sound isolation of the partitions (STC) may be measured/quantified using *ASTM E90: Standard Test Method for Laboratory Measurement of Airborne Sound Transmission Loss of Building Partitions and Elements.*

Impact sound isolation (or IIC) may be measured/quantified using *ASTM E492: Standard Test Method for Laboratory Measurement of Impact Sound Transmission through Floor-Ceiling Assemblies Using the Tapping Machine.*

If the project's basis of design floor assembly has not already been acoustically tested, it is recommended that the floor system be installed and tested in a certified acoustic test laboratory per ASTM E492 and ASTM E90 to confirm that the target acoustic performance can be achieved.

Note that the current ASTM testing standards for impact and airborne sound isolation report only a limited band of frequencies (100 to 3150 Hz in ⅓-octave-band resolution). In order to test the low-frequency performance, the testing laboratory should be able to reliably test down to at least 50 Hz. This may involve the use of testing methods other than the standard ASTM tests noted above.

The "Drop Ball Test" (ISO 140-11) may be utilized to test the low-frequency performance of the floor assembly. The method excites the floor assembly by dropping a standardized 2.7-kg ball from a height of 1 m onto the assembly and measuring the sound produced in the chamber below.

Acoustic Prediction Software

If a prototype of the mass timber assembly cannot be fabricated and tested in an accredited testing laboratory, the acoustic performance may be predicted using first principles and/or commercial software packages (e.g., INSUL, sound flow). These prediction methods typically rely on a combination of theory and empirical data sets for accurate results and therefore should be used carefully given the relativity few empirical data for mass timber. Acoustic predictions should only be conducted by an expert acoustic consultant with mass timber experience given the "garbage in, garbage out" nature of acoustic prediction. Significant variations between predictions and in situ performance often occur, even with traditional constructions/assemblies with well-established acoustic performance.

Limitations of Single Number Ratings

Note that STC and IIC only consider a limited frequency range (ASTM E336 test method is valid only from 100 to 5000 Hz, and the ASTM E1007 test method is valid from 100 to 3150 Hz), while humans can generally hear from 20 to 20k Hz, that is, two octaves above and below the frequency range that STC/IIC even consider. Furthermore, the performance across this limited range is condensed down to a single number rating, resulting in a wide range of performance that might fall within the same single number rating.

To complicate things further, how we experience sound is largely contextual and depends on the holistic acoustic environment. For example, your ability to hear your neighbor's stereo depends on how quiet your apartment is as well as room size and finishes.

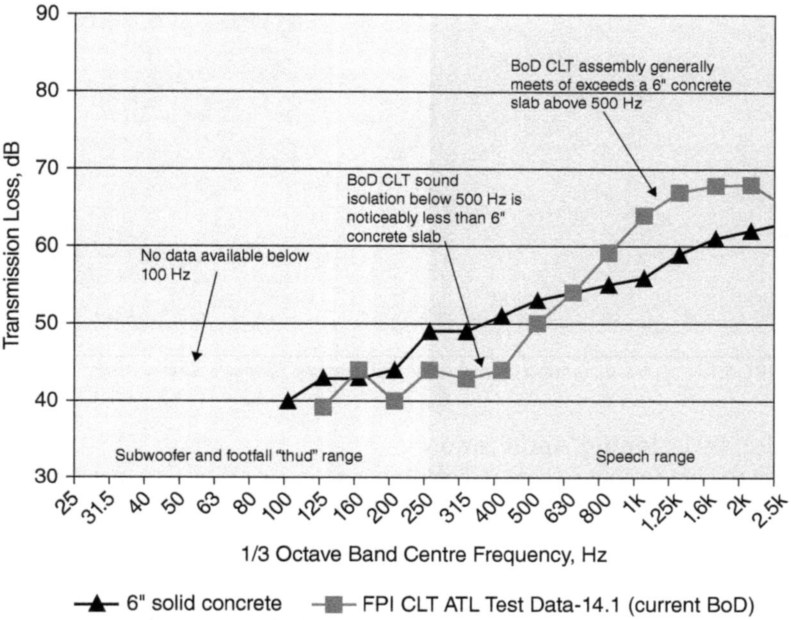

FIGURE 6.13 Airborne sound isolation performance comparison of concrete versus CLT + acoustic topping. (*Image courtesy Arup.*)

All this to say that the single number acoustic rating system does not tell the whole story, nor does it even necessarily correlate to how we experience sound in a particular context.

While mass timber assemblies can readily be designed to meet or exceed the broadband acoustic metrics (e.g., STC, IIC), the low-frequency sound isolation performance is inherently weaker than concrete for both airborne and impact sound isolation given the reduced mass. Note that neither STC nor IIC adequately addresses low-frequency performance, especially below 100 Hz incidentally where contemporary audiovisual media equipment, such as subwoofers, operate. Low-frequency sound, including footfall activity, by residential activity will likely be more audible across mass timber partition constructions than concrete equivalents. For more information, see Figs. 6.13 and 6.14.

FIGURE 6.14 Acoustic testing in the field. (*Images courtesy Katerra.*)

FIGURE 6.15 Acoustic testing in accredited laboratory. (*Image courtesy Arup.*)

Design by Listening/Auralization

Given the limitations of the single number acoustic ratings, more advanced tools may be employed to determine the acoustic performance required for a particular project. "Auralization" is a term introduced to be used in analogy with visualization to describe rendering audible (imaginary) sound fields. Aurlaization may be used to subjectively assess the performance differences between a variety of design options.

Current-day auralization methods can be extremely useful way to communicate the acoustic experience that various design options can provide. As such, owners and design teams do not need to rely on esoteric acoustic metrics (which do not tell the whole story), nor do they need to be an acoustician to understand what level of performance their future will provide to its building occupants. When accurate and precise auralizations are combined with cost and constructability considerations, owners and design teams are enabled to assess the cost benefit of the various acoustic design options (see Figs. 6.15 and 6.16).

FIGURE 6.16 Aurlaization space (Arup SoundLab, New York). (*Image courtesy Arup.*)

6.3.8 Resources for Acoustic Test Data

Government Entities

- NRC-CNRC Apparent sound insulation in CLT: http://nparc.nrc-cnrc.gc.ca/eng/view/fulltext/?id=27e0a78f-5739-4b23-90e1-442155ff4d25

Industry Entities

- CLT Handbook (United States): https://www.thinkwood.com/products-and-systems/clt-handbook
- CLT Handbook (Canada): https://fpinnovations.ca/Extranet/Pages/AssetDetails.aspx?item=/Extranet/Assets/ResearchReportsWP/E4642.pdf#.WwiGXUgvyUl

Manufacturers

- StructureCraft: https://structurecraft.com
- StructurLam: http://www.structurlam.com
- DR Johnson: http://www.drjlumber.com
- Pliteq: http://www.pliteq.com
- Nordic Structures: https://www.nordic.ca/en/home
- FP Innovations: https://fpinnovations.ca/Pages/index.aspx
- Acousti-Tech: https://www.acousti-tech.com

6.4 Future Research and Development

CLT construction is in its nascent stages, particularly with regard to acoustics. There is significant future research and development to be done to encourage the adoption of CLT structures and to make them as well documented and understood as traditional building structures. The following research areas are a short list of examples of the additional work that should be undertaken to help make mass timber a viable building option:

- Develop floor/ceiling constructions that achieve sound isolation and impact noise requirements for multifamily residential, particularly without a drop ceiling below and without a wet-trade mass layer above the CLT slab.
- Increase the quantity of in situ field measurements of sound isolation and impact noise performance to add to an understanding of how these structures perform in practice.
- Conduct cost-benefit analyses of various CLT construction methods to achieve reasonable acoustic performance.
- Incorporate constructability benefits of CLT with acoustic requirements.

6.5 Acoustic Consultants

This chapter has only just scratched the surface of acoustic design for mass timber buildings. Our aim is to reveal some of design challenges and opportunities associated with mass timber construction as well as recent developments that help make building with mass timber a more viable alternative to traditional approaches. Anyone designing a mass timber building should engage with a qualified acoustic consultant to ensure that the project's acoustic needs are considered at the early stage of design and considered holistically with the building design. Qualified consultants may be found via the National Council of Acoustic Consultants (NCAC); however, experience with mass timber buildings should be a key consideration when selecting a consultant.

6.6 Clarifications

This chapter includes information collected from a variety of sources as well as the design experience of the authors. It is recommended that the information and solutions provided herein be validated against the current research, product developments, expert opinion, and the latest design and construction experience given the constantly evolving mass timber market. The authors would like to reiterate that the information contained within this chapter is based on our design and construction experience with mass timber and should be challenged by competing opinions and experience.

Arup is an independent firm of designers, planners, engineers, consultants, and technical specialists, working across every aspect of today's built environment. Together we help our clients solve their most complex challenges—turning exciting ideas into tangible reality as we strive to find a better way and shape a better world.

6.7 A Note from the Authors

At the time of writing, Denis Blount and Matt Mahon were employed by Arup, an independent firm of designers, planners, engineers, consultants, and technical specialists, working across every aspect of today's built environment. Together we help our clients solve their most complex challenges—turning exciting ideas into tangible reality as we strive to find a better way and shape a better world. While the authors understand the challenges that come with mass timber construction, we believe that a shift to building with more renewable resources (such as wood) is an essential shift in the construction industry to help reduce our carbon footprint in the world.

CHAPTER 7

Fire Safety for CLT Projects

David Barber

Arup

7.1 Basics of Timber Reaction to Fire

7.1.1 Wood Structure and Chemistry

Wood is a cellular, composite material harvested from two broad categories of trees known commercially as softwood and hardwood. Overall, the dry weight of wood is made up of approximately 50% cellulose, 25% hemicellulose, and 25% lignin.[1]

Softwoods have a simpler basic structure because they have only two cell types: axial (mainly axial tracheids) and radial (mainly ray parenchyma). Hardwoods are structurally more complex because they have a greater number of basic cell types and variability within those cell types.[2] Both the microstructure variations and the types of extractives present among different species manifest into differences in fire performance. In general, woods of higher density and moisture content have better fire performance.[3] The limited extractive contents of hardwood species reduce their overall variability in flame spread and heat release compared with the softwoods. Most wood used in construction are softwoods.

7.1.2 Wood Pyrolysis and Combustion

The process of heat transfer through a solid is governed by conduction, and when gases are involved, the heat transfer processes are radiation and convection. Since wood is a porous material, all three modes of heat transfer occur. As wood is exposed to elevated temperatures, a complex series of chemical reactions and physical transformations occur, beginning with hydrolysis, then oxidation, dehydration, pyrolysis, and, finally, combustion.[4] Pyrolysis is the thermal decomposition of organic matter at elevated temperatures that leads to the release of volatiles (or flammable gases) and produces tar and char. Once the char is formed, it acts as an insulating layer that controls the pyrolysis kinetics and impedes further degradation of the wood.

At temperatures between 280°F and 390°F (140°C and 200°C), dehydration takes places, and production of acetic acid from the hemicelluloses and noncondensable gases, such as CO_2, occurs.[5] As the wood absorbs energy and the temperature increases, the breakdown of the major polymeric components of wood begins. Hemicelluloses are the most thermally sensitive and begin to break down between 390°F and 500°F (200°C and 260°C).[6] Cellulose thermally decomposes between 460°F and 660°F (240°C and 350°C). Finally, lignin thermally decomposes between 535°F and 930°F (280°C and 500°C) with the dominant result being the formation of char.

7.1.3 Wood Char

Char is a commonly used term when discussing the carbonaceous residue left behind after wood burns. Rowell states that char is a "solid residue that is neither intact substrate nor pure carbon, but a different material at various stages of charring and carbonization."[6] Browne defines char the "non-volatile residue" that "assumes a hexagonal graphitic network in which the carbon-to-carbon bonds are unbreakable by pyrolysis alone to temperatures beyond 3000°C."[7]

In structural fire engineering, an isotherm has become a widely accepted definition of char with the delineation for char being present when the wood reaches a temperature of 570°F (300°C).[8] Fonseca and Barreira have developed an even more general definition of char such that "all pyrolyzing wood can be considered char."[9] Because wood char is largely formed from the cellulose and lignin constituents, the initial chemical composition of the wood will affect the relative thickness of the char layer.[10] Char is highly insulating and protects the remaining wood.[11]

7.1.4 Char Rates

The char rate is usually defined as the depth into the wood at the chosen temperature divided by the amount of time to reach this temperature (see Fig. 7.1). Because of the char's insulating properties, initially, when wood is exposed to pyrolysis temperatures, the char rate is fast, and then, once a layer of char is formed, the char rate decreases. The char layer formed decreases the char rate by limiting the transport of gases involved in combustion, acting as thermal insulation for the underlying wood, and reducing the production of volatiles, impeding the continuation of the combustion cycle.[4] Char rates are especially important for determining the fire resistance of structural timber for buildings and are referenced extensively throughout both prescriptive and performance-based building codes. The method of predicting a charred depth of timber for structural assessment is referred to as the "reduced cross-section" method.[12] Char rates determine how quickly the cross section of a structural member decreases such that a critical design load level can be evaluated. Char, plus a reduced strength "heat-affected zone" directly behind the char layer, termed the "zero-strength" layer, is used in structural capacity calculations.

Fire resistance is a measure of the ability to resist collapse or other failure during exposure to a fire.[13] The fire-resistance rating (FRR) of a structural member or assembly, wood-based or otherwise, has traditionally been obtained by testing the assembly in a furnace. To achieve an FRR, the building element is located within a furnace, appropriate loads are applied, and the element must resist a predetermined time-temperature curve (the "standard fire"[14]). Pass or failure is measured against criteria for load bearing,

FIGURE 7.1 A slice through a Douglas fir glulam column after a standard fire test to 90 minutes, showing the reduced cross-sectional area due to char.

integrity, and insulation. Char rates under the standard fire exposure are well documented and provide a means by which to compare differing engineered timbers. Ignition occurs in about 2 minutes, and charring into the depth of the wood then proceeds at a rate of approximately 1.89 in/hr (0.8 mm/min) for the next 8 minutes, and thereafter the rate decreases to 1.4 in./hr (0.59 mm/min).[15] The char rate measured under the standard fire does vary; however, for practical purposes, the commonly assumed nominal charring rate [1.5 in./hr 0.64 mm/min] is used for all exposed wood members in the absence of other data.

Standard fire testing is important for building construction given the need to provide a means by which architects, engineers, and authorities having jurisdiction can specify and check compliance of a building element against building code required fire ratings. Standard fire testing does have limitations for large timber members, such as CLT floors or walls, given the material is combustible and the amount of energy input required to achieve the standard time-temperature curve will vary with material.[16] Given the limitations in knowledge that standard testing can provide researchers and engineers, charring rates for fire exposures other than the standard fire has become more relevant and prevalent. In 1967, Schaffer determined charring rates for constant temperature exposures at 538°C, 815°C, and 927°C.[10] Many have calculated charring rates and developed pyrolysis models based on fixed heat fluxes between 15 and 100 kW/m^2.[17–20]

More recently, a trend toward using nonstandard fire scenarios has occurred to evaluate the fire performance and determine char rates for mass timber components. These nonstandard fires are often referred to as "natural" fires and often more closely represent a real fire scenario with a decay phase when compared to the standard fire.[20,21]

7.2 Building Stability during Fire

7.2.1 Expected Building Performance When Exposed to Fire

Building primary structural elements are required to perform to an acceptable standard during a fire to enable safe evacuation of the occupants, to prevent fire spread to neighboring buildings, and for firefighters to conduct their search, rescue, and extinguishment operations.[22] Model building codes, such as the International Code Council's International Building Code[23] (IBC), provide guidance on fire protection to buildings and structures, based on use, numbers of occupants, area, and height. Low-rise buildings of up to three stories have minimal fire protection for the structure, and the structure may suffer significant damage in a fire. Buildings are permitted to be larger or taller if they include a fire-rated structure or fire-separated compartments, if they are separated from neighbors, or if sprinkler protection is installed. Medium-rise buildings of up to 85 ft (25.9 m) high are required to have a 1-hour FRR with sprinkler protection and are restricted in area and use. High-rise buildings that are 75 ft (22.9 m) or more to the highest occupied floor are required to have an FRR of 2 hours to the primary structure, sprinkler protection, and a number of additional fire protection features. For buildings over 420 ft (128 m), all fire ratings are 3 hours for the primary structure and 2 hours for floors.

7.2.2 High-Rise Structure Fire Resistance

For high-rise buildings, the level of fire protection to support life safety and firefighting actions is increased. There is also more redundancy within the required fire protection systems for high-rise buildings. The fire protection for the structure is increased such that collapse should be prevented in the event of a fire.[24] Higher fire ratings are required to support the longer evacuation time for occupants and fire department intervention via the fire-protected stairwells and fire service access elevators. The implicit intent of the increased fire-resistance requirements specified by code is for the structure to survive the burnout of a severe, realistic "fully developed fire," being a fire that burns until decay.[25] Hence, there is a significant difference in expected structural performance in fire for a high-rise building when compared to a medium- or low-rise buildings, with high-rise buildings required to survive a fully developed fire (structure survives through to fire decay), assuming an extreme scenario of the sprinklers not activating and firefighting intervention not occurring.

7.2.3 High-Rise Mass Timber Buildings

A significant change has been introduced to the 2021 IBC with the adoption of three new mass timber construction types, including two that permit a high-rise mass timber structures.[26] The code changes will allow mass timber buildings to be constructed up to 12 floors with construction Type "IV-B" and up to 18 stories with construction Type "IV-A." Type IV-B will allow partially exposed mass timber, while Type IV-A requires all mass timber to be protected with fire-rated gypsum board.

For jurisdictions that have not adopted the 2021 IBC, high-rise buildings are required to have noncombustible construction used for the primary structural elements. A high-rise building utilizing mass timber can only be approved by the authority having

FIGURE 7.2 University of British Columbia Brock Commons building, a high-rise building constructed by using an alternative design approach.

jurisdiction through an "alternative materials, design, and methods of construction and equipment" approach to a building permit. Examples of high-rise buildings with mass timber construction, approved through an alternative engineering approach, include the 18-floor Brock Commons building in British Columbia (see Fig. 7.2).

7.3 CLT Fire-Resistance Rating

7.3.1 CLT Manufacturer Compliance Fire Testing

Where CLT is part of the primary structure (load-bearing element) and requires an FRR, as with all construction materials, standard fire testing or calculation methods are used to prove load-carrying capacity under fire exposure. Where fire testing is used, the IBC requires the testing be carried out to ASTM E119[14] or UL 273,[27] with a tested CLT panel compliant with the requirements of ANSI/APA PRG-320.[28]

Standard fire testing of CLT panels occurs primarily for product compliance, and each CLT manufacturer is responsible for carrying out their own fire testing to meet the requirements of ASTM E119 or UL273. By passing a standard fire test on a wall or floor, to achieve either 1- or 2-hour FRRs, the manufacturers are providing authorities having jurisdiction, architects, engineers, and contractors with products that are certified for use and that meet the requirements of the IBC (see Fig. 7.3). Each manufacturer provides their fire test results for a specific panel layup. When CLT is specified as part of the primary structure, the building permit documentation is required to include copies of the panel manufacturer's compliance fire tests and, if required, assessment of fire ratings meeting the NDS[12] or CLT Handbook,[29] where variations to the fire test in CLT layup are present.

FIGURE 7.3 CLT floor after a 2-hour fire test to ASTM E119. (*Image courtesy of DR Johnson.*)

7.3.2 Methodology for Calculating a Panel FRR

To determine the char rate and the heat-affected zone for CLT, standard fire testing was initially carried out in Europe,[30–32] with additional testing in Canada.[33–35] By carrying out standard fire tests on panels, data have been gathered and utilized to better understand how panels respond to fire, which can be compared to the results of other engineered timbers, such as glulam. Fire testing was also aimed at showing consistency in the results between European-produced panels and North American-produced panels given that there was a longer manufacturing and testing history in Europe when compared to the more recently developed North American manufacturing market.

The fire testing has shown the non-charred timber of the CLT panel will retain its strength and continue to support applied forces. The reduced cross-section method allows a depth of char to be calculated easily for design purposes so that the structural capacity can be assessed (see Fig. 7.4). This is documented within the CLT Handbook.[29] The correlation allows the construction industry to assess the fire resistance of panel layups that differ from those fire tested, without needing to undertake additional fire testing, based on the pyrolysis fundamentals. The correlations for assessing fire resistance are appropriately conservative.

The effective residual cross section of the CLT floor is calculated from (see Fig. 7.5)

$$h_{\text{fire}} = h - a_{\text{char}}$$

where h_{fire} = effective cross-section depth (in.)
h = initial cross-section depth (in.)

The char depth can be calculated as

$$a_{\text{char}} = 1.2\ [n_{\text{lam}} * h_{\text{lam}} + \beta_{\eta}\ (t - (n_{\text{lam}} * t_{\text{gi}}))^{0.813}]$$

FIGURE 7.4 CLT floor panel after a standard fire test to ASTM E119, cut to show char depth and full-strength timber.

where a_{char} = effective depth of char
n_{lam} = number of laminations that may burn through
h_{lam} = lamella thickness
β_{η} = nominal charring rate = 1.5 in./hr
t = required fire resistance (h)
t_{gi} = time required to reach a glue laminated interface (h)

FIGURE 7.5 CLT annotated for char depth calculation, as per CLT Handbook, U.S. Edition.

and

$$t_{gi} = (h_{lam}/\beta_{\eta})^{1.23}$$

and

$$n_{lam} = (t/t_{gi})$$

where the number of laminations is rounded up to the lowest whole integer.

The method for assessing fire resistance of panels in North America differs slightly from the European method.[36,37] The European approach reflects the maturity of the European marketplace for panel supply and fire testing research. The two methods for assessing CLT FRR provide similar outcomes.

7.4 Interior Finish and Flammability

Building codes control the flammability of interior finishes (interior linings) to limit early fire development, where occupants may be impacted during their egress. By controlling flammability and smoke development of finishes, fire growth can be slowed, and building occupants can take longer to evacuate in tenable conditions. Interior finishes include floors, ceilings, or walls. Where a combustible product such as timber is used as an interior finish, the influence on fire growth is required to be understood and potentially limited in its use or combustibility.

7.4.1 Test Methods and Results

The flammability of all interior wall and ceiling finish materials is required to be proven through fire testing to ASTM E84[38] (see Fig. 7.6). The test methodology is based on a relative measure of flammability and smoke development. The test allocates a classification, being A through C, where "Class A" is the least flammable material and "Class C" is more flammable.

The IBC limits interior finishes to walls and ceilings within most occupancy types to Class C, when tested to ASTM E84, with sprinkler protection installed to the building. CLT can therefore be used as an interior finish given that most CLT panels will meet Class B or Class C when tested to meet ASTM E84. Data on interior finish classifications for various wood species can be found from the American Wood Council[39] and suppliers of CLT.

7.4.2 Methods to Control Flammability and Smoke Development

Where CLT is being used as an interior finish within a building area that requires a finish classification more restrictive than the panel has achieved, the flammability can be controlled through the application of a clear fire-retardant coating. Once applied, these coatings will normally provide a timber surface with a Class A classification. It should be noted that these topical coatings are effective at reducing the flammability and smoke development of the timber but may not have any significant impact on the fire resistance of the CLT.

FIGURE 7.6 CLT panel having undergone an ASTM E84 test.

7.5 CLT Behavior under Fire Exposure

7.5.1 CLT Panel Reaction to Fire

Bench scale, standardized furnace, and natural fire tests have shown a general consistency in how CLT reacts to fire. When a CLT panel is exposed to the heat of a fire, the reaction will depend on the thickness of the panel, the type of adhesive used in the manufacture, and the thickness of the plys (lamella). Also, to a much lesser degree, the fire performance is influenced by the species of the lumber, the width of the lumber used, the method of adhesive application (face application only or face and edge application), and the stress induced by the application of any load.

When a CLT panel is exposed to a radiative heat flux of more than 20 kW/m^2 for a duration exceeding 60 minutes, produced by a radiant panel or exposure to a compartment fire, the initial charring behavior in the first ply is identical to that of sawn timber or glulam. As the charring reaches the adhesive line, one of two events will occur. Either the charring continues consistently through the adhesive line, or the protective char may dislodge due to lack of adhesive strength under heating, exposing the unburnt wood below. The behavior of char falloff at the adhesive line when a CLT panel is exposed to fire has been referred to as delamination, char debonding, or char stickability. Char falloff is the process where small pieces of mostly charred wood separate from the

FIGURE 7.7 CLT panel that has been fire tested and is showing char falloff.

unburnt CLT base as the charring reaches the ply interface (adhesive line).[30–32,34,40] The separation of the char occurs due to the ply adhesive losing its strength under increased temperatures (see Fig. 7.7). CLT panels located horizontally (underside of floors) are more susceptible to char falloff than panels located vertically (walls). The process of char falloff has been shown to be adhesive dependent (see later section on adhesives).

When char falloff occurs, the heat insulating function of the char is lost. The unburnt timber below the char then becomes exposed to the heat of the fire, and there is a rapid localized increase in the pyrolysis (burning rate) and hence charring rate. This increased burning occurs until a new char layer is formed, which then insulates the remainder of the timber, and the normal char rate returns. Not all CLT panels are susceptible to char falloff, and char falloff only impacts CLT (other engineering timber products have a linear charring rate at the adhesion line). Char falloff is an important issue for a CLT panel that is exposed to a fire, impacting the panel fire resistance and how the compartment fire may behave.

For fire resistance, the load-bearing capacity of the panel is directly impacted by the increased burning rate (char rate) when char falloff occurs. With a faster charring rate, compared to a panel that does not have char falloff, there will be less cross-sectional area of ambient temperature timber to support applied loads for a set fire resistive duration, such as 2 hours. Char falloff is accounted for within the methodology to calculate char rate in panels exposed to the standard fire, within the CLT Handbook.[29]

Delamination or Char FallOff?

Delamination is a term commonly used to describe char falloff for CLT. The term "delamination" was originally defined within a plywood standard[41] as "A visible separation between plies that would normally receive glue at their interface and be firmly contacted in the pressing operation." This definition is based on a bench-scale test by Schaffer[42] for determining failure of plywood at elevated temperatures, where plys are not permitted to separate when exposed to an open flame, as part of the heat durability test for adhesives. Given the term "delamination" is relevant for plywood, the term "char falloff" is more typically used to describe the response that some CLT panels will exhibit when exposed to longer term fires.

7.5.2 Reaction of CLT Adhesives in Fire

Adhesives have been used to laminate wood for well over 100 years, with glulam being the oldest example. Adhesives typically used are phenolic based, being resorcinol-formaldehyde (RF) resin, phenol-resorcinol-formaldehyde (PRF), or a melamine urea formaldehyde (MUF or MF). Due to the concern over formaldehyde-based adhesives, engineered woods are more commonly using polyurethane (PUR) adhesives. PUR is also a phenolic-based adhesive but is formaldehyde-free.

Experimental testing of wood adhesives under fire conditions has been carried out at the Forest Products Laboratory since the 1960s. Schaffer's[42] fundamental work showed that the performance of wood adhesives varies when exposed to fire, depending on the adhesive type. Schaffer showed that some adhesives are more susceptible to failure at the adhesive line than others. Schaffer's work led to MUF and PRF being the preferred adhesives for engineered structural wood products, such as glulam, given their stability under fire conditions. A significant number of fire tests in glulam members have proven the resilience of MUF and PRF adhesives in fire.[43,44]

CLT Panel Fire Testing and Adhesive Reaction

The process of CLT char falloff has been shown to be adhesive dependent.[29,34,45,46] The impact of adhesives on CLT stability in fire was first investigated by Frangi,[47] with testing on a range of PUR and MUF adhesives showing that char falloff only occurred in the PUR-bonded CLT. Frangi observed failure in the PUR-bonded plys between 390°F and 570°F (200°C and 300°C) at the ply interface. Frangi also recorded charring rates of 1.4 in./hr (0.6 mm/min) in the MUF-bonded CLT, indicating the plys were charring through the adhesive line. On the other hand, the PUR-bonded CLT showed an increase in char rate due to the char falloff. Frangi concluded that CLT panel performance in fire is dependent on adhesive type used. It was also noted that for the same overall thickness of CLT, a panel with thicker plys is more fire resistant than the same thickness CLT panel with a greater number of thin layers. The thicker plys minimize the increase in burning rate at the inter layer transition and hence produce a higher fire resistance.

Further evidence of adhesive performance impacting char falloff has been shown in the fire testing by Friquin,[47] which showed no evidence of char falloff with CLT panels bonded with MUF adhesive, with a limited number of tests. Further, a comprehensive and detailed review of CLT fire testing and the influence of adhesives was completed by Klippel.[30] Klippel noted that PRF adhesives have an improved performance at elevated temperatures. He also noted that PUR adhesives perform relatively poorly with a ductile response when heated and that the performance of one PUR adhesive may not represent other PUR adhesives given the wide range of strength characteristics. The instability at the ply interface discussed by Klippel and Frangi is consistent with the findings from Medina[48] and Aguanno.[40] The influence of adhesive type and the relatively consistent result of PUR adhesives failing to adhere at temperatures above 390°F (200°C) are major factors in char falloff. This has also been proven in CLT compartment fire testing.[49,50]

7.6 Influence of Exposed CLT on a Compartment Fire

7.6.1 Why Carry Out Natural Fire Tests on CLT Compartments?

Where CLT is exposed and the fundamental fire behavior needs to be understood, such as within high-rise buildings, the impact of the exposed mass timber on the compartment fire is a key design consideration. The additional fuel introduced by the exposed mass timber will increase the compartment heat release rate (HRR) and duration of the fire in the highly unlikely scenario of a fully developed fire (sprinklers failing and fire department not intervening). The fully developed fire is a scenario required to be assessed for most structural analysis. With the increase in heat release and fire duration, the structural capacity of the load-bearing CLT member needs to be determined based on both heating and cooling of the compartment fire. Once the combustible fuel of the furniture, fixtures, and contents are consumed by the fire, understanding if the radiative heat from the decaying fire will keep the timber pyrolyzing or lead to fire decay is vital. For a high-rise building, the structure is required to remain intact and resist the applied forces while the fully developed fire decays.

To better understand how compartments with areas of exposed CLT panels (not protected) will respond to a fully developed fire, natural fire tests have been carried out. A CLT compartment is constructed with a defined fire load, and the fire can grow to full development and then decay, with temperature and HRR data recorded inside the compartment and temperatures within the panels recorded, with an assumption of no sprinkler activation or fire department action. Small-scale and standard testing provides limited information, whereas natural fire testing is highly informative, as it reflects actual fire conditions.

7.6.2 CLT Compartment Fire Tests—Results and Discussion

Natural fire tests on compartments with exposed CLT have been carried out by various research organizations, including ETH,[51] VTT,[52] Carleton University,[48,53] Delft University of Technology,[54] University of Edinburgh,[49] SP Norway,[55] University of Queensland,[56] FPRF,[50] U.S. Forest Service (USFS) and AWC,[57] Tallinn University of Technology (TUT),[58] and NRCC[59] (see Figs. 7.8 and 7.9). These natural fire tests have shown that a combination of exposed surface area, configuration (relative orientation of exposed surfaces), reliability of the CLT protection, and CLT susceptibility to char falloff can influence fire behavior and whether the compartment fire decays.

Fire Temperatures and Flashover

Where the CLT is exposed, the VTT, Carleton University, Delft University, University of Edinburgh, FPRF, USFS, TUT, and NRCC tests have shown that compartment temperatures were only increased marginally during the early stages of fire development. The peak compartment temperatures were not significantly impacted by the exposed timber, mainly due to the compartment fire tests being ventilation controlled and initial temperature peak due to other compartment fuels. In addition, the tests showed only minor differences in times to flashover (with 600°C room temperature criteria). Results from the FPRF tests have shown that the exposed CLT has some influence on time to flashover, but not significant, with a 6-minute variation between tests.

FIGURE 7.8 CLT compartment fire from the Fire Protection Research Foundation series of tests (undertaken by National Research Council, Canada, with RISE, Sweden, at the National Institute of Standards and Technology).

Impact of Exposed CLT Surfaces and Fire Decay

A number of the test series included partially protected CLT compartments—those by Carleton University, University of Edinburgh (see Fig. 7.9), Delft University, FPRF, USFS, TUT, and NRCC. The area and location of the exposed CLT surfaces will impact the compartment fire, and fire decay may be delayed or not occur if there is char falloff in

FIGURE 7.9 University of Edinburgh and Arup CLT compartment fire test.

the CLT. The key consideration for compartment fire decay is that once all the movable fuel (fixtures and fittings) within the compartment are consumed by the fire, the fire HRR starts to decay, and there is insufficient radiation from the fire to keep the exposed CLT ignited. This decay in the fire occurs even if there is available ventilation, as the fuel and subsequent radiative flux that causes the CLT panels to pyrolyze, have been removed by the fire decaying. This decay in HRR can be adversely impacted by CLT char falloff.

If char falloff occurs, unburnt timber is exposed to the fire, with a resulting increase in burning rate at the freshly exposed timber, until a new layer of char has formed. Where the extent of the char falloff is significant, this can produce an additional growth stage in the fire (termed a "second flashover") as more fuel (exposed timber) becomes available (see Fig. 7.10). The extent of the fire regrowth is influenced by CLT char falloff, the location of the CLT, and the area of exposed CLT. Fire test results have shown consistency in that exposing large areas of mass timber can lead to sustained burning (observed in fire tests by Carleton University, SP Norway, University of Edinburgh, and FPRF). The USFS tests have been the largest CLT compartment test fires to date, with a floor area of 900 ft.^{2} (81 m^2). These tests showed that with very limited areas of exposed CLT that is susceptible to char falloff and good compartment ventilation, a fully developed fire can decay. The NRCC tests are highly relevant, as they included CLT panels that included adhesives to prevent char falloff. These tests included exposed areas of CLT at the walls and ceiling and included an exposed beam and column. The NRCC tests showed that a higher proportion of CLT can be exposed in a compartment and fire decay can occur where char falloff is eliminated.

The conditions for CLT have been the focus of research[54,60,61] that has been aimed at quantifying the conditions for the transition from flaming combustion to smoldering combustion to extinguishment. The results have been generally consistent in addressing the heat flux received for the transition between these combustion phases and correlating these to mass loss. Bartlett and Emberley have shown when the mass loss rate for the pyrolyzing CLT is below a value in the range of 4 to 6 g/m^{2}s, there is a transition from flaming. Crielaard's work showed that the transition from smoldering to self-extinguishment occurred when received heat flux was below 10 kW/m^2. If the CLT experiences char falloff during the transition from flaming combustion through to smoldering combustion, then the exposed unburnt wood can result in fire regrowth. Thus, having CLT panels that are not susceptible to char falloff is critical for a compartment with exposed timber to have fire decay. If fire HRR decay does not occur, the structure is at risk in the highly unlikely scenario of sprinklers failing and no fire department intervention given that for high-rise

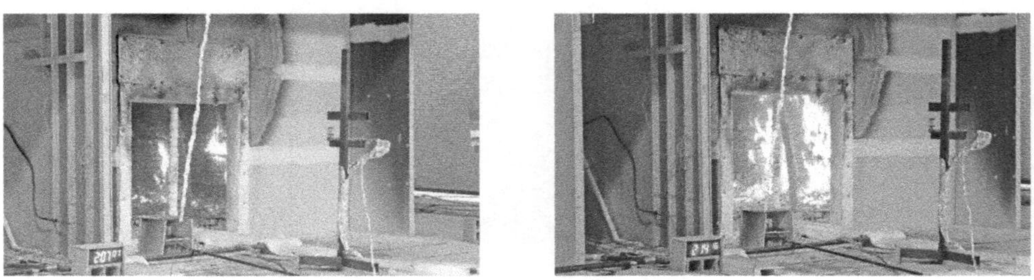

A B

FIGURE 7.10 Regrowth of exposed CLT wall, 2 hours and 7 minutes into a compartment fire (A) and 2 hours and 14 minutes into fire (B); from FPRF compartment fire test.

buildings, the structure is required to survive burn out. The interaction of compartment HRR, received heat flux, CLT susceptibility to char falloff, reliability of CLT protection, and compartment ventilation is important in determining the compartment fire dynamics and whether compartment fire decay occurs.

CLT Protection

Fire testing has shown that limiting the area of exposed CLT will lead to fire HRR decay. To limit the area of exposed CLT, the noncombustible protection must remain in place for the duration of the fire. Both standard and natural fire testing has shown that protection to the mass timber is not always effective. CLT compartment tests involving two exposed CLT surfaces have become tests with three or four CLT surfaces exposed, as the protection (typically fire-rated gypsum board) fails and exposes the CLT to the fire.[48,50,55,62] Test data based on single or two exposed CLT surfaces should be considered carefully, as the tests can evolve to have all CLT surfaces exposed and burning, showing the importance of the protection.

Heat Release outside the Compartment

A factor in determining the compartment fire dynamics is the amount of heat released from a fire outside the compartment by external pyrolysis through flames from windows. The ratio of burning outside the compartment will be dependent on the location of the exposed CLT and ventilation available to the fire, such as breakable windows. Most fire testing undertaken has been on underventilated compartments, where heat release outside the compartment will be high. University of Carleton and TU Delft work shows that the resultant fire from having two walls exposed, with different layouts, does impact the amount of HRR outside of the compartment.

To date, there are no fire tests on well-ventilated compartments, where the ratio of interior to exterior heat release has been recorded. This indicates that when assessing the total heat release from the exposed mass timber, it is conservative to assume all the heat released occurs within the compartment for compartments with limited ventilation. The location of the CLT will also impact the external heat release, with exposed CLT located at the ceiling having a larger impact than exposed CLT located at the walls.

7.6.3 Summary of Outcomes from the Reviewed Fire Tests

From the reviewed fire tests, 29 have been carried out with some CLT exposed (four at Carleton University, five at University of Edinburgh, five at Delft University, one at SP, one at University of Queensland, four for FPRF, four for USFS, one at TUT, and four at NRCC). In summary:

- Of the 27 tests on partially protected CLT compartments (two USFS tests included sprinkler activation and so are not discussed further), 16 tests did not exhibit fire decay (though some were related to protection failures). CLT char falloff was a factor in all tests where fire decay did not occur, except one NRCC test where protection failed for CLT that was not susceptible to char falloff.
- Where fire decay does not occur, the behavior is characterized by either continued burning (nine tests) or a reduction in HRR followed by regrowth in HRR (seven tests).
- In the nine tests where sustained burning occurred, multiple surfaces were exposed (three or more). There was an increase in HRR as the movable fuel load

was consumed, followed by a phase of steady HRR. The temperatures and radiation from burning surfaces within the compartment remain sufficiently high to sustain pyrolysis of the exposed timber until extinguishment by hose streams.

- In the seven tests which exhibited HRR regrowth, at least two adjacent or opposing CLT surfaces were exposed, or a single surface and protection failure occurred to introduce multiple CLT surfaces. FPRF Test 4 (ceiling exposed) appears to be the only test where a single surface was exposed and significant regrowth occurred, with protection remaining intact (see Fig. 7.10).
- In the 11 tests where fire decay occurred, four cases had a single exposed CLT surface. With a single surface exposed, radiation from the protected surfaces and the decaying compartmenting fire is insufficient to maintain pyrolysis of the exposed surface. Char falloff also has a lesser impact, and reliability of CLT protection is imperative. In four cases of fire decay, two opposing or adjacent surfaces were exposed. The susceptibility of CLT char falloff and received heat flux were critical in determining whether significant regrowth in HRR occurs in this configuration. CLT not susceptible to char falloff resulted in two decay scenarios.
- Only the FPRF tests had a load applied. The FPRF test results have concluded that the loading condition had no influence on the compartment fire results.
- From the review of these tests, it is noted that there is no practical definition of compartment fire self-extinguishment. The term is regularly used in research and fire testing and needs more consensus on definition, such as temperature, HRR, or a critical heat flux. For building engineering, fires are normally defined by HRR, and more work is required to link current research on CLT panel self-extinguishment and compartment fire size.

Impact of Mass Timber Compartment Fire Testing

The compartment testing of exposed CLT has been fundamental in influencing code changes that allow for high-rise mass timber construction. Standard fire testing has shown the impact of adhesives that result in CLT char falloff, but the full impact of char falloff was not apparent until compartment fire tests provided conclusive evidence of the fire regrowth issue. The compartment tests are directly applicable to building construction where mass timber and exposed CLT are used. Increased char rates due to char falloff can be accounted for in design correlations, for standard fire exposure but the compartment fire regrowth from char falloff cannot be predicted.

The influence of adhesive on char fall off has resulted in the North American CLT manufacturing standard ANSI/APA PRG-320[29] being revised to exclude adhesives that can lead to char falloff beginning in 2021. This is linked directly to the change in the 2021 IBC with high-rise mass timber construction.

7.7 Building Design with Mass Timber and CLT

7.7.1 Exposed CLT for Low- and Medium-Rise Construction

The IBC permits the use of CLT within Types III, IV, and V construction, which can be built up to 85 ft. The fire performance of CLT for buildings of Type III, IV, and V is addressed through the fire protection specified by the IBC that includes sprinkler protection, two fire-separated means of egress stairs, and access and facilities for firefighting personnel.

Type IV construction (under the 2021 IBC, renamed Type IV-HT for "heavy timber") is based on timber members having a minimum dimension that provides an inherent FRR, and where CLT is used as part of Type IV construction, interior flammability criteria do not need to be met and mass timber can be fully exposed. Type IV construction also permits exterior wall assembles to use CLT, with the exterior side being protected with a noncombustible lining and meeting the requirements of IBC Chapter 14; otherwise, they must be noncombustible.

7.7.2 CLT for High-Rise Construction

Under the 2021 IBC, high-rise buildings are permitted to use mass timber as part of the primary structure, including CLT, for construction Types IV-A and IV-B. For a high-rise building utilizing CLT in a jurisdiction that does not enforce the 2021 IBC or for a building where the area of exposed CLT is more than permitted by Type IV-B or IV-A, an alternative engineering approach (performance-based design) has to be undertaken. The basis of the alternative engineering is for the mass timber primary structure to be assessed to ascertain the load-bearing capacity for not just the fire duration required by the IBC (standard fire) but also to withstand a fully developed fire.

Where CLT (and other elements of mass timber) are used as part of a high-rise building structure and are exposed, the timber will form part of the building combustible fire load. The compartment fire tests undertaken to understand the influence of the exposed CLT on the compartment fire and the testing have shown definite trends in the results.[63] A combination of exposed surface area, configuration (relative orientation of exposed surfaces), reliability of mass timber protection, compartment ventilation, and CLT manufacture related to susceptibility for char falloff behavior influence the predictability of the compartment HRR and resultant time temperature curve so that structural capacity can be determined. These factors also determine whether a compartment fire decays. If the total combustible fuel within the space (furniture, fixtures, and exposed timber) can result in a fire that releases more energy than the primary structure can resist, this may result in excessive structural deflections and localized failure.

Understanding the compartment fire is key to determining accurate char rate and char depth for the CLT so that structural capacity can be assessed. Accounting for large areas of exposed CLT is a significant technical challenge to analyze and one that differs from a building with a noncombustible structure given the interaction of the combustible CLT with the compartment fire. This becomes even more difficult when the compartment fire is impacted by the unpredictable behavior of char falloff.

Application of CLT Compartment Fire Tests Relevant to High-Rise Construction

The following items summarize the impact for fire-safe design for the structure and compartmentation to withstand a fully developed fire when CLT is exposed:

- The area of exposed CLT will need to be evaluated as additional compartment combustible fuel to determine if the fire duration will adversely impact structural fire resistance. Determining a char rate and depth must be based on received heat flux and a fixed char rate cannot be used, such as that determined from standard fire testing.
- The area of exposed CLT needs to be assessed relative to ventilation with areas of exposed CLT not located adjacent to each other (to prevent reradiation).

- Where panels are known to be susceptible to char falloff, the area of exposed CLT will need to be restricted if survivability of the structure in a fully developed fire is a fire safety goal. There is a limited set of empirical evidence to support small areas of exposed CLT in residential compartments and achieve compartment fire decay (USFS compartment tests[57]).
- Where the CLT panels are proven to not have char falloff, the fully developed fire is more predictable, and compartment fire decay can be more readily assessed. Areas of exposed CLT can be increased.
- For nonresidential uses and residential compartments much larger than >1000 ft^2 (93 m^2), the application of the fire test results published to date should be used with caution, understanding the testing limitations.
- Where CLT is not exposed, CLT protection is to be provided to prevent the timber forming part of the fire load, for the full duration of the fire, through to decay (see later section on CLT Protection).

Hazard and Consequence Based Approach to Fire Safety

To determine appropriate fire safety for a mass timber building, an approach based on height, fire protection, and area of exposed timber can be used. First proposed by Buchanan, Frangi, and Östman[64] and further developed by Buchanan,[65] the methodology is based on examining the fundamentals of building height and use and area of exposed mass timber and providing appropriate fire safety measures (see Table 7.1). This includes understanding the impact of exposed timber on the compartment, fire-active measures, fire-resistance ratings, and additional measures to improve the reliability of sprinkler protection. The decision-making process will be informed by the extent of mass timber that can be exposed and whether the CLT being used is susceptible to char falloff.

Methodologies for Assessing Limited Areas of Exposed Timber

Determining the fire performance of exposed mass timber is not a new issue, as there is a limited area of exposed timber that is always going to be acceptable within a compartment, as it will not significantly impact the compartment fire. But there is also a limit to how much timber can be exposed before the mass timber significantly increases the HRR and duration of the fire. To provide an FRR for the timber primary structure to resist the expected HRR and fire duration, the member is increased in sectional area.

Height	Low-rise	Mid-rise	Tall	Very tall	High-rise
Stories	1–2	3–5	6–8	9–15	>15
Likely escape	Quick escape	Slow escape	Assisted escape	Assisted escape	Difficult escape
No sprinklers	Local areas exposed	No exposed wood	Not allowed	Not allowed	Not allowed
Normal sprinklers	Large areas exposed	Local areas exposed	No exposed wood	Full encapsulation	Full encapsulation
Special sprinklers	Large areas exposed	Large areas exposed	Local areas exposed	No exposed wood	Full encapsulation

TABLE 7.1 Possible Decision Basis for Fire Protection Based on Building Height and Area of Mass Timber Exposed. For Complete Description, see Buchanan. Table Replicated from Buchanan.

But increasing the sectional area also introduces more combustible fuel to the room. Thus, the FRR to be provided for the exposed load-bearing timber structure needs to be determined to include the fire load introduced by the structure itself. Determining the maximum amount of energy released by the fully developed fire is key to permitting a safe amount of mass timber to be exposed within a compartment such that the energy released by the fully developed fire does not result in localized failure of the constructed high-rise building elements. For assessment of the mass timber structure, there are several options for the design of a compartment that includes CLT such that the fully developed fire can reliably decay. These are briefly discussed below.

Adhesive Product Small- and medium-scale fire testing has shown that the use of different adhesives can result in a significantly different fire performance of CLT. Tests indicate that CLT with a MUF adhesive and specifically formulated PUR do not exhibit char falloff in compartment fire conditions.[66] Compartment fire tests, standard fire tests, and cone calorimeter tests have led to the same conclusion. Compartment fire tests with CLT that is not susceptible to char falloff can result in a predictable decaying fire, allowing for more timber to be exposed.[59]

Thickness of the Outer Lamella For CLT that is susceptible to char falloff, the falloff behavior does not occur if the char layer does not reach the bond line. An increased thickness of the outer lamella of up to 1.57 in. (40 mm) can therefore potentially prevent char falloff. However, given the variety of fuel within a compartment, variations in available fire ventilation, and the inaccuracies in fire modeling, it is not practical to design with an outer thick lamella for predictable compartment fire performance. This approach is not recommended.

Limitation of Exposed Surface Area The compartment fire can result in reliable fire decay if the area of exposed mass timber is limited. The amount of exposed CLT that would result in a fully developed fire decaying, even after all movable fuel is consumed, is dependent on the timber exposed and ventilation of the compartment. The maximum allowable area of exposed CLT is too complex to accurately model and is dependent on the CLT susceptibility to char falloff.

Protection of CLT A method to prevent CLT char falloff is to protect the CLT with gypsum board that remains in place and prevents charring. The CLT protection must remain in place for the growth, fully developed, and decay phases of the fire. The reliability of the gypsum protection must be based on evidential fire tests with CLT.

Given the need to understand exposed mass timber as the primary structure in high-rise buildings, models are being developed to assess the impact, and existing models are being reevaluated. The exposed mass timber as a fuel source can be accounted for by determining the timber charring, as the char depth is a measure of how much mass has been converted and released as energy (pyrolysis). The volume of timber consumed must be based on understanding the timber properties, compartment ventilation, fire growth, and decay. To accurately account for the exposed mass timber charring, the heat flux needs to be determined at the timber location to then determine a char rate. The char rate can then be calculated based on the expected compartment HRR, and the decay period of the fire can also be accounted for, influenced by the exposed timber. Modeling can be carried out using hand-calculation methods via a spreadsheet or through more complex analysis methods. This approach also provides a means by which to determine the time-equivalence FRR for a timber member based on an approximate natural fire and available ventilation.[67] A constant char rate should not be used and the char rate must be determined based on the received heat flux. Methods are based on CLT that is not susceptible to char falloff, as this behavior can lead to a very unpredictable fire, beyond the capabilities of fire models. Until a predictive model of CLT char falloff and resultant impact on the compartment fire is developed, a very conservative approach to exposed CLT needs to be taken.

7.8 CLT Panel-to-Panel Connections

CLT panel-to-panel connections are required to achieve the same fire resistance as the panels they connect (see Fig. 7.11). To prove an FRR, panel-to-panel half-lap or spline connections are fire tested as part of a standardized test (ASTM E119 or UL273) (see Fig. 7.12). By having the panel connection as part of the fire test, CLT manufacturers can then specify their connections as part of the wall or floor assembly.

The fire-resistant performance of a connection can be improved through the inclusion of concrete or plywood to the top of the CLT or the inclusion of a fire-rated caulk within the connection. By preventing any movement of hot gases through the connection, fire resistance is improved.

To assess the integrity of a CLT floor connection, a methodology developed by FPInnovations[68] has been utilized within the CLT Handbook:

$$t_{\text{int}} = K_j\,(h/\beta_n)$$

where t_{int} = integrity fire-resistance time (h)
K_j = CLT panel-to-panel joint coefficient
h = CLT panel thickness
β_n = nominal charring rate (based on 1-hr exposure, 1.5 in./hr)

The methodology states a 1-hour exposure in the CLT Handbook, but the original research by FPInnovations does not state this limitation and provides test values that are

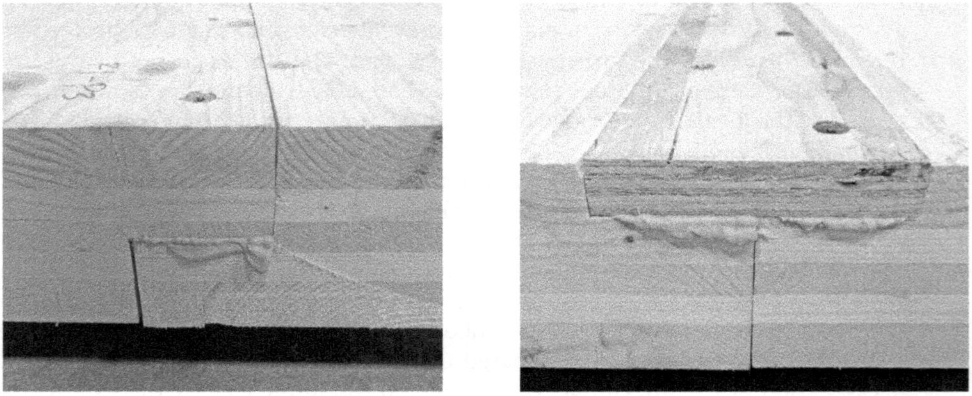

A B

FIGURE 7.11 CLT panel-to-panel connections—half-lap (A) and single surface spline (B).

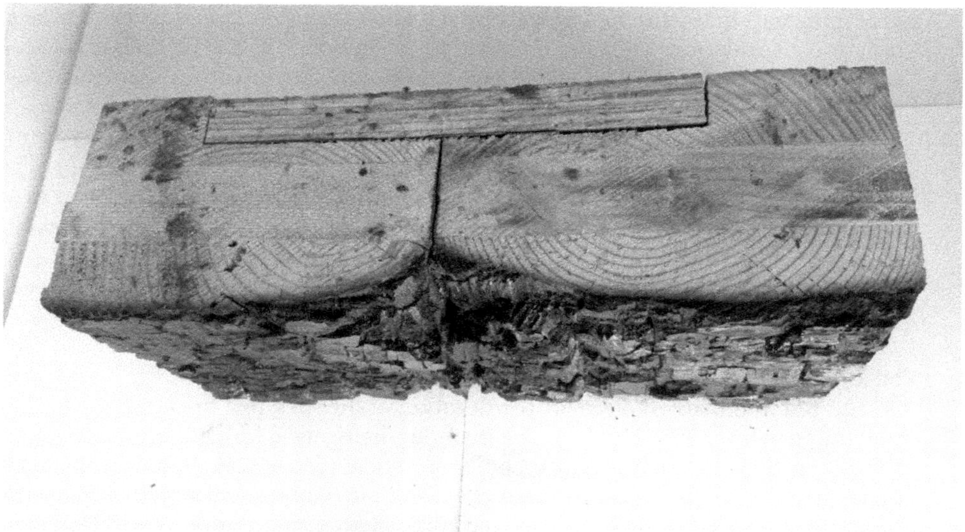

FIGURE 7.12 CLT five-ply panel with single surface spline connection after a 2-hour ASTM E119 fire test.

more than 3 hours. Hence, the methodology is considered acceptable for use in both 1- and 2-hour exposures. The recommended values of K_j are as follows:

- 0.35 for half-lap joint
- 0.6 for a surface spline

The work by Dagenais also recommends that where a concrete slab is in place, the "K_j" factor can be taken as 1.0, showing the effectiveness of a concrete topping in connection performance. CLT panels are not typically butt jointed, as it is not an effective panel-to-panel joint for fire resistance, and so are not typically fire tested.

7.8.1 CLT Floor-to-Wall Connections

An area that has limited testing to date is the connection between CLT walls and CLT floors, especially where the walls are continuous and the floors span onto the walls (referred to as "balloon" framing). In this situation, the construction method is based on the CLT floor being located on a ledger or support. This ledger provides the structural connectivity between the floor and the wall and should also provide an FRR. No standardized fire tests have occurred to prove an acceptable and efficient fire-rated connection. Solutions that provide an FRR are therefore more ad hoc or inefficient, such as adding an extra support bracket to the top of the floor panel.

7.9 CLT Protection Methods and Limitations

For noncombustible structures, FRRs can be achieved by applying products such as fire-rated gypsum board ("drywall") to the primary structure. For mass timber buildings, the noncombustible protection is installed over the exposed timber to prevent its becoming part of the fire fuel (it will also enhance the FRR). This is referred to as protection or encapsulation. The noncombustible protection to the mass timber is required to stay in place for the duration of the expected fire through to decay. For example, a CLT floor may have an inherent FRR due to its thickness of 7 in. (178 mm) to meet a 2-hour FRR but may be required to be protected with multiple layers of fire-rated gypsum board so that the CLT is not exposed to a fire until the fire reaches extinguishment. The noncombustible protection to the mass timber needs to be proven and specified to achieve this goal. The edge details of the protection will also need to be addressed within any specification of the protection. Where the CLT protection butts up against CLT walls or glulam beams or stops midway across an exposed CLT panel, the edge fire seal will need to be considered and appropriately specified.

Potential protection products for CLT include fire-rated gypsum board, noncombustible fiber insulation, and spray-applied fire-resistant materials.[62,69,70] The protection is required to remain in place for the duration of the specified fire, with the CLT surface remaining below 570°F (300°C). The materials being used for CLT protection need to be reliable in their fire performance and be practical for construction. Specifications and products for CLT protection are not prevalent currently, and more testing from manufacturers is needed to provide effective and consistent CLT protection. The protection reliability must be proven through standardized and natural fire testing. Materials will become more widespread in use if they can be proven to be cost effective or allow for installation before panel construction.

7.10 Fire Protection of CLT Through Penetrations and Joints

Through penetrations of CLT walls or floors will occur for metallic and nonmetallic pipes and ducts and for cables. Where these penetrations occur and the floor or wall is providing a required FRR, the through penetration is required to maintain the fire rating. The fire sealing can be provided by various means, such as fire caulk, fire collars, or flexible intumescing materials.

Joints occur where a CLT panel is butting up to another building element and there is the potential for fire to spread through any opening, such as when a CLT floor panel is supported by a CLT wall. Where the CLT panel is required to achieve a fire resistance, the joint must maintain that fire resistance through a sealant product.

The products used to fire seal through penetrations and joints are chosen based on the dimension of the opening and type of material that needs to be sealed, with very different solutions for metallic and nonmetallic pipes. Flexibility in the fire seal may also be required. The annular space between the CLT and a through penetrant is also relevant. As the CLT panel will char during the fire exposure, the fire-sealing materials need to consider this moving interface.

Proven fire-rated solutions are manufacturer based and carried out by fire testing, meeting standards such as ASTM E814[71] or UL 1479.[72] Fire seal suppliers will provide approved solutions based on the fire testing they have completed (see Fig. 7.13). Where a fire test has not been completed for a particular penetrant or opening type, the suppliers may issue an engineering judgment based on similar proven products.

The number of proven fire seals for CLT through penetrations and joints is not extensive, and the information may be proprietary to the suppliers. A higher number of fire-tested through penetration and joint fire seals is required to allow for more efficient design of CLT building elements.

FIGURE 7.13 Fire test of penetration seals after a 2-hour ASTM E119 floor test.

7.11 Use of CLT within Exterior Walls

7.11.1 Fire Spread Fundamentals

Exterior walls have limitations on their construction materials and the number and type of openings to prevent fire spread vertically and between buildings. Codes restrict openings for buildings that are very close to each other and require fire ratings to exterior walls to prevent fire spread. Fire spread between buildings can occur through radiation, which is significantly reduced where an automatic sprinkler system is provided. The intervention of the fire department is also a contributing factor in preventing fire spread.

7.11.2 Fire Spread between Floors

Fire spread between floors is typically prevented through the installation of sprinkler protection. If a fully developed fire were to occur, then the sprinkler protection would have failed, and fire spread vertically by external flaming may occur, regardless of the type of construction. The temperature, height, and width of the flames extending from a window are dependent on the ventilation characteristics of the compartment and the fuel available.

The limited testing information available[53] indicates that where ventilation is low and CLT is exposed, a greater proportion of the compartment heat release occurs outside, compared to a compartment with noncombustible structure and linings. The impact of flame dimensions and temperature for a compartment with a large area of CLT exposed needs to be understood in more detail to predict how fire spread vertically between floors could occur and if there are any differences when compared to a compartment with a noncombustible structure. Available ventilation to the compartment fire will impact the flame projection.

7.11.3 Building with CLT Exterior Walls

The IBC permits CLT to be used within an exterior wall provided it meets the fire-resistance requirements and the exterior is clad with a noncombustible material or fire-retardant-treated (FRT) sheathing. The height to which CLT may be used is limited to 40 ft (12.2 m) above grade plane. The most appropriate method to prevent the CLT panel forming part of the fire is to protect this with a noncombustible and fire-rated board such that the CLT cannot be ignited. CLT used as an exterior wall with the timber exposed on the outside of the building is not permitted above 40 ft.

There are only a limited number of relevant fire tests to understand vertical flame spread up a façade.[73] At the time of this writing, there had been no fire tests on CLT exterior wall assemblies carried out to meet NFPA 285,[74] the reference test standard for exterior wall protection in fire. A test on CLT to the equivalent Canadian standard CAN/ULC S134-13[75] has been completed for CLT and has passed the acceptance criteria of the test. The exterior wall assembly may also need to meet the criteria of NFPA 268[76] to show that the CLT will not ignite when exposed to a radiative flux of 12.5 kW/m^2. The use of CLT within exterior walls has very limited research to date. Full-scale fire testing is required and would be beneficial.

7.12 Construction Fire Safety

There are numerous potential fire hazards associated with construction, and mass timber buildings utilizing CLT may need specific site safety requirements to reduce the likelihood and impact of a fire during construction. The consequences of a fire during construction are significant given that the building will be without many or all fire protection measures.

7.12.1 Methods of Hazard Reduction

The most appropriate approach to address construction fire safety is through a consultative process between the project team, the authority having jurisdiction, fire department, and general contractor to identify potential hazards and implement mitigation strategies. The mitigation strategies then need to be monitored and enforced on-site. This approach is consistent with the overarching framework for site fire safety during construction from NFPA 241.[77]

The approach to fire safety during construction may include the following:

- Site safety and construction fire safety plans.
- Site safety coordinator and/or fire safety manager.
- Establish, promote, and maintain strong communication on the work site regarding the fire risks, with regular training.
- Daily walk-through inspections for construction fire safety plan.
- Maintaining a tidy worksite.
- Limiting or eliminating hot works and enforcing daily permits and fire watch for any hot works.
- Storing materials and equipment safely in a noncombustible enclosure or off-site.
- A no-smoking work site and a no-cooking work site, all workers cooking to be located away from active construction.
- Using companies with timber construction experience (including light frame), as they have a better understanding of the site risks and consequences.

- Improving 24-hour security, with a secure work site with fences to all sides (minimum 8 ft).
- After-hours security with patrols throughout the site.
- Security lighting to the site with motion detectors (combined with surveillance system).
- Standpipe installed once third level is reached, located at stair or construction hoistway, following stair construction. Access to the standpipe is to be documented.

Further Reading

The following documents provide guidance on the fire-safe use of CLT within building projects.

"Fire Resistance of Timber Structures."[64]
"Technical Guide for the Design and Construction of Tall Wood Buildings in Canada."[78]
"Fire Safety in Timber Buildings—Technical Guideline for Europe."[79]
"CLT Handbook, U.S. Edition."[29]
"Calculating the Fire Resistance of Exposed Wood Members, Technical Report No. 10."[80]

Acknowledgment

The author would like to thank Laura Hasburgh, Forest Products Laboratory, for her input and assistance with this chapter.

References

1. Prasad, K., Sangen, E., Visser, P., 1985, "Wood burning cookstoves," Advances in Heat Transfer, No. 17: p. 159–317.
2. Ross, R. (editor), 2010, "Wood Handbook: Wood as an engineering material. General Technical Report FPL-GTR-190," USDA Forest Service Forest Products Laboratory, Madison, Wisconsin.
3. White, R.H., 2000, "Fire Performance of Hardwood Species." Proceedings of XXI IUFRO World Congress, Kuala Lumpur, Malaysia.
4. Shafizadeh, F., 1982, "Introduction to pyrolysis of biomass," Journal of Analytical and Applied Pyrolysis, 3(4): p. 283–305.
5. Hill, C., 2007, Wood Modification: Chemical, Thermal and Other Processes. Vol. 5. John Wiley & Sons.
6. Rowell, R., 1984, "The Chemistry of Solid Wood," in Advances in Chemistry Series. American Chemical Society Washington, DC.
7. Browne, F., 1958, "Theories of the Combustion of Wood and Its Control," Madison, Wisconsin, U.S. Department of Agriculture, Forest Service, Forest Products Laboratory.
8. White, R.H., 1988, "Charring Rates of Different Wood Species," University of Wisconsin–Madison.
9. Fonseca, E., Barreira, L., 2009, "Charring Rate Determination of Wood Pine Profiles Submitted to High Temperatures," WIT Transactions on the Built Environment, No. 108: p. 449–457.
10. Schaffer, E.L., 1967, "Charring Rate of Selected Woods—Transverse to Grain," DTIC Document.
11. Friquin, K., 2011, "Material properties and external factors influencing the charring rate of solid wood and glue-laminated timber," Fire and Materials, 35(5): p. 303–327.
12. American Wood Council, 2015, "National Design Specification for Wood Construction," Leesburg, Virginia.
13. Buchanan, A., 2001, Structural design for fire safety. Vol. 273, Wiley & Sons, New York.
14. American Society for Testing and Materials, 2016, "ASTM E119: Standard Test Methods for Fire Tests of Building Construction and Materials."

15. White, R., Dietenberger, M., 2010, "Fire Safety of Wood Construction," in Wood Handbook, Wood as an Engineering Material R. J. Ross (editor). U.S. Department of Agriculture, Forest Service, Forest Products Laboratory.
16. Schmid, J., Lange, D., Sjöström, J., Brandon, D., Klippel, M., Frangi, A., 2018 "The Use of Furnace Tests to Describe Real Fires of Timber Structures," in Proceedings World Conference on Timber Engineering, Seoul, South Korea.
17. Tran, H., White, R., 1992, "Burning rate of solid wood measured in a heat release rate calorimeter," Fire and Materials, 16(4): p. 197–206.
18. White, R., Tran, H., 1996, "Charring Rate of Wood Exposed to a Constant Heat Flux."
19. Mikkola, E., 1991, "Charring of wood based materials," Fire Safety Science, 3: p. 547–556.
20. Friquin, K., 2010 "Charring rates of heavy timber structures for fire safety design: A study of the charring rates under various fire exposures and the influencing factors," Norwegian University of Science and Technology, Doctoral Thesis, 2010, p. 128.
21. Bartlett, A., Hadden, R., Bisby, L., Law, A., 2015, "Analysis of cross-laminated timber charring rates upon exposure to nonstandard heating conditions." Proceedings. Interscience Communications Ltd, p. 667–681, 14th International Conference and Exhibition on Fire and Materials 2015, San Francisco, California, February 2–4.
22. Custer R., Meacham B., 1997, "Introduction to Performance-Based Fire Safety," Society of Fire Protection Engineers.
23. International Code Council, 2018, "International Building Code."
24. Department of Commerce, 1942, "Building Materials and Structures, Report BMS92, Fire Resistance Classification of Structures," National Bureau of Standards, Report of Subcommittee on Fire Resistance Classifications of Central Housing Committee on Research, Design and Construction.
25. American Society of Civil Engineers, 2016, "ASCE/SEI 7–16, Minimum Design Loads for Buildings and Other Structures."
26. https://www.iccsafe.org/products-and-services/i-codes/code-development/cs/icc-ad-hoc-committee-on-tall-wood-buildings.
27. Underwriters Laboratories, 2014, "UL 263—Standard for Fire Tests of Building Construction and Materials."
28. ANSI/APA, 2015, "PRG-320 Standard for Performance-Rated Cross-Laminated Timber."
29. Karacabeyli, E., Douglas, B., 2013, "CLT Handbook, U.S. Edition" FPInnovations SP529-E.
30. Klippel, M., 2014, "Fire Safety of Bonded Structural Timber Elements," ETH Zurich, Institute of Structural Engineering, Thesis Number 359.
31. Frangi, A., Fontana, M., Knobloch, M., Bochicchio, G., 2008, Fire Behaviour of Cross-Laminated Solid Timber Panels. Institute of Structural Engineering, ETH Zurich.
32. Klippel, M., Fahmi, R., 2017, "COST Action FP 1404: Expert Meeting CLT, The Way to Standardization—Definition of Gaps (Joint Meeting WG1/TG4 and WG2/TG1)," ETH Zurich.
33. Osborne, L., Dagenais, C., Benichou, N., 2012, "Preliminary CLT Fire Resistance Testing Report," FPInnovations and NRC-CNRC, Project No. 301006155.
34. Craft, S.T., Desjardins, R., Mehaffey, J., 2011, "Investigation of the Behaviour of CLT Panels Exposed to Fire," Proceedings of the Twelfth International Conference Fire and Materials.
35. Embury-Williams, L., Karacabeyli, E., 2009, "Application of Analysis Tools from NEWBuildS Research Network in Design of a High-Rise Wood Building," NSERC Strategic Research Network on Innovative Wood Products and Building Systems (NEWBuildS) in Partnership with FPInnovations.
36. Klippel, M., Schmid, J., Frangi, A., 2016, "Fire Design of CLT," paper submitted for joint event of COST Actions FP1402 and FP1404, KTH Stockholm.
37. Klippel, M., Schmid, J., 2017, "Design of Cross-Laminated Timber in Fire," Structural Engineering International No. 2/2017.
38. American Society for Testing and Materials, 2015, "ASTM E84: Standard Test Method for Surface Burning Characteristics of Building Materials."
39. American Wood Council, 2015, "Design for Code Acceptance 1—Flame Spread Performance of Wood Products Used for Interior Finishes," Leesburg, Virginia.

40. Aguanno, M., 2013, "Fire Resistance Tests on Cross-Laminated Timber Floor Panels: An Experimental and Numerical Analysis." Department of Civil and Environmental Engineering, Carleton University, Ottawa, Canada.
41. National Institute of Standards and Technology, "Product Standard PS1-83, Construction and Industrial Plywood," 1983, Standard Services Division, Gaithersburg, MD (supersedes VPS PS I-66 "Softwood Plywood, Construction and Industrial," NBS Prod. Std. 1–66).
42. Schaffer, E., 1968, "A Simplified Test for Adhesive Behavior in Wood Sections Exposed to Fire," Research Note FPL-175, USDA Forest Service, Forest Product Laboratory, Madison, Wisconsin.
43. White, R., 2000, "Charring Rate of Composite Timber Products," FPL.
44. Klippel, M., Frangi, A., Fontana, M., "Influence of the Adhesive on the Load-Carrying Capacity of Glued Laminated Timber Members in Fire," Fire Safety Science-Proceedings of the 10th International Symposium, 2011.
45. Frangi, A., Fontana, M., Hugi, E., Jobstl, R., 2009, "Experimental analysis of cross-laminated timber panels in fire," Fire Safety Journal, 44: p. 1078–1087.
46. Hasburgh, L., Bourne, K., Peralta, P., Mitchell, P., Schiff, S., Pang, W., 2016, "Effect of Adhesives and Ply Configuration on the Fire Performance of Southern Pine Cross-Laminated Timber," Proceedings from the World Conference of Timber Engineering, August 22–25, Vienna.
47. Friquin, K. L., Grimsbu, M., Hovde, P. J., 2010, "Charring Rates for Cross Laminated Timber Panels Exposed to Standard and Parametric Fires."
48. Medina-Hevia, A., "Fire Resistance of Partially Protected CLT Rooms," Thesis, Carleton University, 2014.
49. Hadden, R., Bartlett, A., Hidalgo, J., Santamaria, S., Wiesner, F., Bisby, L., Deeny, S., Lane, B., 2017, "Effects of Exposed Engineered Timber on Compartment Fire Dynamics," Proceedings of the 12th International Association of Fire Safety Science Symposium, Lund, June 10–16.
50. Su, J., LaFrance, P., Hoehler, M., Bundy, M., 2018, "Fire Safety Challenges of Tall Wood Buildings—Phase 2: Task 2 & 3 Cross Laminated Timber Compartment Fire Tests," Fire Protection Research Foundation, Report No. FPRF-2018–01.
51. Frangi, A., Bochicchio, G., Ceccotti, A., 2008, "Natural Full-Scale Fire Test on a 3-Storey XLam Timber Building," Fire Safety Science—Proceedings of The Ninth International Symposium, p. 1279–1290.
52. Hakkarainen, T., 2002, "Post-flashover fires in light and heavy timber construction compartments," Journal of Fire Sciences, p. 133–175.
53. McGregor, C., 2013, "Contribution of Cross Laminated Timber Panels to Room Fires," Thesis, Carleton University, Ottawa, Canada.
54. Crielaard, R., 2015, "Self-Extinguishment of Cross-Laminated Timber," Master's Thesis report, Delft University of Technology, Delft, Netherlands.
55. Hox, K., et al., 2015, "Branntest av massivtre," SP Fire Research, Report A15101 (in Norwegian, translated by Arup).
56. Emberley, R., Gorska, C., Bolanosa, A., Lucherinia, A., Solartea, A., Soriguera, D., Gutierrez, M., Humphreys, K., Hidalgoa, J., Maluka, C., Law, A., Torero, J., 2017, "Description of small and large-scale cross laminated timber fire tests," Fire Safety Journal, 91: p. 327–335.
57. Zelinka, S., Hasburgh, L., Bourne, K., Tucholski, D., Ouellette, J., 2018, "Compartment Fire Testing of a Two-Story Mass Timber Building," General Technical Report FPL–GTR–247, May 2018.
58. Just, A., Brandon, D., Nele Mäger, K., Pukk, R., Sjöström, J., Kahl, F., 2018, "CLT Compartment Fire Test," Proceedings of World Conference on Timber Engineering, August 20–22, 2018, Seoul, South Korea.
59. Su, J., Leroux, P., Lafrance, P., Berzins, R., Gratton, K., Gibbs, E., Weinfurter, M., 2018, "Fire Testing of Rooms with Exposed Wood Surfaces in Encapsulated Mass Timber Construction," National Research Council, Canada, Report No. A1-012710.1.
60. Bartlett, A., Hadden, R., Hidalgo, J., Santamaria, S., Wiesner, F., Bisby, L., Deeny, S., Lane, B., 2017, "Auto-extinction of engineered timber: Application to compartment fires with exposed timber surfaces," Fire Safety Journal, 91: p. 407–413.

61. Emberley, R., Do, T., Yim, J., Torero, J., 2017, "Critical heat flux and mass loss rate for extinction of flaming combustion of timber," Fire Safety Journal, 91: p. 252–258
62. Hasburgh, L., Bourne, K., Dagenais, C., Osborne., L., Roy-Poirier, A., 2016, "Fire Performance of Mass-Timber Encapsulation Methods and the Effect of Encapsulation on Char Rate of Cross Laminated Timber."
63. Barber, D., Crielaard, R., Li, X., 2016, "Towards Fire Safe Design of Exposed Timber in Tall Timber Buildings," Proceedings of the World Conference of Timber Engineering, Vienna, August 22–25 2016.
64. Buchanan, A., Östman, B., Frangi, A. (editors), 2014, "Fire Resistance of Timber Structures (draft)," National Institute of Standards and Technology White Paper.
65. Buchanan, A., 2015, "Fire resistance of multi-storey timber buildings," invited keynote lecture, in Proceedings of the 10th Asia-Oceania Symposium on Fire Science and Technology, October 5–7, 2015, Tsukuba, Japan.
66. Brandon, D., Dagenais, C., 2018, "Fire Safety Challenges of Tall Wood Buildings—Phase 2: Task 5—Experimental Study of Delamination of Cross Laminated Timber (CLT) in Fire," Fire Protection Research Foundation, Report No. FPRF-2018-05.
67. Barber, D., 2016, "Fire Safety Engineering of Tall Timber Buildings in the USA," Proceedings of the World Conference of Timber Engineering, Vienna.
68. Dagenais, C., 2015, "Fire Performance of Cross-Laminated Timber Panel-to-Panel Joints," FPInnovations Project Report No. 301009649.
69. Western Fire Center, 2017, "Fire Resistance Testing of CLT Floor/Ceiling Assemblies to Establish Contribution of Gypsum Protection," Conducted for American Wood Council, WFCi Report No. 17091r1.
70. Berzins, R., LaFrance, P., Leroux, P., Lougheed, G., Su, J., Benichou, N., 2014, "Report to Research Consortium for Wood and Wood-Hybrid Mid-Rise Buildings—Solutions for Mid-Rise Wood Construction: Intermediate-Scale Furnace Tests with Encapsulation Materials," National Research Council, Canada, Report No. A1-100035-01.2.
71. ASTM, 2017, "E814—13a Standard Test Method for Fire Tests of Penetration Firestop Systems."
72. Underwriters Laboratories, 2015, "UL 1479—Standard for Fire Tests of Penetration Firestops."
73. White, N., Delichatsios, M., 2014, "Fire Hazards of Exterior Wall Assemblies Containing Combustible Components—Final Report," Fire Protection Research Foundation, Report No. EP142293.
74. NFPA 285, 2019, "Standard Fire Test Method for Evaluation of Fire Propagation Characteristics of Exterior Non-Load-Bearing Wall Assemblies Containing Combustible Components."
75. CAN/ULC S134-13, "Standard Method of Fire Test of Exterior Wall Assemblies."
76. NFPA 268, 2017, "Standard Test Method for Determining Ignitability of Exterior Wall Assemblies Using a Radiant Heat Energy Source."
77. NFPA 241, 2013, "Standard for Safeguarding Construction, Alteration, and Demolition Operations."
78. Karacabeyli, E., Lum, C., (editors), 2014, "Technical Guide for the Design and Construction of Tall Wood Buildings in Canada, First Edition," FPInnovations Special Publication SP-55E.
79. Ostman, B. (editor), 2010, "Fire Safety in Timber Buildings—Technical Guideline for Europe," SP Report 2010:19.
80. American Wood Council, 2015, "Calculating the Fire Resistance of Exposed Wood Members, Technical Report No. 10," Leesburg, Virginia.

CHAPTER 8

Environmental Aspects of Wood as a Construction Material

Jim L. Bowyer

Professor Emeritus
Department of Bioproducts and Biosystems Engineering,
University of Minnesota

8.1 Introduction

Proposals for increased use of wood for any purpose invariably raise questions about the wisdom of doing so. Some wonder about the overall environmental impact of using wood versus available alternatives. Others question whether use of wood, especially whether *increased* use of wood, is sustainable. Still others question wood durability over the long term.

These issues are addressed in this chapter, first by examining environmental implications of wood as a construction material, with comparisons to alternative materials based on systematic assessment of a range of impact estimators. Next, the current state of North American forests and recent and historical trends in forest cover and growth-harvest relationships are discussed. Forest conditions and trends in other world regions are also examined. Consideration of long-term durability of wood structures concludes the chapter.

8.2 Environmental Impacts of Construction Materials

8.2.1 Determination of Impacts

Life Cycle Assessment

Environmental life cycle assessment, or LCA, has become the tool of choice for leading organizations in both the public and the private sector. Sometimes referred to as "cradle to grave" analysis, LCA provides a mechanism for systematically evaluating the environmental impacts linked to a product or process and in guiding process or product improvement efforts. LCA-based information also provides insights into the environmental impacts of raw material and product choices and maintenance and end-of-product-life strategies. Because of the systematic nature of LCA and its power as an evaluative tool, the use of LCA is increasing as environmental performance becomes more and more important in society.

A key part of LCA is the life cycle inventory, or LCI. An LCI involves a careful accounting of all the measurable raw material inputs (including energy), product and coproduct outputs, and emissions to air, water, and land. Examination of energy use is particularly revealing since energy generation from fossil fuels—far and away the leading source of energy in the United States—is linked to a wide array of impacts in extraction and combustion. Carbon dioxide, sulfur dioxide, nitrogen oxide, and methane are but a few of the emissions from fossil fuel combustion.

Impact measures may be determined for manufacturing of a particular building product or subassembly or even for an entire building. In any case, assessment begins with examination of raw material extraction (or collection of materials for recycling) and all steps in subsequent processing through conversion to final product, including transportation. Quantities of inputs—raw materials, water, and various forms of energy—are measured, as are product and coproduct outputs and all emissions and wastes. Energy directly consumed in industrial processes is accounted for, as is energy needed to provide heating of manufacturing plants. If LCA involves assessment of impacts in creating a total structure, all activities involved in the building construction process are also considered, and again, all emissions, effluents, solid wastes, and other releases associated with consumption of energy and all other materials are accounted for. Finally, when an assessment boundary includes building operation and maintenance through the expected life of a building, all materials and processes involved in the use and maintenance of the building are considered, as are processes involved and/or materials recovered at the end of the useful life of the structure.

When impacts of a total building are assessed, energy consumption is divided into two categories: (1) embodied energy, or the total quantity of energy consumed in raw material extraction, transportation, and processing through every step in manufacturing from primary to finished products, transport to building site, and incorporation in the finished structure, and (2) operational energy, or the energy required to heat, cool, and light, that is otherwise used in operating a building.

An LCA can be conducted for a period of time that is less than the full product life. For instance, a number of recent analyses have examined all of the steps involved up to the point that construction of a building is completed. In these analyses, use and maintenance and recycle/waste management stages are not considered.

A set of international protocols, as published by the International Organization for Standardization (ISO), specify a uniform set of guidelines for the planning, conduct, and reporting of an LCA. These protocols help to ensure comparability of like assessments and to minimize the potential influence of bias on the part of analysts.

Third-Party Certification

Given the rigor and expense of gathering and analyzing LCI data, assessments are typically made only periodically (i.e., every five years or more frequently if there are significant changes to production processes), with data often collected on an industry-wide basis for particular products. LCA does not report impacts which require subjective evaluation or those which are likely to vary widely from site to site. Thus, LCA assessment of impacts linked to raw material extraction focuses on those things that can be more or less precisely measured (e.g., consumption of fuel and lubricants, emissions resulting from machinery use and transportation) and not on factors such as esthetics and impacts to biodiversity, outdoor recreation, and local communities. In these situations, and where third-party verified evaluation of impacts to landscapes is desired, certification programs have been developed.

For wood, there are today a number of forest certification programs in operation. While there are differences between them, they all are similar, featuring independent boards of directors, engagement of independent auditors who make periodic on-the-ground assessments of forest management and harvest sites, and chain-of-custody tracking of wood harvested throughout market channels. These programs are also defined by freely available documentation which clearly explains program requirements and systems for verifying compliance. Subjective judgment is often needed in verification to certification standards. For wood products, the combination of LCA results and third-party certification provides strong assurance of responsible production. As of this writing, there are no programs for third-party certification of mining activity.

What LCA Reveals about Relative Impacts of Building Materials

There are three primary structural materials used in construction: steel, reinforced concrete, and wood. In all cases, relative impacts depend upon the specific products involved. Embodied energy and carbon for various materials, as determined by LCA and on both a mass and a volume basis, are shown in Table 8.1. While various materials are not used in equal mass and volume when constructing functionally equivalent buildings or components, the figures nonetheless provide insights into relative impacts of materials.

Material	MJ/kg	kgCO$_2$e/kg	MJ/m^3	kgCO$_2$e/m^3
Concrete	0.75	0.107	1805	258
Steel-reinforced concrete	1.79	0.184	4550	837
Brick	3.00	0.240	6150	492
Softwood lumber	7.40	0.20 + 0.39 (bio)	3774	102 + 199 (bio)
Steel (100% recycled)	9.40	0.470	73,790	3690
LVL	9.50	0.33 + 0.32 (bio)	4845	168 + 163 (bio)
Glulam	12.00	0.42 + 0.45 (bio)	6120	214 + 230 (bio)
Steel (structural)	13.10	0.720	102,835	5652
Steel (galvanized rolled)	22.60	1.540	180,800	12,230
CLT	27.20	0.80 + 0.93 (bio)	12,792	373 + 438 (bio)
Steel (virgin)	35.40	2.890	277,890	22,687
PVC	77.20	3.100	11,940	4495

Source: ICE Database, Bath University (2010).

*Note that while measures are presented in metric units only, the relative margin of difference for all categories is the same regardless of whether metric or imperial units are used.

TABLE 8.1 Embodied Energy and Carbon in Common Construction Materials*

Steel

The environmental impact of steel construction materials is highly dependent upon recycled content since energy consumption in making steel is more than three times higher if producing from iron ore versus from steel recovered for recycling. Recycled content depends upon the specific product. Light-gauge steel used in making steel studs, for example, has an average recycled content of about 30%. Large beams, on the other hand, incorporate a larger percentage of recycled steel. Consequently, energy requirements and overall environmental impacts associated with producing light-gauge steel products are relatively high in comparison to large section modulus steel products.

Concrete

Impacts linked to concrete depend upon designed strength, which in turn is defined by the proportion of cement; the greater the quantity of cement, the greater the impacts. Structural concrete, which is reinforced with steel, is also a relatively high-impact product.

Some find the high emissions levels associated with concrete production surprising. The high impacts are attributable to cement, the active ingredient in concrete that binds sand, gravel, and other aggregate material together to form the finished product; cement typically accounts for 12% to 15% of the dry weight of concrete. Although made of natural ingredients such as limestone (calcium carbonate), the environmental impact of cement production is relatively high due to both the quantity of energy needed to reduce calcium carbonate to lime (or calcium oxide) in a cement kiln where temperatures above 2700°F are maintained and the substantial release of CO_2 inherent in the process. Chemically, reduction of limestone looks like this:

$$\underset{\text{limestone}}{CaCO_3} \xrightarrow[\Delta]{} \underset{\text{lime}}{CaO} + \underset{\text{carbon dioxide}}{CO_2}$$

Thus, in addition to the mining activity needed to obtain limestone, sand, and gravel (and sometimes clay and other ingredients), cement production includes significant environmental impacts from combustion of coal or natural gas and associated emissions to air as well as direct release of carbon dioxide in production of lime. In general, the production of a ton of cement results in the release of about a ton of carbon dioxide.

Wood

Wood is produced within growing trees using solar energy, and relatively little additional energy is required to convert raw wood into useful products, including building materials. Moreover, virtually all wood product manufacturers in North America use woody biomass—sawdust, bark, and trimmings generated in the manufacturing process itself—as a source of energy for some or all of their energy needs, meaning that they are 60% to 100% energy self-sufficient. Consequently, relatively little fossil energy is needed to produce wood building materials, translating to low fossil fuel emissions in comparison to other materials. This is a significant advantage in a world seeking to reduce fossil carbon emissions.

The lowest-impact wood products are those requiring the least processing. In producing lumber, for example, wood needs only to be cut from logs, shaped, and dried in producing a final product. As a result, quantities of energy required to produce lumber are low in comparison to wood products requiring more manufacturing steps and use of adhesives. Greater energy input is needed in producing various engineered wood products—parallel strand lumber, glulam beams, laminated veneer lumber, and cross-laminated timber (CLT)—and overall environmental impacts are greater than when producing solid-sawn lumber (Table 8.1).

It is important to note that whereas the hierarchy of carbon emissions linked to various forms of construction is sometimes expressed in terms of the quantity of emissions per ton of material, the mass of material actually used in construction varies widely depending upon the material used. For instance, the mass of a concrete wall is far greater than the mass of steel or wood used in constructing functionally equivalent walls. Consequently, when environmental impacts linked to whole buildings are determined, relative impacts are determined by the mass of various materials used. Structures built dominantly of concrete invariably result in greater carbon emissions than those made of steel, even though energy requirements in steel construction are greater. And while wood structures invariably show lower impacts across a range of measures than those built primarily of concrete or steel, the differences between wood and steel construction are less than is suggested by considering impacts linked to an equal mass of materials.

8.2.2 Environmental Assessment of Whole Buildings

Over the past several decades numerous life-cycle-based studies of the environmental impacts of various types of construction have been conducted. Differences in impact resulting from construction using alternative materials are often substantial and consistently point to higher embodied energy and life-cycle impacts of concrete construction than for comparable wood or steel-framed structures. These embodied energy and life-cycle impact differences primarily relate to raw material extraction and production of construction materials. When otherwise equivalent buildings (i.e., buildings of equal size, configuration, thermal insulation, orientation on-site) are constructed of different materials, there is little to no difference in the quantity of energy required to heat and cool them once the construction process has been completed. What this means is that the environmental implications of building material selection are immediate, with the initial advantage of one building material over another persisting throughout the life of the structure. However, the relative magnitude of difference tends to narrow with each year of building operation even when operational energy for two buildings is the same. A hypothetical example of two buildings, one of which requires one-half the energy to construct as the other, illustrates this point (Table 8.2). Note that the large (100%) difference in energy consumption at the point of construction between buildings A and B becomes progressively smaller with progressive heating/cooling seasons even though energy consumption for building operation remains the same.

Despite a diminution of low-impact initial advantage over time, differences in environmental impact at the point of construction are often substantial and increasingly important. As buildings become more and more energy efficient, the impact at time zero (i.e., the time of building completion) rises in significance. For example, in the case of a net-zero building, embodied energy would account for 100% of building energy and all related impacts.

Table 8.2 shows how an advantage of low initial energy (embodied energy) is eroded over time, even when operational energy requirements of two buildings are the same. It is perhaps obvious, then, that if two buildings are compared in a situation in which one of the buildings is designed to have a higher thermal rating, that building will eventually demonstrate superior environmental performance (lower environmental impact) even if

		Building A	Building B	% Difference
1	Units of energy consumed in extraction of raw materials, production of building products, and building construction	500	1000	100
2	Units of energy consumed in heating and cooling the structure over a 20-year period	500	500	
3	Units of energy consumed in construction of building *and* heating and cooling over a 20-year period. (line 1 + line 2)	1000	1500	50
4	Units of energy consumed in heating and cooling the structure over a 40-year period	1000	1000	
5	Units of energy consumed in construction of building *and* heating and cooling over a 40-year period. (line 1 + line 4)	1500	2000	33
6	Units of energy consumed in heating and cooling the structure over a 60-year period	1500	1500	
7	Units of energy consumed in construction of building *and* heating and cooling over a 60-year period. (line 1 + line 6)	2000	2500	25

TABLE 8.2 Relative Differences in Energy Consumption for Two Functionally Equivalent Buildings Constructed of Different Materials

embodied energy of that building is higher than the one with which it is being compared. For this reason, standard practice in LCA, when the goal is to compare the effect of construction material on environmental impacts, is to compare thermally equivalent buildings. This is an essential part of the definition of "functionally equivalent."

8.2.3 Environmental Impacts of Concrete versus Wood Construction

A number of LCAs of concrete construction are outlined below. These assessments encompass construction of a wide range of buildings, from large commercial structures to multistory apartment buildings to single-family homes. Many of the examples come from Europe, where there is a longer history of consideration of life-cycle impacts in building construction. Examples include both traditional wood construction and construction using CLT and other engineered wood products.

Analysis of a Single-Family Residential Structure

A study conducted by the Consortium for Research on Renewable Industrial Materials (CORRIM) examined concrete block and wood-frame construction options for a typical home built in the Atlanta metropolitan area (Lippke et al. 2004, Perez-Garcia et al. 2005). The 198-m^2 (2135 ft^2) house was a concrete block slab-on-grade design. An LCA examined two virtual structures, one of wood, the other of concrete block. There were material differences only in the exterior walls; both were designed to sit on a concrete slab, and both featured identical wood roof systems and wood-framed partition walls (Fig. 8.1, Table 8.3).

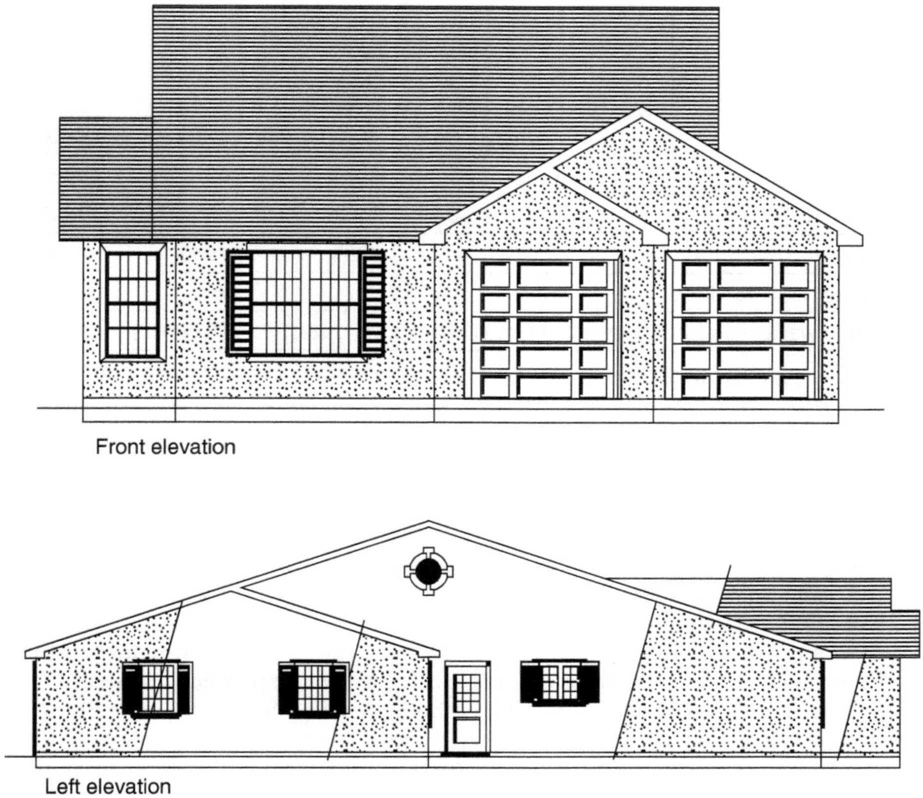

FIGURE 8.1 Atlanta house design analyzed for concrete block versus wood construction. [*Source: Kasal and Huelman (2010). Image used with permission of the Consortium for Research on Renewable Industrial Materials.*]

Wood Frame	198 m² (2153 ft²), 1 story	Concrete Block
	On-slab	
38 × 89 mm studs@406 mm o.c. 2 × 4 (nominal) studs @ 16 in. o.c.	Exterior walls	Concrete blocks
38 × 89 mm studs@406 mm o.c. 2 × 4 (nominal) studs @ 16 in. o.c.	Interior walls and sheathing	38 × 89 mm studs@406 mm o.c. 2 × 4 (nominal) studs @ 16 in. o.c.
	Roof all wood	
	Concrete floor slab	
	Extraction (primary materials)	
11,498 kg (12.7 short tons) of wood		8397 kg (9.3 short tons) of wood
8388 kg (9.2 short tons) of limestone		14,487 kg (16.0 short tons) of limestone

TABLE 8.3 Design Differences for Atlanta House, Concrete Block versus Wood Construction

Atlanta house	Concrete construction	Wood frame	Difference	Concrete environmental impact versus wood—total structure
Embodied energy—GJ	461	398	63	+16%
GWP (CO_2)—kg (short tons)	28,004 (30.9)	21,367 (23.6)	6637 (7.3)	+31%
Air emission index	6007	4893	1060	+23%
Water emission index	7	7	0	0%
Solid waste—kg (short tons)	11,269 (12.4)	7442 (8.2)	3827 (4.2)	+51%

Source: Lippke et al. (2004).

TABLE 8.4 Environmental Impact Comparison—Wood to Concrete: Total Structure

A life-cycle comparison of the two designs from material extraction through building construction showed concrete construction to result in significantly higher consumption of energy and emissions of greenhouses gases [indicated by global warming potential (GWP)], emissions to air, and generation of solid wastes than wood construction) (Table 8.4). Because mill residues provide the bulk of energy used in manufacturing of wood components used in these buildings, the difference in consumption of non-biomass energy (primarily fossil energy) is far greater than indicated by embodied energy values. For instance, while the total embodied energy in the concrete block house was determined to be only 16% greater than for the wood house, the concrete block structure used 250% more non-bioenergy than the wood structure.

A comparison of above-grade portions of the structure only (which factors out the common slab foundation but which retains common elements in the interior walls and roof systems) shows a greater difference, with impacts 38% to 164% greater for concrete construction than for wood (Table 8.5). However, the most meaningful comparison is one in which only the different elements are compared—in this case the exterior walls

Atlanta house	Concrete construction[b]	Wood frame[c]	Difference	Concrete environmental impact versus wood
Embodied energy—GJ	231	168	63	+38%
GWP (CO_2) —kg (short tons)	14,982 (16.5)	8345 (9.2)	6637 (7.3)	+80%
Air emission index	3373	2313	1114	+46%
Water emission index	2	2	0	0%
Solid waste—kg (short tons)	6152 (6.8)	2325 (2.6)	3827 (4.2)	+164%

[a]Roofs of both buildings wood trusses with OSB sheathing.
[b]Walls constructed of concrete blocks with furred-out wood-stud walls with sheetrock.
[c]Walls constructed of 38 × 89 mm (2 × 4 nominal) kiln-dried SYP, with sheetrock and vinyl siding.
Source: Lippke et al. (2004).

TABLE 8.5 Energy Efficiency Comparison Wood to Concrete: Above Grade Only[a]

	Concrete block wall MJ/m³ MJ/ft²	Lumber-framed wall MJ/ft² (MJ/m³)
Structural[a]	816.9 (75.89)	67.5 (6.27)
Insulation[b]	91.6 (8.51)	91.6 (8.51)
Cladding[c]	87.7 (8.09)	240.1 (22.31)
Total	995.6 (92.49)	399.2 (37.09)

[a]Includes studs and plywood sheathing for the lumber wall design and concrete blocks and studs (used in a furred-out wood-stud wall) for the concrete wall design.
[b]Includes fiberglass and six mil polyethylene vapor barrier for both warm climate designs.
[c]Includes interior and exterior wall coverings. Exterior wall coverings are vinyl (lumber wall design) and stucco (concrete wall design). Interior wall coverings gypsum for both warm climate designs.
[d]Includes subtotals from Structural, Insulation, and Cladding categories.
Source: Edmonds and Lippke (2005).

TABLE 8.6 Energy Efficiency Comparison Wood to Concrete: Exterior Walls Only

(Table 8.6). The embodied energy in the concrete building is 38% higher than for the wood design, and the global warming potential is 80% higher. Emissions to air and solid wastes were also found to be substantially greater for concrete block walls versus wood-framed and sheathed walls.

Assessment of a Multistory Apartment Building

An LCA focused on energy and greenhouse gas (GHG) emissions associated with construction of a four-story apartment building in Växjö, Sweden, examined concrete and wood alternatives (Börjesson and Gustavsson 2000). Known as the Wälludden study, the initiative focused on evaluation of a building contained 16 apartments, comprising 1189 usable m^2 (12,800 ft^2) in total (Fig. 8.2). This study was conducted by analyzing the actual

FIGURE 8.2 Building types considered in life-cycle assessment of reinforced concrete versus wood construction, Växjö, Sweden. (*Photo courtesy Växjö Kommun.*)

Material	Wood	Concrete
Lumber	59 (65)	33 (36)
Particleboard	18 (20)	17 (19)
Plywood	21 (23)	20 (22)
Concrete	223 (246)	1352 (1490)
Plasterboard	89 (98)	25 (28)

Source: Based on data from Gustavsson (2008).

TABLE 8.7 Materials Used in Construction of the Two Structures [metric tons (short tons)]

construction of a wood-frame building and then designing and analyzing (but not constructing) the same structure from reinforced concrete. Included in the analysis were raw material extraction, manufacturing, all transportation, and building construction activity.

Wood products used in these buildings (actual and virtual) included lumber, particleboard, and plywood (Table 8.7). Both buildings had concrete foundations. The concrete design used wood for the roof system.

Findings revealed significantly higher primary energy consumption in producing the building materials needed for concrete construction than for wood construction (Fig. 8.3). This is the case even given the many common elements in the two structures. The difference was due to the much greater use of concrete in the concrete frame building. Use of crushed gravel in the concrete mix increased energy requirements 20% to 30% over use of natural gravel.

As little difference in energy use was found in materials transport and building construction, the energy use differential also applies to the completed structures.

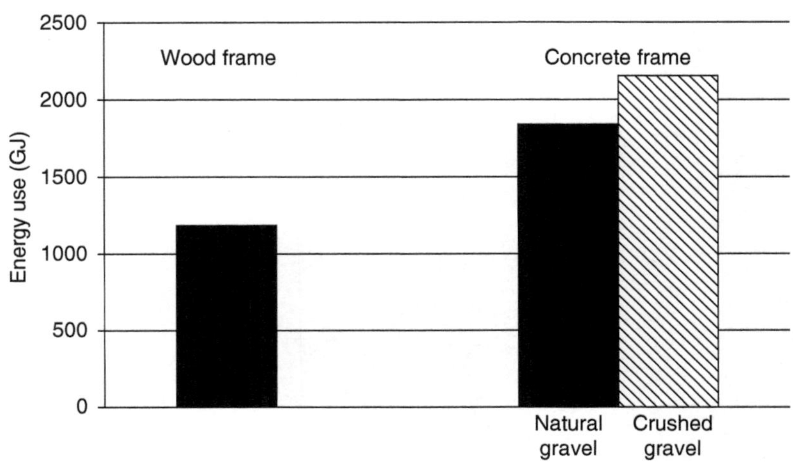

FIGURE 8.3 Primary energy use in the production of building materials—concrete versus wood frame. [*Source: Börjesson and Gustavsson (2000).*]

GHG emissions at the point of building construction were found to be far greater for the concrete frame structure than indicated by the difference in energy requirements, with this attributed to substantial carbon emissions in the cement production processes. When analyses were extended through the full building life and variously considering useful lives of 50, 100, and 300 years, large GHG emission differences were also found.

Use of renewable bioenergy (commonly used in wood products manufacture) versus nonrenewable (fossil) energy (commonly used in cement manufacture) gives wood a distinct advantage relative to carbon emissions. In this analysis, relative emissions were determined assuming bioenergy use for production of all building materials, including cement and concrete components. Nonetheless, the average GHG mitigation potential over a 100-year perspective was determined to be 2 to 3 times better for the wood building than the concrete building. It was also found to be better over 50-year and 300-year building life cycles, with the wood GHG mitigation advantage at 300 years 3 to 5 times greater for wood than concrete.

The carbon emission advantages of wood building materials are highly influenced by the fate of materials at the end of useful life of a building. If wood building materials are landfilled and with no methane subsequently collected at the disposal site, the GHG mitigation advantage of wood disappears. Mitigation advantages are greatest when wood is reused and/or used for energy production to replace fossil fuels.

A key finding of this study was the following:

- The use of wood building materials in place of concrete, coupled with the greater integration of wood by-products into energy production, would be an effective means of reducing fossil fuel use and net CO_2 emissions to the atmosphere.[1]

This and related studies led to initiation of the Växjo Wooden City, part of an effort initiated in 1991 by the city of Växjo to become the world's first fossil fuel-free city and the "greenest city in Europe." Results of the Wälludden Project were the basis for a focus on wood construction as part of this effort. The building pictured in Fig. 8.4 is part of what is known as the Limnologen project. Constructed over a first floor of concrete, the load-bearing structure of the top seven floors of the building consists of CLT. CLT was used for both floors and most walls. Traditional timber framing was used in building walls that separate apartments.

Assessment of a Mid-Rise Office Building (Concrete vs. CLT)

An assessment by the University of British Columbia and the Canadian Wood Council (Robertson et al. 2012) examined relative environmental impacts of building a mid-rise office building with concrete versus CLT and engineered wood. In this case, the office building evaluated—a five-story 14,215-m^3 (153,000 ft^2) structure with three levels of underground parking, constructed of cast-in place reinforced concrete, was an actual building, constructed in British Columbia, Canada, in 2009.

For purposes of comparison, a functionally equivalent structural system and building envelope was designed and virtually constructed but using a combination of glulam

[1]A subsequent study (Lenzen and Treloar 2002) found that Börjesson and Gustavsson had substantially underestimated both energy consumption and carbon emissions in production of building materials for both building designs. However, recalculation of energy and carbon emissions nonetheless confirmed the energy and GHG mitigation advantages of the wood-frame alternative.

FIGURE 8.4 Limnologen project, Växjo, Sweden. (*Photo courtesy Växjö Kommun.*)

and CLT for the vertical and horizontal force-resisting systems in the aboveground portion of the building, in conjunction with a reinforced-concrete shear core. The roof system, sheathing, and exterior insulation were also different in the redesigned building (Table 8.8). Following redesign, LCA was employed in determining the full range of environmental impact estimates for each building.

Key findings of this assessment (Table 8.9) were the following:

- Impacts associated with the concrete design were significantly greater in every category measured, except for fossil fuel depletion. Greater fossil fuel consumption in the wood design was attributable to the fossil fuel-based adhesives used.
- Global warming potential of the concrete design was almost four times greater than for the wood design.

Very similar results to those reported above have been obtained by virtually every research entity around the world which has studied these issues. For instance:

- A U.S. study of various wall configurations, including wood 2 × 4 construction, steel stud construction, autoclaved cellular concrete, and EPS insulating concrete form, found that wall systems made of concrete had poorer energy performance when considered both at the point of completion of building construction and over the long term. The study evaluated wall systems with equivalent R-values used in cold climate regions of the United States and Canada and included an evaluation of energy required for wall construction and for subsequent building operation over the long term (Pierquet et al. 1998).
- A Canadian LCA of a generic three-story office building constructed alternatively of wood, steel, and reinforced concrete found significant differences in the aboveground portion of the buildings (above the common foundation and parking structure). Embodied energy requirements in the concrete and steel

Material group	Unit of measurement	Concrete design	Timber design
Foundation—No differences in concrete and timber design			
Superstructure			
Primary shear walls and cores	m^3 of concrete	No differences	
Vertical load-bearing walls	m^3 (ft^3) of concrete/CLT	181 (6392)	128 (4520)
Above-grade floors and roof	m^3 (ft^3) of concrete/CLT	3628 (128,122)	2950 (104,178)
Above-grade columns	m^3 (ft^3) of concrete/glulam	268 (9464)	122 (4308)
Beams and roof parapet	m^3 (ft^3) of concrete/glulam	166 (5862)	947 (33,433)
Wood sealer	m^2 (ft^2)	—	1586 (17,072)
Building Enclosure			
Curtain wall	m^2 (ft^2)	No differences	
Cedar siding	m^2 of 13-mm thickness (ft^2 of 0.5-in. thickness)	—	13,374 (143,957)
ccSPF insulation	m^3 (ft^3)	2134 (75,362)	—
R-13 (RSI-2.3) insulation	m^3 (ft^3)	258 (9111)	115 (4061)
Steel stud framing	m^2 @ 406 mm o.c. (ft^2 of 16 in. o.c.)	1617 (17,405)	—
Wood stud framing	m^2 @ 406 mm o.c. (ft^2 of 16 in. o.c.)	—	1617 (17,405)
Gypsum wallboard	m^2 of 13-mm (or 0.5 in.) thickness	No differences	

Source: Robertson et al. (2012).

TABLE 8.8 Design Details for Mid-Rise Office Building (Shaded cells indicate differences in concrete and timber design)

Impact category	Concrete design	Wood design
Global warming potential (GWP)	100	27
Ozone depletion	100	58
Human health effects	100	67
Criteria air pollutants	100	70
Water intake	100	71
Eutrophication	100	71
Ecological toxicity	100	82
Smog	100	82
Acidification	100	87
Fossil fuel depletion	94	100

Source: Robertson et al. (2012).

TABLE 8.9 Environmental Impact Comparisons, Mid-Rise Office Building (Highest impact estimate in each category expressed as 100) structures were determined to be 27% and 61% higher, respectively, than for the wood structure (Cole and Kernan 1996).

- An LCA (Damberger 1995, as reported by Scharai-Rad and Welling 2002) of single-family houses in Germany found concrete block construction to require 1.58× the energy overall and 2.85× the energy for the above-grade portion of structures. When considered over an 80-year life cycle, overall energy requirements for a concrete structure (considering total energy needed for construction and building operation) were found to be 21% greater than for a functionally equivalent wood structure. Global warming potential was similarly higher for the concrete structure.
- A Swedish study of concrete and wood-framed buildings (Gustavsson and Sathre 2006, Gustavsson et al. 2006) found higher energy and CO_2 balances for concrete structures (with differences in the range of 30 to 130 kg C per m^2 of floor area) and concluded that reducing the proportion of concrete building materials relative to wood building materials would be an effective means of reducing fossil fuel use and CO_2 emissions.

While the vast majority of assessments have found that the concrete construction process (including all steps from raw material extraction through completion of construction) results in the consumption of more energy and greater emissions of carbon dioxide than wood construction, a seemingly conflicting conclusion was reached in two 2008 assessments (Marceau et al. 2008a, b). Reported results were that while wood construction results in lower environmental impacts initially, cumulative environmental impacts linked to building operation over the life of a structure shift the advantage to concrete. However, a review of methodology (Bowyer et al. 2008) revealed that the buildings compared were designed to different thermal standards, with concrete designs insulated to a higher standard—a fundamental violation of LCA protocols. In other words, the comparisons were not of functionally equivalent buildings and therefore were irrelevant.

8.2.4 Energy and Related Differences between Wood and Steel Framing Systems

Analysis of a Single-Family Residential Structure

A 2004 assessment of relative impacts of wood and steel construction examined two virtual houses (neither was actually constructed) designed to code standards in Minneapolis, Minnesota; one was built of wood, the other with steel framing. The two buildings had identical building envelope designs (Fig. 8.5) with matched thermal performance. The wood design featured solid 2 × 6 wood framing members, except for wood composite I-joists. Oriented strand board (OSB) was used for roof and wall sheathing and for floors, and preengineered wood trusses were used for the roof system. Substitution of plywood for OSB and solid timbers for I-joists was also evaluated. In the steel design, steel floor joists and wall studs were substituted for wood, and extra insulation was used in order to meet code requirements. Both designs incorporated a concrete foundation/basement.

Comparison of the steel and wood designs from material extraction through building construction showed steel construction to result in greater consumption of energy and emissions to air, including GHG, and far greater emissions to water (Table 8.10). Wood construction, however, was estimated to generate about 1% more waste.

FIGURE 8.5 Minneapolis house design analyzed for steel framing versus wood construction. [*Source: Kasal and Huelman (2010). Image used with permission of the Consortium for Research on Renewable Industrial Materials.*]

A comparison of above-grade portions of the structure only (which factors out the common basement and foundation but which retains common elements in the wall and roof systems) shows greater differences, with impacts as much as 867% greater for steel construction than for wood (Table 8.11). The waste difference was magnified as well, with the wood building estimated to yield 9% more on-site waste than the steel building.

As noted previously, lumber manufacturers in the United States and Canada are largely energy self-sufficient, using scraps of bark and what was once waste wood to produce their own energy in large industrial boilers and cogenerating equipment. As a result, the lumber industry is 50% to 60% energy self-sufficient overall. In contrast, the

Minneapolis house	Steel construction	Wood frame	Difference	Steel environmental impact versus wood—total structure
Embodied energy (GJ)	764	651	113	+17%
GWP (CO_2)—kg (short tons)	46,826 (51.6)	37,047 (40.8)	9779 (10.8)	+26%
Air emission index	9729	8566	1163	+14%
Water emission index	70	17	53	312%
Solid waste—kg (short tons)	13,641 (14.8)	13,766 (15.2)	−125 (−0.4)	−0.9%

Source: Lippke et al. (2004).

TABLE 8.10 Environmental Impact Comparison—Wood to Steel: Total Structure

Minneapolis house	Steel construction[b]	Wood frame[c]	Difference	Steel environmental impact versus wood
Embodied energy (GJ)	296	250	46	+18%
GWP (CO_2)—kg (short tons)	17,262 (19.0)	13,009 (14.3)	4253 (4.7)	+33%
Air emission index	4222	3820	402	+11%
Water emission index	29	3	26	867%
Solid waste—kg (short tons)	3181 (3.5)	3496 (3.9)	−315 (−0.4)	−9%

[a]All walls designed with 11-mm (7/16 in.) plywood sheathing and vinyl siding.
[b]Walls constructed of 20-gauge (0.9525 mm), 38 × 139-mm (2 × 6 nominal) galvanized studs containing average recycled content for steel framing in North America.
[c]Walls constructed of 38 × 139-mm (2 × 6 nominal) kiln-dried SPF.
Source: Lippke et al. (2004).

TABLE 8.11 Environmental Comparison Wood to Steel: Above-Grade Only[a]

North American steel industry is fueled almost entirely by fossil fuels, namely, coal and natural gas. From an environmental point of view the difference is both significant and important, since energy generation from wood and other biomass is close to carbon neutral[2] and less polluting in general than energy generation from fossil fuels. This results in large differences in emissions of fossil carbon. For instance, although the completed steel-frame house required only 17% more energy than the wood-framed house (Table 8.11), the steel-frame design would have used 281% more nonbioenergy than the wood-frame design.

When all common elements in the buildings are eliminated from consideration and only the different elements compared—in this case the floor and roof assemblies (Table 8.12)—the magnitude of impact resulting from substitution of steel for wood can be clearly seen. The embodied energy difference in this case is 67% and the global warming difference 157%. Emissions to air and solid wastes were also found to be substantially greater for concrete block walls versus wood-framed and sheathed walls.

Assessment of Alternative Roof Structures for an Airport Terminal

A 2002 assessment of steel beams versus glue-laminated spruce wood beams for the roof of the Oslo International Airport terminal (Fig. 8.6) found that the manufacturing of steel beams for that project would have required 2 to 3 times more energy and 6 to 12 times more energy from fossil fuels than would functionally equivalent glulam beams. Analysts

[2]There is some debate over whether harvesting of forests, which removes carbon stored in trees from the landscape and some release of carbon stored in forest soils, can be considered carbon neutral or nearly so. Those questioning neutrality cite periods of several decades or more following harvest to replenish carbon lost from forests. An opposing and currently dominant view is that as long net growth in a given forest equals or exceeds net removals, thereby maintaining or increasing forest carbon stocks over time, harvesting can be considered carbon neutral. The fact that wood, when used in long-term applications such as building materials, continues to store carbon throughout its useful life, coupled with the reality that wood product manufacture typically results in lower carbon emissions than potential substitutes, bolsters the carbon neutrality argument.

Environmental performance index	Floor and roof assemblies [a]		
	Wood	Steel	Difference
Embodied energy (GJ)	109	182	67%
Global warming potential (CO_2)—kg (short tons)	3763 (4.1)	9650 (10.6)	157%
Air emission index (index scale)	981	1813	85%
Water emission index (index scale)	17	70	312%
Solid waste—kg (short tons)	13,766 (15.1)	13,641 (15.0)	–0.9%

[a]Floor joists are 38 × 235 mm (2 × 10 nominal), for steel and wood, with steel of 18-gauge (1.27 mm).
Source: Perez-Garcia et al. (2005).

TABLE 8.12 Environmental Performance Indices for Steel and Wood Floor and Roof Assemblies for a Home Built to Minneapolis Code Standards

noted that if virgin rather than recycled steel were used, the differences as indicated above would be substantially greater. In the most likely scenario, steel beam manufacture was estimated to result in a fivefold increase in GHG emissions in comparison to glulam beams. The structure was subsequently built using spruce glulam beams.

A subsequent (2011) similar evaluation led to use of glulam beams in the roof system of the Raleigh-Durham Airport terminal in North Carolina. With options of using steel, reinforced concrete, or engineered wood for the structural support system, designers turned to LCA to assess environmental impacts of design choices. Results indicated that wood glulam beams rather than traditional materials for this application would result in energy savings of 20,160 gigajoules (5600 MWh), equivalent to the electricity use of 500 homes over 1 year or 23 days of operational energy use for the terminal, and GWP savings of 1000 t CO_2e, equivalent to 1 year of taking 200 cars off the road or 85 ha (210 acres) of forest carbon

FIGURE 8.6 Roof of Oslo International Airport terminal. (*Photo Courtesy Novatop, AGROP NOVA a.s., Petnti, Czech Republic.*)

sequestration potential. Assessment again led to selection of glulam beams in this project (ARUP, as reported by Athena Sustainable Materials Institute 2011).

8.2.5 Wood and Environment—Additional Considerations

As indicated in numerous assessments, production of wood building materials results in far lower carbon emissions (CO_2 equivalents) than those made of concrete or steel. In virtually every case, wood has been shown to have a substantial advantage in relation to other materials in terms of energy consumption per unit of finished products and to generate lower emissions to air and water in the process of raw material to product conversion. But wood has an additional advantage in this regard: It stores massive quantities of carbon. All wood products store carbon throughout their life. Dry wood is one-half carbon by weight, with the result that wood use translates to significant carbon storage.

Residential houses, which in North America are dominantly wood construction, store large quantities of carbon over the long term. As reported by the Idaho Forest Products Commission, a new single-family home built in the United States in 2011 had an average floor space of 223 m^2 (2400 ft^2) and contained about 38 m^3 (16,000 board ft) of lumber and 1300 m^2 (14,000 ft^2) of wood panels. This quantity of wood incorporated almost 11,800 kg (26,000 pounds) of carbon, with a carbon dioxide equivalent of over 42,500 kg (94,000 pounds). With over 70 million single-family homes in the United States and almost 50 million additional housing units contained within townhouses and multiple-occupant residences and a growing number of commercial/industrial and other structures, a massive quantity of carbon is stored.

Consequently, it is not only the trees growing within the nation's forests and neighborhoods that store carbon for long periods of time but also the homes that make up neighborhoods. Similarly, the industrial and commercial buildings, including high-rise buildings, which are increasingly built of wood, create additional carbon pools across the landscape while at the same time reducing the overall environmental impacts of building material production.

8.3 The Renewability of Wood—An Underappreciated Reality

Forests are renewable, a fact that, unfortunately, is often not fully appreciated. An example of forest renewability is provided by the 62-year period ending in 2013, during which over 24,806 million m^3 (876 billion ft^3) of timber were harvested from U.S. forests, a volume approximately equivalent to a bulk pile of wood (no spaces) measuring $1.6 \times 1.6 \times 9.5$ km (1 mile × 1 mile × 5.9 miles) in height! This wood was used in building some 93 million homes and in producing countless other products. And what was the impact on domestic forests? Because forests are renewable, the volume of wood within domestic forests *increased* by more than 60% during those 62 years. Accompanying the massive use of wood was an increase in the volume of carbon stored within U.S. forests, long-term storage of billions of tons of carbon within residential structures and other buildings, and avoidance of even greater quantities of carbon through use of wood rather than other more fossil energy-intensive products.

The differences between renewable and nonrenewable materials are fundamental and dramatic. These differences are sometimes overlooked or discounted in discussions of environmental policy. Maintaining access to a versatile raw material that, if properly managed, will not only maintain but actually increase its volume is a gift that should not be taken lightly.

8.4 Wood—More Than Enough

Proposals for increased harvesting activity in forests evoke strong emotions in some people who tend to equate harvest with deforestation such as seen in the world's tropical forests. However, the extent of forested land in the United States (and in Canada) is today within 1% of the same area as in the early 1900s despite ongoing harvest activity. This is consistent with the relatively recent finding that the lowest rates of deforestation and net forest carbon emissions occur in global regions with the highest rates of industrial wood harvest and forest product output. Conversely, global regions with the highest rates of deforestation and forest carbon emissions rank lowest in harvest of wood for use in building and production of other forest products (Ince 2010). In short, forest landowners who are unable to realize ongoing financial returns from forests have an economic incentive to convert their lands to agriculture or some other use that will yield income. An obvious and perhaps counterintuitive implication is that an effective way to maintain or increase forest carbon stocks on private lands is to ensure the existence of a strong market for forest products.

But what would be the impact of a large increase in production of CLT buildings and use of engineered wood products? Is the wood resource base sufficient for this kind of shift in building construction? This discussion focuses on those two questions, first examining wood resources and the state of forests in North America and Europe—the source of the vast majority of wood used for industrial purposes, including construction. Thereafter, wood supply considerations for the near and longer terms are discussed.

8.4.1 Forest Resources

North America

Only the forests of the United States and Canada are included in this discussion. Together, the two countries account for 90% of forest cover in North America and over 95% of forests managed for timber production. Slightly under one-half (48%) of the world's certified forests are located within Canada and the United States (FAO 2016).

United States

The United States has 310 million ha (766 million acres) of forests which cover one-third of the land area of the country and which account for 8% of the forested area globally. U.S. forests are mostly privately owned, with individuals, families, investment groups, and the forest industry controlling 58% of forestland. Almost two-thirds of privately owned forestland is in the hands of individuals and family forest owners. The remaining 42% of forestlands are public, with three-fourths of these owned by the federal government and the rest by state, municipal, and local governments. Periodic harvesting to produce forest products is heavily concentrated in privately owned forests; these forests account for 90% of the annual harvest.

Surveys of college students and others over the past three decades have revealed a pervasive belief that forests of the United States are disappearing, a perception apparently fueled by ongoing loss of forest cover in the world's tropical regions. Reality does not match that perception. In fact, the area covered by forests in the United States has remained remarkably constant over the past 100 years (Fig. 8.7), with a recent trend toward *increasing* forest area. The area of timberland (forests that are available for periodic timber harvest) has similarly remained largely unchanged over an extended period. Timberland makes up about two-thirds of the forest area of the United States.

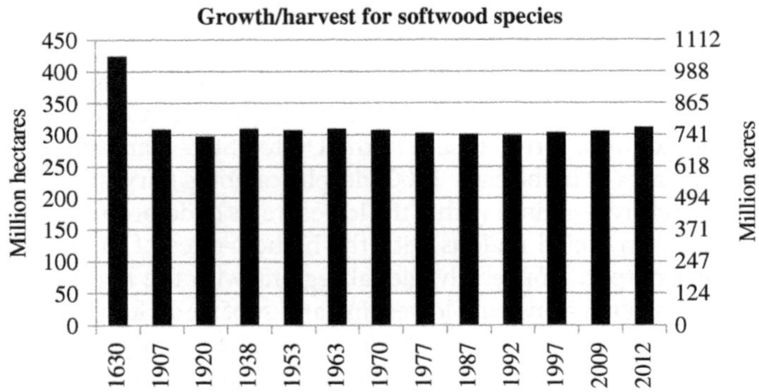

FIGURE 8.7 Area of forestland in the United States, 1630–2012. [*Source: Based on data from Oswalt et al. (2019).*]

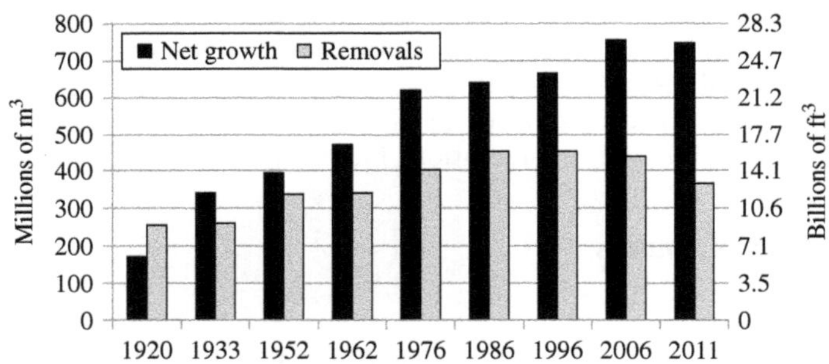

FIGURE 8.8 Net growth versus annual removals on U.S. timberland, 1952–2011. [*Source: Oswalt et al. (2019).*]

The extent of forests in the United States today is about the same as in 1900 and about two-thirds of that estimated to have existed at the time of European settlement.

Another widely held misconception is that more timber is harvested each year than is grown. Despite rising population, wood consumption, and harvest levels, net forest growth in the United States has exceeded removals for at least seven consecutive decades,[3] with an increasing margin between growth and removals (Fig. 8.8). For softwood species which provide essentially all building materials for the core and shell of structures, there has been a similar pattern in net growth and removals (Fig. 8.9). A marked decline which occurred in forest harvest rates during the sharp recession of 2007–2010 is reflected in 2011 numbers in both Fig. 8.8 and Fig. 8.9. In 2011, the rate of tree growth and wood formation for all species was more than double the rate of removals; for softwoods, net growth exceeded removals by 89%. Because the volume of wood added annually through new growth exceeds the volume of wood removed, the volume of trees (and of wood) in U.S. forests is increasing. Trees within the one-third of U.S.

[3]Net forest growth is the sum of all new growth minus all losses due to decay, insects, fire, wind, and other such events. Removals include volumes extracted in harvesting, land clearing, and similar activities.

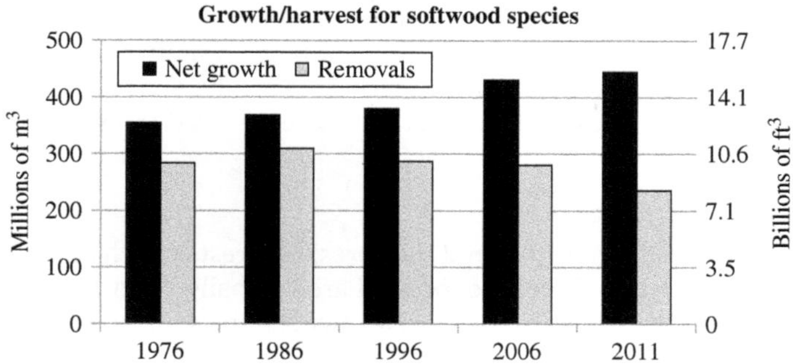

FIGURE 8.9 Net growth versus annual removals of softwood species on U.S. timberland, 1976–2011. [*Source: Oswalt et al. (2019).*]

forests that are off-limits to periodic harvesting, such as forests within wilderness areas, are not counted when determining timber volume.

As earlier indicated, the volume of wood contained within U.S. forestlands designated as timberland has increased by over 60% over the past 62 years. The record of forest growth in relation to forest removals across the United States over the past six decades provides a testament to the accomplishments of forestland managers. This is not, however, the whole story. Another reason that the volume of trees has increased in the face of expanding wood consumption is that the United States has increasingly turned to imported wood, softwood in particular, to fulfill wood consumption needs. Net softwood lumber imports increased markedly after 1990 when government actions largely halted forest harvesting. Were the United States providing all of its own wood needs, the nation would still have a positive net growth to harvest ratio for softwoods, but by a narrower margin than indicated by net growth-to-harvest ratios.

Canada

Canada has 347 million ha (857 million acres) of forests which cover 42% of the land area of the country and which account for 9% of the forested area globally. In sharp contrast with the United States, 94% of Canada's forests are publicly owned, with almost all of these under the control of provincial and territorial governments. Only 4% and 6% of forests are owned by the federal government and private interests, respectively. About one-tenth of Canada's annual timber harvest comes from privately owned lands (Natural Resources Canada 2017).

Canada accounts for 10% of the world's forests and retains 91% of the forest cover that existed prior to European settlement. The forestland base in Canada has remained quite stable since the beginning of the 20th century. In recent decades there has been a very slight trend toward forest conversion to other uses; during the 25-year period 1990 through 2014, forest loss amounted to 0.03% of the forestland base.

Two-thirds of Canada's forests are managed, and the country has a greater area of third-party certified forests than any other country. As in the United States, there is a long history of greater net growth than removals nationwide. Data from 2014 show a net growth-to-harvest ratio for softwoods of about 1.3:1.0 and for hardwoods of about 2.2:1.0. Overall, net forest growth exceeded harvests in 2013 by over 50% (Natural Resources Canada 2016).

Despite favorable growth-to-harvest statistics, Canada's forests have recently become a carbon source due to a devastating mountain pine beetle outbreak in the western provinces that began in the early 1990s. The effect has been a marked reduction in net growth, a loss of forest carbon stores as dead trees decay, and a reduction of harvest levels.

Europe

Europe has 182 million ha (450 million acres) of forests which cover 43% of the land area and which account for 5% of the forested area globally. Over 60% of the EU-28's forests were privately owned in 2010. Private ownership is most common in western and northern European countries, where private ownership often accounts for 70% or more of forestland. In contrast, public ownership is more common in eastern Europe (Eurostat 2011).

The forestland base throughout Europe has been expanding slowly since the late 1800s, following centuries of rampant forest exploitation. In the period 2000–2010 the forested area expanded about 2%.

Net forest growth within the EU significantly exceeds removals, with a steady increase in the growth-to-removals ratio in recent decades. The most recent statistics (2010) show net forest growth over 60% higher than removals. Consequently, the volume of wood within EU forests is increasing (Eurostat 2017).

8.4.2 Potential for Expanded Wood Use

Wood Supply Considerations

While wood was used to a greater extent in construction of commercial buildings in 2017 than in most of the last century, there remains considerable potential for greater wood use in multifamily residences and in nonresidential buildings of all sizes. As part of an initiative to gauge the potential for greater wood use building construction and specifically of tall buildings, the impact on U.S. forests of greater harvest levels was estimated.[4] This analysis examined the potential for greater wood use based on the effect on wood demand of greater wood use in low-rise nonresidential buildings, multifamily residences of 3 to 6 stories, and taller buildings of all kinds in the 7- to 15- story range. Findings indicate the potential for increased annual wood consumption of 16.1 to 17 million m^3 (6.8 to 7.6 billion board ft) (Table 8.13). The near-term potential increase anticipated from use of wood in constructing buildings to heights over 6 stories is 3.8 to 5.7 m^3 (1.6 to 2.4 billion board ft).

With regard to differences between annual growth and removals, the most recent estimate (2016) shows net annual growth-to-removal ratios for all species at 1.9 and for softwood species at 1.7. Expressed differently, the difference in net annual growth and annual removals for all species is 340 million m^3 [12.0 billion ft^3, or 97 billion board feet (Bbf)], and for softwoods it is 187 million m^3 (6.6 billion ft^3 or 53 Bbf). The estimates of wood volumes associated with CLT potential equate to 7% to 8% of net annual wood growth in the United States and approximately 14% of the current available growth that occurs each year in excess of existing market demands. In addition, production of mass timber products such as CLT and nail-laminated timber can utilize standing dead and

[4]Forest Climate Group (2015).

	Approximate additional wood volume needed	
	Million cubic meters (m^3)	Billion board ft (Bbf)
Low-rise nonresidential[a]	10.6	4–5
Multifamily[b]	1.7	0.7
U.S. buildings 7–15 stories	3.8–5.7	1.6–2.4
Total	16.0–17.9	6.8–7.6

[a]Nonresidential includes nonresidential low-rise buildings such as offices, stores, schools, public buildings, and hotels. For simplicity, 100% conversion of only the near-term opportunity to wood was assumed, a figure expected to increase over time.

[b]Multifamily includes multifamily residential buildings such as 3- to 5- story apartment buildings and condominium-style buildings. Market penetration for wood in this market was about 80% at the time of the analysis (2015), and increased wood volume potential was based on an assumption of 100% conversion of near-term opportunities to wood.

Source: Forest Climate Group (2015).

TABLE 8.13 Potential Near-Term Annual Increase in Wood Use in U.S. Construction

dying timber, such as that impacted by the mountain pine beetle over vast areas of western North America, for which there are few other markets, further expanding the near-term potential wood supply. Thus, when viewed from this perspective, a potential increase in harvest volume of 16.0 to 17.0 million m^3 (6.8 to 7.6 Bbf) provides a considerable margin within the bounds of sustainability (Table 8.13).

Environmental Implications

The previously discussed estimate of the potential for increased use of wood in construction also examined impacts regarding avoided carbon emissions. It was estimated that if increased quantities of wood as defined in Table 8.14 were realized, and the near-term carbon benefit could be as high as 32.7 million metric tons (mmt) (36 million short tons) of CO_2e per year in the United States and 7 to 10 mmt (7.7 to 11 short tons) specifically resulting from construction of buildings over six stories in height.

	Approximate additional wood volume needed annually million cubic meters (Bbf)	Million metric tons (million short tons)		
		Stored CO_2e	Avoided CO_2e	Total carbon benefit
Low-rise nonresidential	9.4–11.8 (4.5)	6 (6.6)	13 (14.3)	19 (21)
Multifamily	1.65 (0.7)	1 (1.1)	2 (2.2)	3 (3.3)
U.S. buildings 7–15 stories	3.8–5.7 (1.6–2.4)	2–3 (2.2–3.3)	5–7 (5.5–7.7)	7–10 (7.7–11)
Total	16.0–17.9 (6.8–7.6)	9–10 (9.9–11)	20–22 (22–24)	29–32 (32–35)

Source: Forest Climate Group (2015).

TABLE 8.14 Potential Near-Term Annual Increase in Wood Use in U.S. Construction

As shown above, the effect of using an additional 3.8 to 5.7 million m^3 of wood (1.6 to 2.4 Bbf) in construction of 7- to 15- story buildings would be avoided emissions of 7 to 10 mmt (7.7 to 11 short tons) of carbon dioxide (CO_2e). Of this, 2 to 3 mmt (2.2 to 3.3 million short tons) would be stored within the wood products, and 5 to 7 mmt (5.5 to 7.7 million short tons) of emissions would be avoided due to use of wood in place of more energy intensive (and particularly fossil energy intensive) building materials. The effect would be significant, equivalent to removing 1.5 to 2.2 million vehicles from U.S. streets and highways or shutting down two to three coal fired power plants for 1 year. Greater wood use across all multistory and nonresidential construction would result in an even greater carbon benefit. Estimates of potential carbon benefits of greater wood use in construction are confirmed by recent comparative LCAs of buildings made of various materials.

8.5 Longevity of Wood Structures

The use of lower useful service life estimates for wood in comparison to other materials appears to be reinforced by a pervasive view of wood inferiority among many in the building industry. A survey of 683 architects, structural engineers, builders, and developers in the United States and Canada (Gaston et al. 2001 as reported by O'Connor 2004) regarding building durability revealed that survey participants expected nonresidential wood buildings to last for far shorter periods than buildings constructed of other materials. Respondents indicated an average expected life for wood buildings of 46 years, whereas useful lives of steel, masonry, and concrete buildings were estimated at 77 to 87 years. It is a perspective that subsequent research has found has no factual basis.

Only one definitive study of building durability by primary type of material used in construction has been conducted; this involved an examination of 227 building demolitions in the Minneapolis/St. Paul metropolitan area for the years 2000–2003 (O'Connor 2004). About two-thirds of the 227 buildings for which there were public records (which included both residential and commercial structures) were wood, one-fourth were concrete, and the remainder were steel or various combinations of wood, steel, and concrete. The largest concentration of buildings fell into the 76- to 100- year age group at the time of demolition. While the ages of demolition for concrete and steel structures were as long as 100 years+, the majority of these buildings reached the end of their useful lives at about 25 to 30 years of age. The wood structures were, on average, far older at the time of demolition (Fig. 8.10), a result influenced by the fact that many of these were houses.

About half (105) of the buildings studied were nonresidential. Of these, the structural systems of 54 were concrete, 10 were steel, and 30 were wood, for a total of 94 buildings. The structure of one building was aluminum, and the rest had structural systems of various combinations of concrete, steel, and wood. Comparing the age of demolition by type of structural material with the impressions of design professionals indicated previously revealed a wide gap between perception and reality. Only one-fifth of the steel buildings were more than 50 years old at the time of demolition, with half of these no more than 25 years old. Similarly, only a third of concrete structures reached an age of 50 years or more prior to being demolished. In contrast, over 60% of the wood buildings were older than 50 at demolition, with the largest group demolished at 76 to 100 years of age (Fig. 8.10).

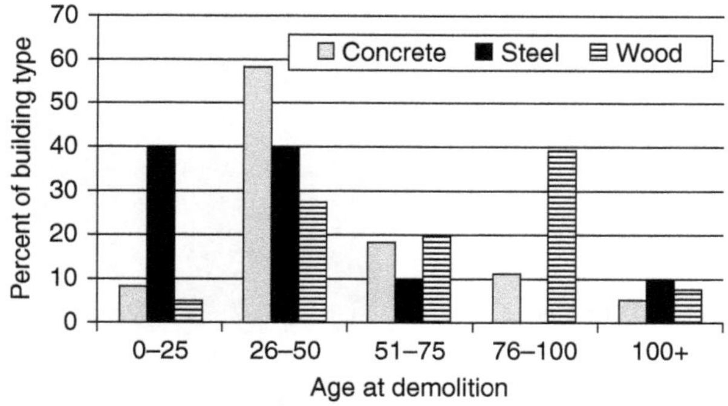

FIGURE 8.10 Age of 94 nonresidential buildings by structural material at the time of demolition. [*Source: O'Connor (2004).*]

Interestingly, upon delving into the reasons for demolition, it was found that most buildings in this study were demolished for reasons that had nothing to do with the physical state of the structural systems. About 60% of structures were removed because the buildings no longer fit the needs of the owner or tenant due to changing land values, socially undesirable use, or an inability to economically bring a building up to code. For only 8 of the 227 buildings studied, structural failure was identified as the primary reason for demolition, and in all but one of these, the problem was foundation failure. In one building, wood decay was identified as the problem.

Fire damage was reported as the reason for demolition of 3.5% of the buildings studied. A greater percentage of steel buildings were demolished for this reason than of buildings constructed of wood or concrete (Fig. 8.11).

Examination of the data from Minneapolis/St. Paul led to the conclusion that despite a widely held perception that the useful life of wood structures is lower than other buildings, no meaningful relationship exists between the type of structural material and average service life. Results also showed that wood structural systems are fully capable of meeting longevity expectations, a reality that is conclusively illustrated by the Butler Building—an eight-story, brick-clad, 46,500-m^2 (500,000 ft^2) building in Minneapolis which was built of heavy timbers in 1906 (Fig. 8.12).

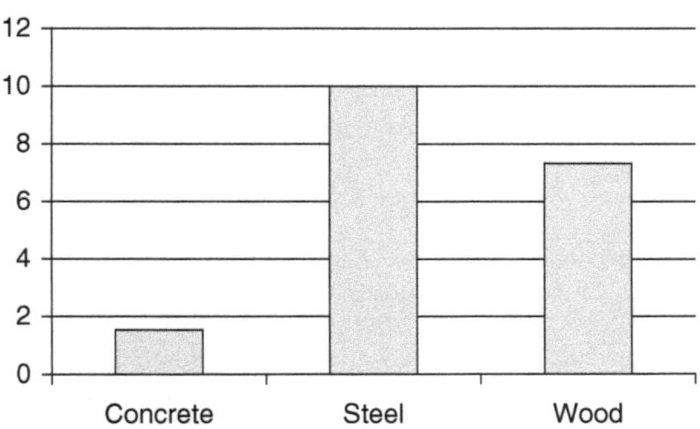

FIGURE 8.11 Percent of each building type demolished because of fire damage. [*Source: O'Connor (2004).*]

FIGURE 8.12 Butler Building, Minneapolis, Minnesota. (*Photo: Tom Wallace. Permission for use from McGough Facilities Management, Minneapolis.*)

8.6 Summary

As long as forests from which wood is obtained are managed sustainably, structures constructed of wood almost invariably result in lower environmental impacts across a wide array of impact categories than those built of nonwood materials. Moreover, wood structures are equally as durable as those constructed of steel or concrete.

Forests in those world regions from which structural wood materials are commonly obtained are generally stable or expanding, with net growth rates greater than removals. Sustainable management of forests, coupled with the fact that forests and the wood they produce are renewable and traditional structural materials are not, makes wood products unique among the array of materials available to building designers.

Recognition of the environmental superiority of sustainably produced wood products does not mean that every part of every building should necessarily be built of wood. Rather, wood should be properly viewed as one of many materials available to the design community for construction of buildings of all types and sizes, which, when used appropriately, can markedly improve environmental performance.

References

Athena Sustainable Materials Institute. 2011. Raleigh-Durham Airport Terminal-2. (http://www.athenasmi.org/resources/case-studies/raleigh-durham-airport-terminal-2)

Börjesson, Päl, and Leif Gustavsson. 2000. Greenhouse Gas Balances in Building Construction: Wood versus Concrete from Life-Cycle and Forest Land-Use Perspectives. *Energy Policy* 28(9): 575–588. (https://www.sciencedirect.com/science/article/pii/S0301421500000495)

Bowyer, Jim, Steve Bratkovich, Kathryn Fernholz, Jeff Howe, and Alison Lindberg. 2008. Concrete as a Construction Material: How Does It Compare to Wood, Steel? Dovetail Partners, Inc., August 25. (http://www.dovetailinc.org/report_pdfs/2008/dovetailconcrete0808a.pdf)

Cole, Raymond, and Paul Kernan. 1996. Life-Cycle Energy Use in Office Buildings. *Building and Environment* 31(4): 307–317. (http://citeseerx.ist.psu.edu/viewdoc/download?doi=10.1.1.453.8646&rep=rep1&type=pdf)

Damberger, B.F. 1995. Comparison of "Brick Buildings" with "Wooden Buildings" in Relation to the Global and Regional CO_2 Budget and the Generated Goods and Energy Flows. Diploma Thesis carried out at the Institute for Water Quality and Waste Management of the TU, Vienna. [In German]

Edmonds, Lucy, and Bruce Lippke. 2005. Reducing Environmental Consequences of Residential Construction through Product Selection and Design. Consortium for Research on Renewable Industrial Materials (CORRIM), Fact Sheet #4. (https://corrim.org/wp-content/uploads/2017/12/CORRIM_Factsheet_September_2005.pdf)

Eurostat. 2011. Forestry in the EU and in the World: A Statistical Portrait. Accessed October 24, 2017. (http://ec.europa.eu/eurostat/documents/3217494/5733109/KS-31-11-137-EN.PDF)

Eurostat. 2017. Forests, Forestry, and Logging. Accessed October 24, 2017. (http://ec.europa.eu/eurostat/statistics-explained/index.php/Forests,_forestry_and_logging)

FAO North American Forest Commission. 2016. Analysis of Results of the Global Forest Resources Assessment 2015 for North America. (http://www.fao.org/3/a-mp390e.pdf)

Fernanda Laguarda Mallo, Maria, Urs Buehlmann, and Omar Espinoza. 2014. Outlook for Cross-Laminated Timber in the United States. *Bioresources* 9(4): 7427–7443. (http://www.forum-holzbau.com/pdf/58_IHF_2015_Espinoza.pdf)

Forest Climate Group. 2015. Expanding the Use of Wood in Buildings—Including Tall Wood Buildings—Helps to Support Climate Preparedness and Mitigation. January 14. (http://www.sustainabilitypractice.net/sites/default/files/pdfs/Expanding_the_use_of_wood_in_buildings.pdf)

Gaston, Christopher, Robert Kozak, Jennifer O'Connor, and David Fell. 2001. Potential for Increased Wood-Use in North American Non-Residential Markets. Forintek Canada Corp.

Gustavsson, Leif. 2008. Substitution Effects of Wood-Based Construction Materials: Harvested Wood Products in the Context of Climate Change Policies. Palais des Nations, Geneva, September 9–10. (https://www.unece.org/fileadmin/DAM/timber/workshops/2008/hwp/pdf/05_Gustavsson.pdf)

Gustavsson, Leif, Kim Pingoud, and Roger Sathre. 2006. Carbon Dioxide Balance of Wood Substitution: Comparing Concrete- and Wood-Framed Buildings. *Mitigation and Adaptation Strategies for Global Change*, Volume 11, Number 3. Springer, Netherlands. (http://www.task38.org/finswe-fullreport.pdf)

Gustavsson, Leif, and Roger Sathre. 2006. Variability in Energy and Carbon Dioxide Balances of Wood and Concrete Building Materials. *Building and Environment* 41: 940–951. (http://cat.inist.fr/?aModele=afficheN&cpsidt=17598440)

Idaho Forest Products Commission. 2011. How Much Wood Goes into a House? Accessed September 7, 2017. (http://www.idahoforests.org/woodhous.htm)

Ince, Peter. 2010. Global Sustainable Timber Supply and Demand. In: Rowell, Roger., Caldeira, e Fernando. and Rowell, Judith (eds.). Sustainable Development in the Forest Products Industry. Universidade Fernando Pessoa, Portal, Portugal. pp. 29–41.

Kasal, Bo, and Pat Huelman. 2010. CORRIM: Phase I Interim Report, Appendix F: Design of Residential Building Shells—Minneapolis and Atlanta. (https://www.researchgate.net/publication/271910129_Design_of_Residential_Building_Shells_-_Minneapolis_and_Atlanta_-_CORRIM_Phase_I_Interim_Report)

Lenzen, Manfred, and Graham Treloar. 2002. Embodied Energy in Buildings: Wood versus Concrete—Reply to Börjesson and Gustavsson. *Energy Policy* 30(3): 249–255. (http://dro.deakin.edu.au/view/DU:30008540)

Lippke, Bruce, Jim Wilson, John Perez-Garcia, Jim Bowyer, and Jamie Meil. 2004. CORRIM: LifeCycle Environmental Performance of Renewable Building Materials. *Forest Products Journal* 54(6): 8–19.

Marceau, Medgar, and Martha VanGeem. 2008a. Comparison of the Life Cycle Assessments of a Lightweight Concrete Masonry House and a Wood Frame House. Portland Cement Association Publication SN3042. (http://www.cement.org/bookstore/profile.asp?itemid=SN3042)

Marceau, Medgar, and Martha VanGeem. 2008b. Comparison of the Life Cycle Assessments of an Insulating Concrete Form House and a Wood Frame House. Portland Cement Association Publication SN3041. (http://www.polysteel.us/GreenGuide/PDF_files/tab8/06_Life%20Cycle%20Comparison%20ICFs%20vs.%20Wood.pdf)

Natural Resources Canada. 2016. Forest Resources, Statistical Data. Accessed September 8, 2017. (https://cfs.nrcan.gc.ca/statsprofile)

Natural Resources Canada. 2017. The State of Canada's Forests 2016. (http://cfs.nrcan.gc.ca/pubwarehouse/pdfs/37265.pdf)

O'Connor, Jennifer. 2004. Survey on Actual Service Lives for North American Buildings. Presentation at Woodframe Housing Durability and Disaster Issues Conference, Las Vegas, October. (http://cwc.ca/wp-content/uploads/2013/12/DurabilityService_Life_E.pdf)

Oswalt, Sonja, W. Brad Smith, Patrick Miles, and Scott Pugh. 2019. Forest Resources of the United States, 2017: A Technical Document Supporting the Forest Service Update of the 2020 RPA Assessment. USDA-Forest Service, General Technical Report WO-97. (https://www.fs.fed.us/research/publications/gtr/gtr_wo97.pdf)

Penaloza, Diego, Joakim Norén, and Per-Erik Eriksson. 2013. Life Cycle Assessment of Different Building Systems: The Wälludden Case Study. SP Technical Research Institute of Sweden. (http://www.diva-portal.org/smash/get/diva2:962737/FULLTEXT01.pdf)

Perez-Garcia, John, Bruce Lippke, David Briggs, Jim Wilson, Jim Bowyer, and Jamie Meil. 2005. The Environmental Performance of Renewable Building Materials in the Context of Residential Construction. In: *Wood and Fiber Science*, Volume 37, CORRIM Special Issue, December, pp. 3–17. (https://wfs.swst.org/index.php/wfs/article/view/2071/2071)

Petersen, Ann, and Birger Solberg. 2002. Greenhouse Gas Emissions, Life Cycle Inventory and Cost Efficiency of Using Laminated Wood Instead of Steel Construction. Case: Beams at Gardenmoen Airport. *Environmental Science and Policy* 5(2): 169–182. (http://www.sciencedirect.com/science/article/pii/S1462901101000442)

Pierquet, Patrick, Jim Bowyer, and Pat Huelman. 1998. Thermal Performance and Embodied Energy of Cold Climate Wall Systems. *Forest Products Journal* 48(6): 53–60. (https://www.researchgate.net/publication/276287585_Thermal_Performance_and_Embodied_Energy_of_Cold_Climate_Wall_Systems)

Robertson, Adam, Frank Lam, and Raymond Cole. 2012. A Comparative Cradle-to-Gate Life Cycle Assessment of Mid-Rise Office Building Construction Alternatives: Laminated Timber or Reinforced Concrete. *Buildings* 2(3): 245–270. (https://www.researchgate.net/publication/276034037_A_Comparative_Cradle-to-Gate_Life_Cycle_Assessment_of_Mid-Rise_Office_Building_Construction_Alternatives_Laminated_Timber_or_Reinforced_Concrete)

Scharai-Rad, Mohammad, and Johannes Welling. 2002. Environmental and Energy Balances of Wood Products and Substitutes. University of Hamburg/Federal Research Centre for Forestry and Wood Products/Food and Agricultural Organization of the United Nations. (http://www.fao.org/Docrep/004/Y3609E/y3609e00.htm)

University of Bath. 2010. Inventory of Carbon and Energy (ICE). Sustainable Energy Research Team. Accessed August 17, 2017. (http://www.ghgprotocol.org/Third-Party-Databases/Bath-ICE)

CHAPTER 9
Sustainability Related to CLT

Kathryn Fernholz
Dovetail Partners, Inc.

Introduction

Cross-laminated timber (CLT) is a wood product and therefore depends upon the availability of forest resources to be produced. All materials, including wood products, have environmental impacts and considerations of sustainability that are relevant when evaluating the use of CLT as well as other building materials. This chapter discusses the concept of sustainability, what it means, and what is considered. The specifics of forest sustainability are also addressed, including historic factors, international conditions, and North American forest resources. The modern methods of forest protection, management, and restoration are described along with current concerns for forest health, carbon storage, sustainable livelihoods, and other benefits associated with sustainability. Finally, this chapter outlines the various ways that CLT and the use of CLT in construction can contribute to sustainability goals for forests, communities, and our built environment.

9.1 Defining Sustainability

Sustainability is a holistic concept that embraces many dimensions. At its heart, the concept refers to creating the conditions that can allow something to continue in perpetuity (i.e., to last forever). In practice, sustainability is an ongoing pursuit, an objective, and a goal. Sustainability can be achieved, but efforts must be ongoing to maintain the conditions that support sustainability. A goal of sustainability is widely held for many things—sustainable energy systems, sustainable economic models, and sustainable food production are just a few examples. But sustainability is not feasible for all things. To put it bluntly, an individual human life cannot be sustainable given that each person eventually perishes. However, the concept of sustaining the human existence through the creation of a sustainable environment that supports our needs and the needs of future generations is a goal to be pursued in our policies and actions.

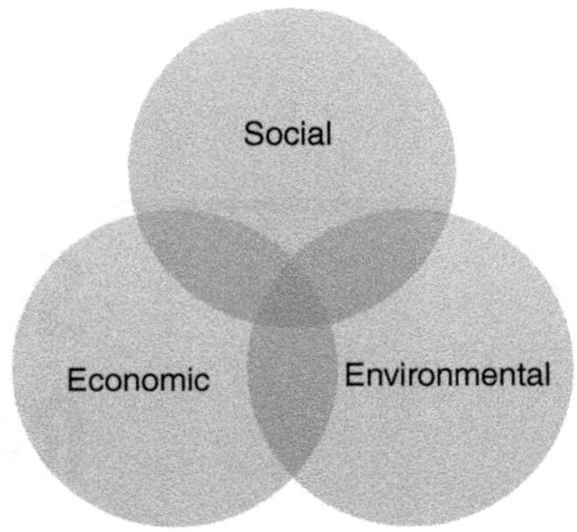

FIGURE 9.1 Sustainability components.

If we image what a sustainable environment for human existence might require, we can categorize the needs into the areas of social, environmental, and economic considerations. Sustainability occurs where these areas intersect and operate in balance (Fig. 9.1). Our needs and the needs within many systems include specific requirements in each of these categories. We can measure our success in achieving sustainability by imaging and evaluating outcomes across these categories.

9.1.1 Social

The category of social considerations for sustainability includes basic concepts of livability, quality of life, freedoms, and rights. Examples of social sustainability goals and measures include the Bill of Rights within the U.S. Constitution[1] and the Universal Declaration of Human Rights[2] proclaimed in 1948. These documents include ideals for human existence and sustaining our ability to freely express ourselves and pursue our own interests and happiness. The Bill of Rights puts limits on the power of government to hinder freedoms and individual actions. In many ways, social sustainability relies upon enabling policies, good governance, and the rule of law. These conditions also depend upon an environment that provides the resources to support human achievement, including the basic needs of clean water, healthy food, and safe shelter. Quality education, job and career opportunities, health care, and social systems throughout all life stages further expand upon these basic needs.

Social sustainability can be challenging to define clearly or consistently. At least three unique definitions for social sustainability can be used to illustrate the concept:

- Maintaining human well-being,
- Human actions to sustain natural resources, and/or
- Ensuring benefits for future generations[3]

[1]https://billofrightsinstitute.org/founding-documents/bill-of-rights
[2]http://www.un.org/en/universal-declaration-human-rights/index.html.
[3]Adapted from Borrini-Feyerabend and Buchan (1997), 115–117.

FIGURE 9.2 Sustainable development goals. (*Source: United Nations, 2015. Available at: http://www.un.org/sustainabledevelopment/blog/2015/12/sustainable-development-goals-kick-off-with-start-of-new-year. Instructions for using UN SDG image and page to access downloadable files: http://www.un.org/sustainabledevelopment/news/communications-material.*)

The definition(s) that may apply to any given situation may vary due to changing circumstances and over time. Given the diversity of human needs, cultures, and other factors, a flexible and inclusive approach to defining and measuring social sustainability is important.

In recent years, social sustainability has been identified as being closely linked to business operations. In this context, the focus is on understanding and influencing the impacts that businesses have on people, including employees as well as other community members and stakeholders. It is recognized that businesses can impact many people both positively and negatively, and, in turn, the company is also impacted. Businesses directly impact their employees, contractors, suppliers, neighbors, and other community members. Businesses indirectly impact other members of the community as well as people more generally as their actions have effects on other activities in society. The conditions in society also impact businesses and their sustainability, including the availability of a quality workforce, education services, housing, legal protections, and financial systems. These interactions can be evaluated through a lens of social sustainability. Efforts to strength social sustainability are recognized as offering benefits to businesses and people.

In 2015, the United Nations defined 17 sustainable development goals (SDGs) (Fig. 9.2).[4] The first six SDGs focus closely on social sustainability concepts—poverty, hunger, health, education, equality, and clean water.

[4]http://www.un.org/sustainabledevelopment/sustainable-development-goals/; https://www.unglobalcompact.org/sdgs.

Individual governments clearly have an important role to play in supporting sustainability locally and globally; however, collaboration with businesses and other interests is necessary to expand the benefits and support greater success. More than 9,000 companies and over 4,000 other organizations have committed to the SDGs as participants in the United Nations Global Compact.[5]

To support social sustainability, governments, businesses, and other organizations can work together to improve peoples' lives, provide employment and career opportunities, make investments and enforce policies that enhance security, and protect food, water, and environmental systems that are essential to basic needs.

In the context of sustainable forestry, these social considerations can be made more specific. Forests play a role in meeting basic human needs, as they provide clean water, offer sources of food, and provide materials to build homes and other forms of shelter and materials, including heat and cooking fuels. Forests also support jobs and education and research opportunities and are linked to human health benefits. In general, social sustainability goals focus around the holistic needs of the human experience; however, even from this brief discussion it is clear the social considerations overlap significantly with environmental and economic areas of interest.

As shown in Fig. 9.3, forests and use of sustainable wood have the potential to contribute to all of the SDGs through the various facets of the forest sector. Several of the SDGs are addressed in more than one area of the forest sector, including SDG 13 addressing climate action, which is addressed throughout the sector in manufacturing innovations, forest management, use of wood as a renewable and low-impact material, and forest products within construction and a sustainable economy. Forests and forest products are unique in the many benefits they offer throughout their life cycle. This uniqueness stems from the natural way in which forest products are grown and carries through to their versatility in uses and their global availability.

9.1.2 Environmental

Within sustainability efforts there are also significant environmental considerations. Some of these overlap with social considerations. For example, clean water is good for meeting the needs of people and also for healthy aquatic habitats and associated fish and wildlife populations. However, there are some environmental sustainability considerations where the shared benefits to people are less direct or may even be undefined. For example, significant efforts are made to sustain rare or endangered species even in situations where the species is not of direct interest and where human interests may in fact be diminished. An illustration of this complexity is the restoration of wolf species. There are environmental sustainability benefits to restoring prey species within an ecosystem; however, these species can also create conflict and even harm economic interests by their impact on people and livestock. At the same time, there are also cultural connections with these species and social sustainability considerations that are relevant. This type of scenario begins to illustrate how interconnected sustainability goals can be and the necessity of a good understanding of these connections and the possible impacts of alternative actions.

Environmental sustainability is often the category that we think of first; however, it is complicated by the challenges to understand our environment and work effectively to

[5]https://www.unglobalcompact.org.

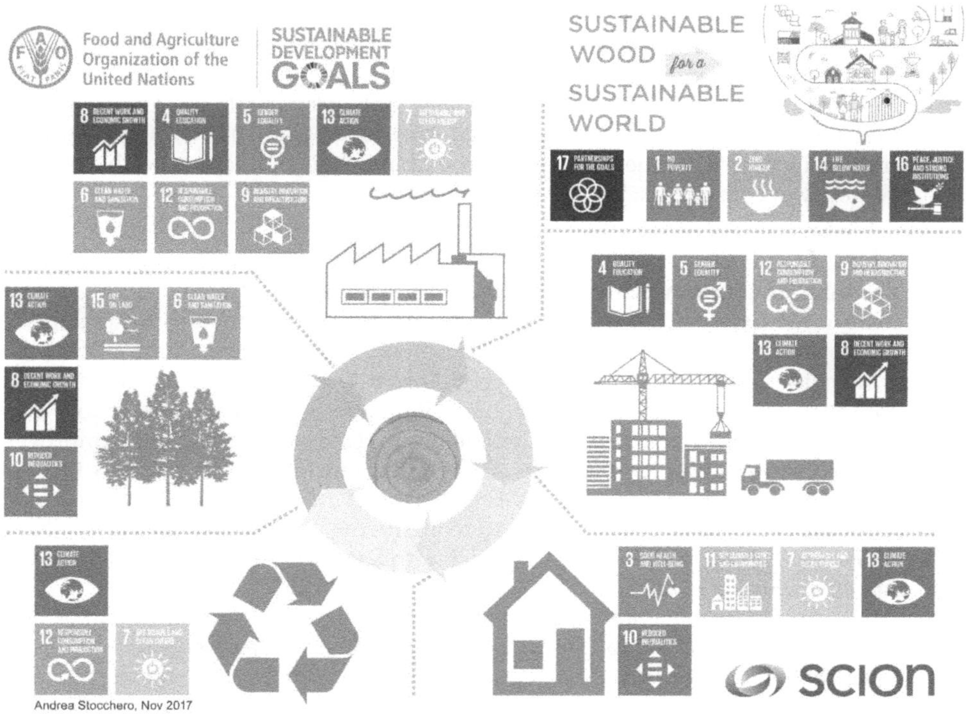

FIGURE 9.3 Sustainable wood use contribution to the sustainable development goals. [*Source: Scion (2017). Available at: http://www.scionresearch.com/about-us/about-scion/corporate-publications/scion-connections/scion-connections-2/scion-connections-issue-26,-december-2017/sustainable-wood-for-sustainable-development.*]

influence its health and productivity. With environmental sustainability efforts there are often many unknowns about the species or systems that we are trying to protect, restore, or enhance. Our ability to achieve environmental sustainability is often limited by the available information and our knowledge. To address these limits, it is often necessary to integrate knowledge from diverse sources, including scientific research as well as the experiences of practitioners, community members, and indigenous cultures. In a pursuit of sustainable forestry, engagement with indigenous peoples is often important to understand changes that have occurred in the systems and traditional practices that support ecosystem resiliency.

Achievement of environmental sustainability is also challenging because of the continuous change. Strategies for environmental sustainability include efforts to protect, restore, and/or enhance. In some situations, a natural forest ecosystem may be largely intact with native species, healthy populations, no significant disease threats, or other damage. This type of forest may be a candidate for protection through special management limits or other methods that support stability (e.g., low-impact logging, prescribed fire, or monitoring). However, this situation can change very quickly if the forest is impacted by a devastating wildfire, hurricane, insect outbreak, or invasive species. When these impacts occur, it may be necessary to take management action to restore or enhance the forest and reestablish the desired conditions or combat the threats. Many areas of the world, including areas throughout North America, have been impacted by past human and natural disturbances which lead many land managers to prioritize efforts related to restoration and enhancement to increase resiliency and reduce threats. The need for active strategies to support environmental sustainability may be increased as the impacts from climate change create new challenges to the resiliency of ecosystems.

9.1.3 Economic

The role of economics in sustainability is also important and closely tied to the achievement of social and environmental outcomes. Economic sustainability can be supported through job creation, improvements in productivity and profitability, and growth in trade. Social benefits can follow if these economic gains are linked to improved quality-of-life outcomes. Environmental benefits can also follow if productivity gains are linked to more efficient processes, energy system improvements, reduced waste, or avoided impacts. Economic gains may also be directly invested into social and environmental initiatives. Even more fundamentally, in many situations economic sustainability is essential to realizing the other benefits. Without long-term economic viability, it may not be possible to support the systems that deliver social and environmental needs.

Economic sustainability and the concept of a sustainable economy can be applied at a local scale as well as more broadly. Achieving a sustainable economy often requires collaboration between private and public interests, including government, business, and nongovernmental organizations.

Approaches to economic sustainability may be tailored to specific parts of the economy, including agriculture and food systems, manufacturing and industry, and energy or technology. The considerations for sustainability may vary across these parts of the economy due to the scale of operations, cycle times, broader sector trends, supporting policies, and other factors. For example, sustainability within agriculture and food systems has many long-term considerations but is also impacted by an annual cycle for most production. Periodic events such as drought can create significant short-term challenges to the sustainability of agriculture.

Economic sustainability can be supported by community-led efforts as well as business initiatives and enabling policies. For example, establishing local farmer's markets can be part of a community's efforts to enhance food system sustainability, but it may rely on a new ordinance or policy that allows for direct food sales and a local business may be recruited to sponsor the market or provide a space for the vendors.

One of the challenges to achieving economic sustainability is balancing the near-term financial potentials with long-term benefits that may or may not be realized or monetized. Paying a higher wage or investing in workforce training has associated costs and will impact an organization's financial performance in the near term. It is not always easy to demonstrate that these investments will produce significant future benefits or economic returns. For businesses, there is often a legal requirement to put the interests of the company first, and this can be interpreted as a requirement to maximize financial benefits. There are many examples of organizations that have broadened their corporate missions to include other measures of success; however, the legal precedence of holding business accountable to maximizing shareholder return can be a hindrance to sustainability initiatives. Recent expansion of for-benefit corporation legislation and for-benefit certification programs can help support corporate behaviors beyond financial return to include economic sustainability considerations and even social and environmental interests as well. Between 2010 and 2015, 38 states pursued Benefit Corporation legislation as a legal structure available to for-profit businesses that are committed to conducting business in a socially and environmentally sustainable manner. Businesses under this corporate structure may conduct business in a way that reflects their mission and goals in addition to financial targets.

Measures of economic sustainability go beyond the effort of an individual company to include a review of the systems and conditions throughout a community. The presence of diverse employment opportunities, small and large businesses, service and technology sectors, health systems, and education can also contribute to economic sustainability within a community or region. The availability of natural resources, including forest products, can also be a foundational component of a region's sustainable economy.

9.2 Understanding Forest Sustainability

While social, environmental, and economic considerations define the scope of sustainability interests, measuring progress in these areas can be accomplished in several more precise ways. Many measurable indicators have been identified for monitoring progress on comprehensive forest sustainability.

Common indicators used to measure forest sustainability include the following:

- Amount of forest cover (e.g., total forest area)
- Amount of protected forest area
- Amount of forest area under sustainable forest management
- Rate of deforestation (e.g., conversion of forest areas to agriculture and other land uses)
- Occurrence of illegal logging

Sustainability can also measure indicators that include the landscape beyond forests, to include the following:

- Land tenure (e.g., ownership and use rights)
- Land use (current conditions and/or changes)
- Land cover (current conditions and/or changes)
- Biodiversity and ecological conservation (including voluntary and legal protections)

The specific indicators that are used to evaluate and measure forest sustainability depend upon several factors. Specific stakeholders may identify indicators that are related to their needs or interests. For example, government agencies with responsibility for legal compliance may focus on indicators related to illegal logging and land use rights. Organizations concerned with climate change impacts may be most interested in land use change or the rate of deforestation associated with conversion of forest areas to agriculture and other land uses. Many national and international efforts have identified widely held objectives for sustainable forestry and have also defined the indicators that help track progress toward the shared goals.

In 2006, the United Nations Forum on Forests identified four global objectives on forests:

- Reverse the loss of forest cover worldwide
- Enhance forest-based economic, social, and environmental benefits
- Increase the area of protected forests
- Increase resources for implementation of sustainable forest management

Although this is a short list of goals, these four goals cover major concepts for assuring sustainability. For example, the first goal is fundamental to forest sustainability. By definition, something cannot be sustainable if it is being lost and decreasing in health and capacity. The prioritization of the stabilization of forest cover worldwide is key to being able to achieve the other objectives. However, although these four global objectives are comprehensive and very appropriate, they are also complex and challenging. It is important to translate broad sustainability objectives into more actionable strategies that can be clearly connected to governmental and nongovernmental opportunities.

The United Nations Forum on Forests identified several key strategies for enabling sustainable forest management:

1. Improving governance at local and regional levels, including the following:
   a. Strengthening the rule of law
   b. Improving accountability and transparency
   c. Local participatory planning and decision making
2. Forest ownership and tenure
   a. Public, private, formal, and customary rights
   b. Ensuring clarity and long-term security of ownership and rights
   c. Efficient markets and market access
3. Promoting forest-based economic development
   a. Role of government
   b. Reducing uncertainty
   c. Promoting assets, cooperation, and market building
4. Addressing forests and conflicts
   a. Forest access and disputes
   b. Policies and mitigations
   c. Negotiations and measures to address inequities

Throughout these broad and global efforts, there is widespread recognition of the importance of sustainable forest management as a tool for protection, restoration, and expansion of forests and the prevention of forest loss or degradation. There is also growing recognition of the value of local engagement, the special need to improve the livelihoods of forest-dependent people, and the role of indigenous peoples. The link between sustainably managed forests and supporting markets for the products and services these forests provide is also recognized as a key strategy, along with governmental and nongovernmental investments in these systems of management.

For the forests of North America, there is an additional specific sustainability effort that began in 1994. The Montréal Process and the Montréal Process Working Group include the governments of Argentina, Australia, Canada, Chile, China, Japan, Korea, Mexico, New Zealand, the Russian Federation, the United States, and Uruguay. These countries represent the majority (83%) of the world's temperate and boreal forest regions, about half of the world's total forest area, and 40% of the world's wood production.[6]

Within the Montréal Process, criteria and indicators have been identified to support data collection, monitoring, and country-level reporting. This information contributes to understanding and evaluating the condition and sustainable management of temperate and boreal forests around the world. The results of this work inform policy initiatives and other government actions as well as the activities of businesses in the forest sector and beyond.

The Montréal Process provides the following broad criteria for evaluating forest sustainability:

Criterion 1: Conservation of Biological Diversity.

Criterion 2: Maintenance of Productive Capacity of Forest Ecosystems.

Criterion 3: Maintenance of Ecosystem Health and Vitality.

Criterion 4: Conservation and Maintenance of Soil and Water Resources.

Criterion 5: Maintenance of Forest Contribution to Global Carbon Cycles.

Criterion 6: Maintenance and Enhancement of Long-Term Multiple Socioeconomic Benefits to Meet the Needs of Societies.

Criterion 7: Legal, Institutional, and Economic Framework for Forest Conservation and Sustainable Management.

Source: National Report On Sustainable Forests—2010. Available at: ***https://www.fs.fed.us/research/sustain/criteria-indicators.***

As with many other efforts related to forest sustainability, the criteria identified within the Montréal Process provide important metrics to evaluate conditions, identify trends, and influence actions throughout the forest sector. It is also clear that these criteria aim to comprehensively consider sustainability goals by including environmental, social, and economic conditions. By applying these criteria, areas of need can be identified and prioritized. Where these criteria are being effectively achieved and maintained, there is confidence that forests and their associated benefits are being sustained.

At the start of this chapter, Fig. 9.1 illustrates a basic approach to understanding sustainability. A weakness of the approach shown in Fig. 9.1 (and also illustrated in Fig. 9.4) is that it implies only small or weak intersections between environmental, economic, and social considerations. The illustration leaves the impression that there is only a small area of intersection where sustainability can occur. It also implies that the three areas of consideration are not very closely connected to each other. There are other ways to illustrate sustainability and understand the interconnections. As shown below (Fig. 9.4, right side), these considerations of the economy, society, and the environment can also be shown as concentric circles with complete overlap and interconnectedness. In this type of illustration, sustainability becomes the only mechanism by which growth or balance is maintained. For example, in this illustration, if the environment is degraded and thereby decreased in size and capacity, that shrinking circle puts pressure and limitations on society which may also decrease in capacity due to conflict and resource constraints which thereby put pressures and limitations on the economy. Similarly, within this model, if

[6]https://www.fs.fed.us/research/sustain/organizations.php.

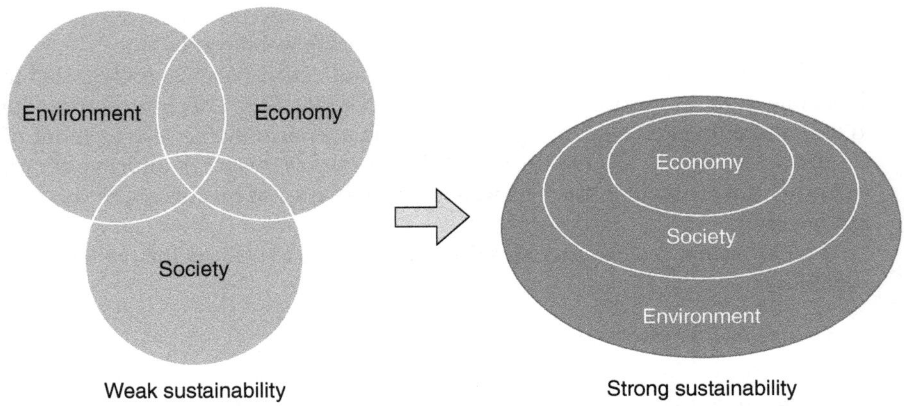

FIGURE 9.4 Models of weak and strong sustainability. (*Source: National Report on Sustainable Forests—2010. Available at: https://www.fs.fed.us/research/sustain/what-is-sustainability.php.*)

economic growth occurs at an aggressive rate, the expansion may put pressures on society which may also translate into pressures on the environment and natural resources.

The concentric circle model of sustainability also illustrates the fundamental reliance on the environment as the basis for our economic and social systems. Our natural resources, including food and water, come from the environment, as do the materials that are used to build our homes, develop our infrastructure, and support our economy. The role of society is also strengthened in this model as a central element that influences the linkages between the economy and our environment as demonstrated in the policies and laws of our societies as well as cultural values and traditions. The economy is at the core of the model and may be reflective of the conditions, practices, and priorities that surround it. By viewing the three sustainability considerations—environmental, social, and economic—as concentric rather than intersecting circles, we can imagine the relationships in new ways and better understand the operational systems.

There are many local, regional, and global concerns related to forest sustainability today. These concerns range from threats to biodiversity to widespread impacts from climate change as well as more localized land use changes and economic worries. While these threats may be characterized as forest sustainability issues, the reality is that their causes and their ultimate solution are related to choices and conditions that operate well beyond the forest sector. Policies and other choices made in almost any area of the economy or society have the possibility of positively or negatively impacting our environment and the world's forests. For example, policies that address energy resources will influence the production of these resources, which may be positive for forests if responsible renewable biomass energy is encouraged or negative for forests if the policy favors trees being cleared to make way for extraction of fossil fuels. Similarly, a food system policy that incentivizes agricultural investment may be negative for forests if it favors annual crops and results in the clearing of forestlands to expand areas of production or if it includes systems to integrate forests into farm productivity and diversification. These trade-offs are also evident in private sector actions, including the sourcing policies and production practices of local, national, and multinational organizations.

The recognition of these trade-offs and the interconnectedness of our policy and consumption choices has led many companies to establish their own programs for sustainable supply chains and reduced impacts. Most notably in recent years is a trend from some companies to pursue zero deforestation goals and even take the step to own and manage their own forestlands to have direct control over their role in forest sustainability. For example, in 2016, Apple invested in approximately 36,000 acres of land in the eastern United States with the intention to use wood responsibly harvested from these lands to reduce the environmental impact of their products and packaging. Similarly, in 2018, IKEA, the Swedish furniture and home goods retailer, invested in 25,000 acres in the southern United States. The company already owned about 250,000 acres of forestland in Europe.

There are clear roles for everyone within sustainability objectives, from individual actions at home, work, or school to the collective actions or our institutions, the policies of government, and the initiatives of small and large organizations of every type. One finding from a recent global review of sustainable forest management was that "forestry... issues are best tackled as part of broader sustainable development frameworks covering natural resources management as a whole—given that many sustainability issues in the Forestry Sector are driven by non-forestry factors (e.g., agriculture and food security, climate change, energy)" (Ross, 2015).

9.2.1 Historic Conditions, Global Forestry, and North American Forests

Forests are living systems, and as such they are continually changing. In many parts of the world, the development of forested regions can be traced back to formative events, including periods of glaciation, volcanic, or seismic activity and the expansion or contraction of oceans and deserts. There are a large number of natural events that influence historic and current forest conditions. In addition to the examples of formative events, there are also disruptions that occur as a result of storm events such as hurricanes or tornadoes, tsunamis, flooding, or fire. These events impact the environment and also threaten human lives and public safety. Insect outbreaks and diseases can also change forests. It isn't always clear if changes caused by nature help or hinder sustainability goals. While natural disturbance can be very destructive, it may also create vibrant new growth or result in greater habitat variability that can support increased biodiversity. One way to understand or predict the outcome of natural disturbance is to try to place them within an historic context and to apply an evaluation of how current events compare to what might have occurred in the past. In this way, ecologists and land managers can make an educated guess about what the outcome of a new event might be given what is known to have occurred as a result of a past event. For example, if a past insect outbreak lasted for three years and resulted in 20% of the trees dying and now a new outbreak occurs that does not last as long, then perhaps the mortality rate will be less. However, it is not always that easy. For example, the prior outbreak may have corresponded with seasons of average or above-average rainfall, so mortality may have been mitigated by available moisture allowing more trees to survive. In contrast, if the new insect outbreak corresponds with a period of drought, mortality may be higher even if the outbreak has a shorter duration. Understanding the impact of natural events and predicting the immediate and long-term outcomes is complicated by the fact that events overlap and conditions have multiple dimensions. These conditions and their interactions are a key part of achieving forest sustainability because the forest sector has to consider the impacts of

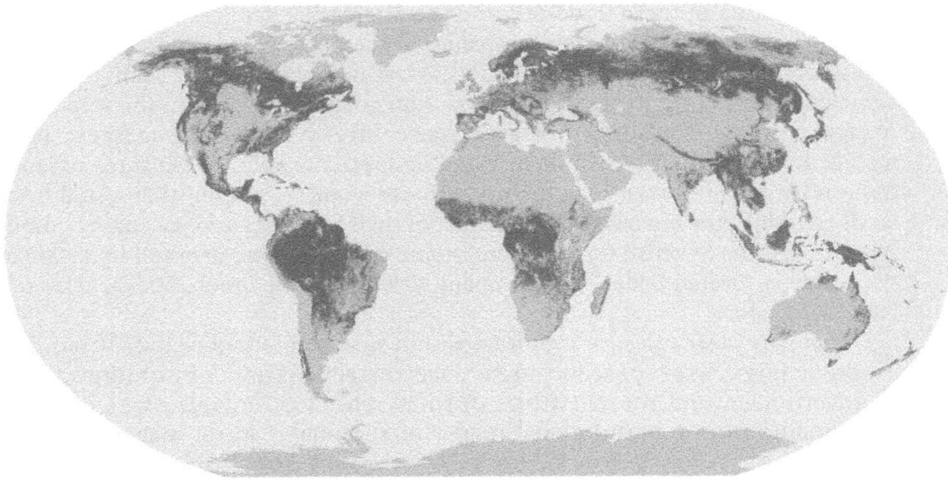

FIGURE 9.5 The world's forests. (*Source: FAO, 2006. Global Forest Resources Assessment. FAO. Rome. Available at: http://www.fao.org/forestry/fra/41256/en.*)

both human actions and natural disturbances in predicting and evaluating forest conditions around the world.

Globally, about 30% of the land area is currently in forest cover (Fig. 9.5). The growth and total area of forests is largely limited to regions that have the ecological conditions that can support forests, including sufficient precipitation, accommodating climate, and suitable soils. The extent of forests is also limited by human activities the change forestlands to other types of land use, including development, urbanization, and agriculture. The difference between the current amount of forest and the maximum amount that could be ecologically possible is a measure of both environmental change and land use change activities. Some areas that had forests in the past do not have forests today because deserts have expanded, water levels or water bodies have expanded, or other ecological conditions have changed and resulted in forest growth no longer being viable. In other areas, lands that were forested in the past are now part of towns and cities, roads and railways, and farms and pastures. These types of land use developments can result in reductions in forest area which can have a negative environmental impact, but they are also important parts of economic development and improvements in quality of life and sustainability for individuals and communities. There are also regions of the world where forests are expanding due to changes in temperature and moisture and due to policies and economic incentives that favor forest cover. Today there are parts of the world where forest areas are expanding as well as areas where forest area is declining (Fig. 9.6). When policymakers and stakeholders ask questions about how much forest we have today and how much forest we want to have in the future, it is important to consider the ecological conditions necessary to grow forests as well as the other land uses that societies require. It is not possible to grow forests everywhere because not all locations can grow trees and also because people need to use land for other purposes.

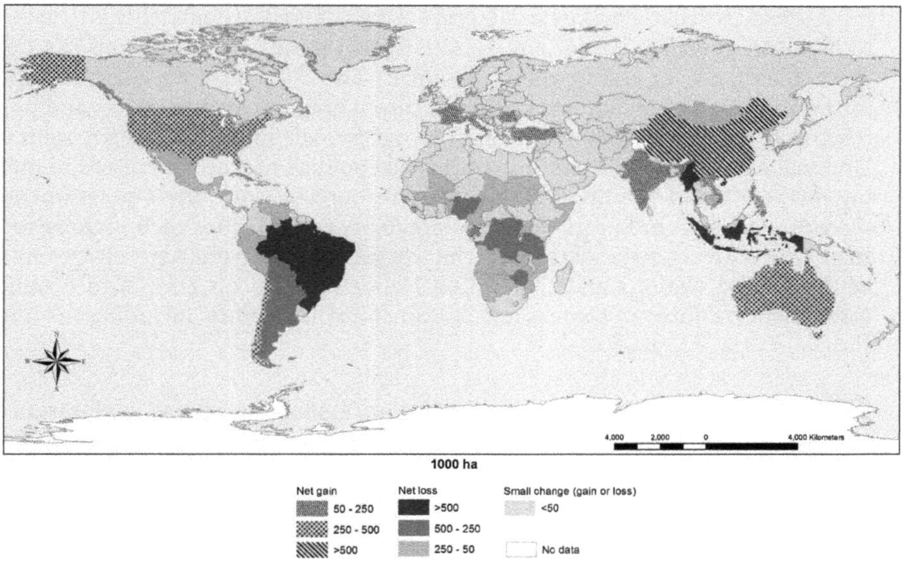

FIGURE 9.6 Annual change in forest area (2010–2015). (*Source: FAO, 2015. Global Forest Resources Assessment. FAO. Rome. Available at: http://www.fao.org/forest-resources-assessment/current-assessment/maps-and-figures/en.*)

As discussed in Chap. 8 regarding the Environmental Aspects of CLT, the renewability of wood and the environmental potential to grow wood around the world are significant and fundamental considerations when examining current and historic forest conditions, global forestry, and the forests of North America.

The forests of the United States and Canada are discussed in detail in Chap. 8. Some of the major conclusions from the information presented in that chapter include the following:

- *Forest area:* The United States has 766 million acres (310 million ha) of forest (about one-third of the land area of the country).
- *Forest ownership:* The forests in the United States are predominantly privately owned (58%).
- *Wood harvest:* The privately owned forests in the United States provide most of the annual harvest of wood products (90%).

In terms of the sustainability of forests in the United States:

- *Forest area change:* The area of forest in the United States today is about the same as it was in 1900 (which is also equivalent to about two-thirds of the forest area that is estimated to have existed at the time of European settlement).
- *Wood harvest rate:* The volume of wood grown annually in the United States has been greater than the volume harvested each year for at least 70 years in a row (in recent years, growth has been more than double the rate of removal).
- *Protected area:* Approximately one-third of U.S. forests are on lands where timber harvesting is not allowed.

As discussed earlier, a key strategy in ensuring forest sustainability is to use indicators to monitor and track changes in conditions and trends so that actions can be taken to protect, enhance, or restore forest resources. The U.S. Forest Service has responsibility for monitoring the conditions of forests throughout the United States and periodically releases sustainability reports. These reports provide important information to inform land managers, policymakers, and other stakeholders that are interested in understanding the health and condition of our forest resources. Many other governments around the world conduct monitoring and sustainability reporting for their nations' forests, and their efforts are aggregated through the work of the Food and Agriculture Organization of the United Nations and the United Nations Forum on Forests to produce global assessments of forest resources and sustainability indicators, including the information shown in Figs. 9.5 and 9.6.

9.2.2 Modern Methods for Forest Protection, Management, and Restoration

Forestry is a profession that involves the growing, tending, and harvesting of trees and management of other products and services from the forest. Like many modern professions, the practice of forestry has evolved significantly from generation to generation. This evolution is the result of ongoing research to inform the practice of forestry. Change in forestry practice also occurs as a result of policy developments as well as environmental, social, and economic expectations.

Today's forestry practice includes strategies to protect and conserve diverse and unique forest types. There is also management on working forestlands where activities are undertaken to meet landowner objectives for providing forest-based products and services. Across North America and in many other parts of the world, there are extensive efforts related to forest restoration. These efforts include reforestation and replanting of trees as well as treatments that combat invasive species, improve biodiversity, and enhance forest health and productivity. The work of forest restoration helps ensure that these lands are more resilient in the face of growing and diverse threats, including climate change.

A key concept of modern methods of caring for forest resources is *ecosystem management*, which involves taking an ecological approach to achieving management goals. In the past, land managers may have focused on a narrow set of goals, such as increased productivity and timber yields. With ecosystem management, a goal for productivity or timber must be considered within other objectives for wildlife habitat, water quality, recreation, and additional functions and values. This shift in management has the potential to provide greater benefits and is increasingly important as the impacts of climate change and other sources of management uncertainty expand.

Forests are managed through a variety of techniques within the practice of *silviculture*. Silvicultural practices influence the structure and growth of the forest and the individual trees within the forest. Silviculture can be applied to increase tree growth, diversify species within a forest, enhance wildlife habitat conditions, and provide other benefits associated with different forest structures and composition. To help foresters, land managers, and landowners make decisions about which practices to apply, researchers have identified the silvical characteristics of many trees species, including about 200 types of trees that are native to North America. These characteristics describe the growth and development patterns of each type of tree so that managers know how they might respond to management actions. These characteristics also help with understanding how specific trees respond to other events that disrupt the forest, like wildfires or windstorms.

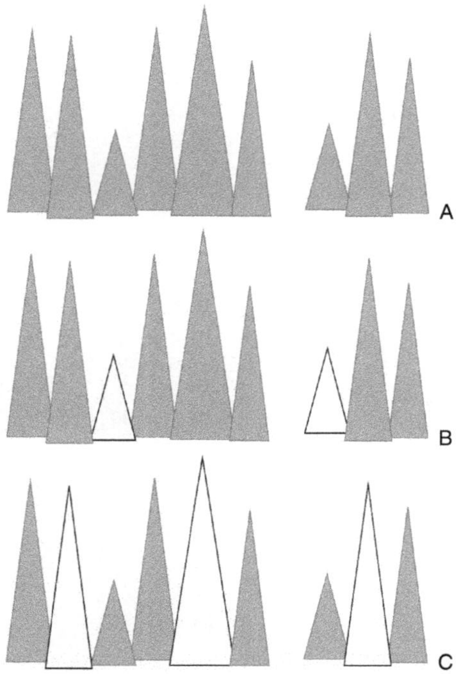

FIGURE 9.7 Forest management examples.

The techniques of silviculture offer many different approaches to achieve desired outcomes, including diverse harvesting systems to selectively remove trees. As shown in Fig. 9.7, there are many different approaches that can be used in each forest. The top image in Fig. 9.7, image A, illustrates a simplified example of a forest. In this forest, as in most forests, there are some large trees as well as midsized and smaller trees. Some of the trees are crowded together, and others have more space between them. When a forester is interested in understanding what management options might work in a specific forest, a first step is to examine the forest and measure the trees to understand how they are interacting in terms of their size, spacing, and composition. After evaluating the current conditions, the forester will then decide what actions (if any) are needed to achieve goals.

You can imagine that if no trees are removed from the forest shown in image A, over time, the trees would continue to grow and become more crowded with less space and openings in the forests. These changes can create benefits for some types of wildlife that prefer dense, closed-canopy forest types, while other species will lose habitat as the grassy openings and meadows fill in with trees. To influence the growth of the trees and the habitat conditions, the forester may choose to remove the smaller trees, as shown in image B (removed trees shown in white). By removing the smaller trees, the larger trees will have additional space to grow, openings or meadows may be expanded, and the wood from the harvested smaller trees may be available to use in small-dimensional lumber (including CLT manufacturing), paper, fiber, or energy products. Alternatively, the forester may

FIGURE 9.8 Hardwood forest (top) and softwood forest (bottom). (*Photos courtesy Dovetail Partners, Inc.*)

choose management that includes removing larger trees, as shown in image C. Through this approach, the smaller trees can be given more space to grow, more sunlight may reach the forest floor and allow for a greater diversity of understory plants to develop, and the harvested wood may be available for products like small- and larger-dimensional lumber, again including potential use in CLT. There are any number of other possibilities that can be applied to any given forest, and each application comes with its own trade-offs.

Trees may be removed because of insect or disease issues, to reduce wildfire risks, to eliminate invasive species, or to favor trees that have more social or economic value. For example, a forester managing for oak timber may remove competing maple trees. Alternatively, a forester managing a forest for maple syrup production might remove the oak trees! These kinds of diverse possibilities can be imagined when looking at all kinds of forests, such as the hardwood and softwood forests shown in Fig. 9.8.

Like the examples in Fig. 9.7, these forests have a mixture of smaller and larger trees and could be left to grow for some period of time, or some trees could be removed to create changes in the conditions that favor specific desired outcomes. Within the hardwood forest on the left in Fig. 9.8, there is a developing understory of shrubs and other plants that might respond to more sunlight, and in the softwood forest on the right, there is a patch of young pine trees in the understory that could be given the opportunity to grow if some trees were removed to provide more space. In almost all forests, the landowner and the land manager have the choice to do nothing or to take management action to create desired conditions. However, it is important to remember that the forest is always changing as the trees and plants are growing and going through their life cycles, so even a choice to not take any management action will result in changing conditions over time.

9.2.3 Current Concerns with Forest Health, Carbon Storage, Sustainable Livelihoods, and Multiple Benefits

There are many current concerns about forests. These include concerns related to sustainability as well as threats to forest health, a desire for increased forest carbon storage, a need for quality jobs and economic opportunities, and many other products and services that depend upon forests. Many governmental and nongovernmental organizations are actively working to address these concerns and reduce the risks and threats to forests. The sustainability programs and monitoring strategies discussed in this chapter are important activities that help identify and respond to forest concerns. At the same time as there are many concerns, there are also many opportunities related to our forests. It is increasingly recognized that forests and the products and services they provide are unique in many important ways, and these unique features can support innovative new products, new engineering and design possibilities, and new ways to reach sustainability goals.

9.3 How CLT Contributes to Sustainability Goals

Trees and forest provide incredible environmental benefits and contribute to achieving many sustainability goals. Many of these benefits are also provided in the wood products that come from trees and forests, including CLT.

Trees filter the air and absorb carbon dioxide. The wood products made from trees continue to store captured carbon (Fig. 9.9). Half of the weight of wood is composed of carbon. The wood products will continue to store carbon as long as they exist. The manufacture and use of CLT can result in significant carbon storage in buildings for long periods of time.

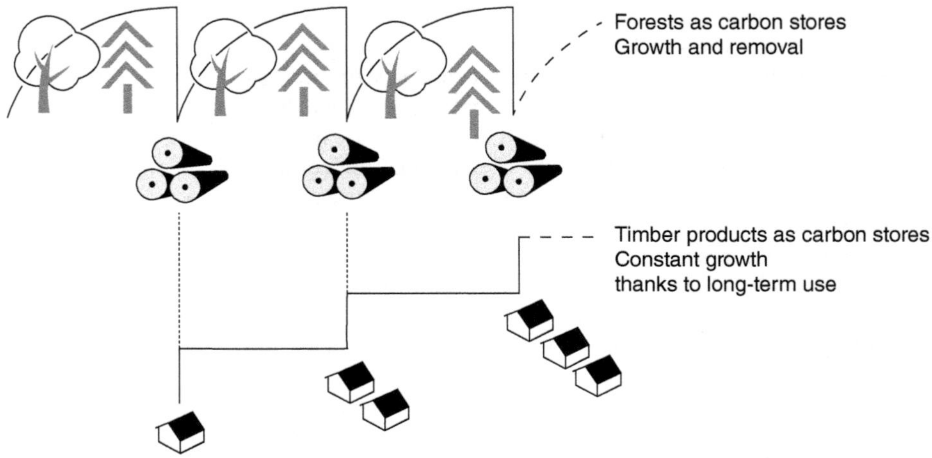

FIGURE 9.9 Forests and timber products as carbon stores. (*Source: Building with Wood = Proactive Climate Change Protection. 2015.*)

The use of wood products and CLT helps to store carbon. These products also contribute to sustainability goals in other ways. Because the wood used to make CLT comes from the growth of trees, it takes less fossil fuel energy to produce than other types of building materials. Chapter 8 detailed these benefits and differences that can be quantified through the use of life-cycle assessment.

It is extremely important to fully recognize the carbon storage benefits of wood products and CLT as the world works to address the threats of climate change. In addition, there are many other benefits of wood products and CLT that also contribute to our quality of life and a sustainable future, and these benefits are unique to wood and are not provided by other products. Wood products come from forests, and they come from trees growing in forests for several years to many decades. At first, the use of trees that have been growing for some time may seem like a bad thing for sustainability; however, when trees grow in a forest for many years, they are providing environmental, social, and economic benefits throughout that time period. The growing trees filter the air and capture carbon and provide homes for wildlife and nesting birds, and the forest is a place for recreation and relaxation for families and communities. These benefits are provided at the same time that the trees are growing and producing wood that will build future homes and provide other products. When the trees are harvested to make those products, the owner of the forest is paid for that wood and can use that money to grow the new forest so that the cycle of providing benefits while the trees are growing and producing wood can continue over and over again. The regeneration and renewability for forests creates incredible unique benefits and is a living model of what sustainability strives to achieve.

9.3.1 Innovation and High-Quality Design

The use of CLT and other innovative products creates new opportunities for high-quality design and construction that meets multiple purposes (Fig. 9.10). As discussed in other chapters, there are possibilities to utilize the strength, flexibility, and longevity

FIGURE 9.10 Wood in construction. (*Photo courtesy Dovetail Partners, Inc.*)

of wood to create buildings that are resilient in diverse situations and perform well when tested against threats such as earthquakes, hurricanes, or fires. These are exciting opportunities for the engineering and architecture professions as well as for agencies and organizations that have responsibilities for community planning and public safety considerations.

9.3.2 Utilization of Materials, Improved Forest Health, and Carbon Storage

Production of products such as CLT can utilize standing dead and dying timber, including areas of forest impacted by the mountain pine beetle in the western regions of North America, for which there are few other markets. Small-diameter timber (e.g., small trees) can be used in producing CLT, which can support forest management practices that remove these trees and create diverse benefits. Among the climate change–related benefits are low carbon emissions and long-term carbon storage within these products.

9.3.3 Job Creation and Rural/Urban Connections

Expanding the use of CLT, including its use in high-rise and more commercial structures, is being supported by a variety of interests—government, architects, engineers, industry, and researchers—and for a range of reasons—economic development, design innovation, climate change mitigation, and positive implications for forest health. In many regions, significant areas of land are dedicated to growing trees, and the local economies depend upon these trees being valued for their many benefits, including markets that value the wood products they can produce. By having markets for trees from these local forests, the local community can create job and business opportunities ranging from work in the forest to jobs in manufacturing to design and construction occupations. These diverse economic and employment opportunities create connections from the rural forest regions through to the homes, communities, and urban areas where the products are made, installed, and enjoyed.

9.3.4 Working Collaboratively

Throughout the United States and around the world, there are many collaborative efforts that support forest sustainability. These efforts range from community-based projects and the work of land trusts to multinational and global initiatives to address illegal logging threats. Many tools have been developed to work at these different scales and to support success in achieving sustainability goals.

Forest Certification

Since the 1990s, there have been global efforts to engage forest owners in responsible forest practices that can be recognized through eco-labels on products in the marketplace. These forest certification efforts have established international standards for responsible forestry and resulted in millions of acres around the world being inspected to verify compliance with the requirements. Today there are thousands of different products in the marketplace that carry labels from forest certification programs and help consumers identify materials coming from well-managed forests. These efforts have helped improve forest management and increase transparency in the forest products supply chain.

State Programs

Many U.S. states have significant forest management responsibilities and work in partnership with state and local organizations as well as the federal government to care for these forests. The National Association of State Foresters (http://www.stateforesters.org) is an important organization for understanding the many diverse needs and opportunities related to forest resources across the nation. Each state has worked with partners and stakeholders to develop a Forest Action Plan, which collectively represents a strategic plan for forests throughout the United States. The Forest Action Plans include assessments of current conditions, identification of priority forest landscape areas, and strategies to address threats and improve forest health (http://www.forestactionplans.org).

Tribal, First Nations, and Indigenous Forestry

Within North America and in many parts of the world, forest resources are cared for within indigenous communities, including Tribes and First Nations in the United States and Canada. The United Nations estimates that 1.6 billion people—including 2000 indigenous cultures—depend on forests for their livelihoods. Many initiatives are focused on protecting and supporting these cultural heritages and reducing threats. These efforts are an important component of sustainability forestry and help ensure that social values are maintained.

9.4 Conclusions

9.4.1 Needs and Opportunities for Supporting CLT and Ensuring Goals for Sustainability Are Addressed

The recent emergence of CLT and other innovations in wood products is an exciting development in the forest sector and offers the possibility of contributing significantly to forest sustainability goals. To ensure that these new opportunities continue to provide positive impacts, it is important that the many organizations and stakeholders that are committed to forest sustainability remain committed to the many systems and programs that support success. This includes local, national, and international efforts to monitor our forest resources to understand conditions, threats, and changes. It also includes governmental and nongovernmental initiatives that aim to support healthy and productive forests that are fundamentally linked to local communities and economies. Finally, there is the need for ongoing research to continually improve our understanding of forests, trees, and the many ways they touch and improve our lives.

References

Aplet, G. H., N. Johnson, J. Olson, and V. A. Sample (eds.). 1993. *Defining Sustainable Forestry.* The Wilderness Society. Island Press.

AppleInsider Staff. 2015. Apple Buys 36,000 Acres of Forest to Create Sustainable Eco-Friendly Product Packaging. April 16. Available at: http://appleinsider.com/articles/15/04/16/apple-buys-36000-acres-of-forest-to-create-sustainable-eco-friendly-product-packaging.

Borrini-Feyerabend, G. and D. Buchan (eds.). 1997. *Beyond Fences: Seeking Social Sustainability in Conservation.* Vol. 2. International Union for Conservation of Nature, 115–117.

Bowyer, J., et al. 2016. *Modern Tall Wood Buildings: Opportunities for Innovation.* Dovetail Partners. January 15. Available at: http://www.dovetailinc.org/report_pdfs/2016/dovetailtall woodbuildings0116.pdf

Burns, Russell M. and Barbara H. Honkala (technical coordinators). 1990. *Silvics of North America,* Vol. 1. Conifers. U.S. Department of Agriculture (USDA), Forest Service, Agriculture Handbook 654.

Burns, Russell M. and Barbara H. Honkala [Technical coordinators]. 1990. *Silvics of North America,* Vol. 2. Hardwoods. U.S. Department of Agriculture, Forest Service, Agriculture Handbook 654.

Dangel, U. 2017. *Turning Point in Timber Construction—A New Economy.* Birkhauser Verlag GmbH.

FAO. 2015. Global Forest Resource Assessment. Available at: http://www.fao.org/forest-resources-assessment/documents/en.

Hanessian, S., J. Howe, J. Bowyer, S. Bratkovich, K. Fernholz, M. Frank, H. Groot, and E. Pepke. 2015. *Benefit Corporations and Certified B Corps: New Opportunities for Business.* Dovetail

Partners. February 17. Available at: http://www.dovetailinc.org/report_pdfs/2015/dovetailbcorps0215.pdf.

Howe, J. and K. Fernholz. 2015. *Building with Wood = Proactive Climate Change Protection*. Produced by Dovetail Partners and State University of New York. Funded by Binational Softwood Lumber Council. Available at: http://www.dovetailinc.org/report_pdfs/2015/building_with_wood.pdf.

IKEA. 2018. IKEA Group Acquires 25,000 Acre Forest in Alabama Latest Investment Furthers the IKEA Group Expansion in the U.S. January 9. Available at: http://www.ikea.com/us/en/about_ikea/newsitem/020118-IKEA-Group-aquires-forest-in_alabama.

MacDicken, K. G., P. Sola, J. E. Hall, C. Sabogal, M. Tadoum, and C. de Wasseige. 2015. Global Process toward Sustainable Forest Management. *Forest Ecology and Management* Vol. 352, 47–56. Available at: https://doi.org/10.1016/j.foreco.2015.02.005Get

Packham, J. R., D. J. L. Harding, G. M. Hilton, and R. A. Stuttard. 1992. *Functional Ecology of Woodlands and Forests*. Chapman & Hall.

Panwar, R., R. Kozak, and E. Hansen. 2016. *Forests, Business and Sustainability*. Routledge. The Earthscan Forest Library. Available at: http://www.routledge.com/books/series/ECTEFL.

Ross, K. 2015. Measuring Sustainable Forest Management: A Report on On-Going and Emerging Global Initiatives to Develop Results Frameworks and Performance Indicators for Sustainable Development, Agriculture, and Natural Resource Management. Forest Economics, Policy and Products (FOE) Division of the Food and Agriculture Organization of the United Nations (FAO). January 15. Available at: http://www.fao.org/forestry/42575-0ee3fc1e9d0f9619b8adfb-c78f836d604.pdf.

Smith, D. 1986. *The Practice of Silviculture*. 8th ed. John Wiley & Sons.

United Nations Forum on Forests. 2007. *Highlights from Enabling Sustainable Forest Management: Strategies for Equitable Development, for Forests, for People*. United Nations. Available at: http://www.un.org/esa/forests/pdf/publications/Enabling_SFM_highlights.pdf

USDA Forest Service. 2011. *National Report on Sustainable Forests—2010*. June. FS-979. Available at: https://www.fs.fed.us/research/sustain/docs/national-reports/2010/2010-sustainability-report.pdf.

Viana, V. M., J. Ervin, R. Z. Donovan, C. Elliott, and H. Gholz (eds.). 1996. *Certification of Forest Products: Issues and Perspectives*. Island Press.

Index

Note: Figures are indicated by *f*; tables are indicated by *t*.

www.ingramcontent.com/pod-product-compliance
Ingram Content Group UK Ltd.
Pitfield, Milton Keynes, MK11 3LW, UK
UKHW050805230325
456577UK00001BA/2